BUSINESS MATHEMATICS

BUSINESS MATHEMATICS

ROBERT J. HUGHES

Dallas County Community College District
Dallas, Texas

IRWIN

Chicago · Bogotá · Boston · Buenos Aires · Caracas
London · Madrid · Mexico City · Sydney · Toronto

Senior sponsoring editor: Richard T. Hercher, Jr.
Developmental editor: Gail Korosa
Project editor: Denise Santor-Mitzit
Production supervisor: Ann Cassady
Designer: Heidi J. Baughman
Art coordinator: Eurnice Harris
Compositor: J.M. Post Graphics, a division of
                                  Cardinal Communications Group, Inc.
Typeface: 10.5/12 Helvetica
Printer: Webcrafters, Inc.

Library of Congress Cataloging-in-Publication Data

Hughes, Robert James
   Business mathematics / Robert J. Hughes.
       p.   cm.
      Includes index.
      ISBN 0-256-16514-9
      1. Business mathematics.   I. Title
   HF5691.H789   1995
   650'.01' 513—dc20   94–46606

Printed in the United States of America
1 2 3 4 5 6 7 8 9 0 WC–W 1 0 9 8 7 6 5 4

PREFACE FOR THE STUDENT

"Math courses have always been difficult. The textbooks always seem to discuss abstract concepts and then apply those same concepts to apples and oranges."

—Jackie Marston
Math Student

Unfortunately, the feelings expressed by Jackie Marston are all too common today. Many students feel the same way when they walk into a math class. That's why it is important to begin this text with a basic idea: *Business Math Doesn't Have to Be Difficult.* In fact, we have done everything possible to eliminate the typical problems that students encounter in a business math class. Let's begin with an overview of how this text is organized.

ORGANIZATION OF THE TEXT

Take a moment and look at the Table of Contents on page XI. Notice that the first four chapters provide review material on our number system, addition, subtraction, multiplication, and division, fractions, and percentage. Even if you have never been successful in math, the material in these early chapters will help you build a foundation for the business applications that follow in Chapters 5 through 15.

CHAPTER FORMAT

All of the features in each chapter have been evaluated and recommended by instructors with years of teaching experience. In addition, business math students were asked to critique each chapter component. Based on this feedback, the following features are included in each chapter.

Chapter Preview

Each of the text's 15 chapters begins with a concise overview of the math topics that follow. You should look at the preview because it provides a capsule summary of each topic and the sequence in which topics are covered.

Math Today Opening Case

Chapter opening vignettes entitled "Math Today" introduce the content of the chapter through real-world situations. Take a moment and look at the opening case from Chapter One on page 1 that highlights how numbers can be used to describe Blockbuster Video. Other Math Today opening vignettes highlight Pepsi, Nike, Automated Data Processing (ADP), AT&T, The U.S. Government, and Ford.

Learning Units

The content in each chapter in *Business Mathematics* is broken down into smaller learning units. Generally, each learning unit covers a basic math concept and has the following components:

1. *Learning Objectives.* If you have a purpose for studying the material in each learning unit, you will learn more than if you just wander aimlessly through the text. Therefore, each learning unit in *Business Mathematics* contains clearly stated learning objectives that signal important concepts to be mastered.

2. *Clearly Written Explanations.* To help you understand the math concepts in this text, we have given special attention to word choice, sentence structure, and the presentation of business terms.

3. *Step-by-Step Instructions.* Every chapter in *Business Mathematics* contains numerous examples to illustrate important concepts. It is our belief that even the most difficult business application problems are easier to solve when a step-by-step approach is used. Take a moment and look at any learning unit in the text for examples of how this approach can help build success.

4. *End-of-Unit Problems.* As soon as a concept is presented, both drill and word problems are used to reinforce each concept. Special effort has been made to choose problems that allow you to apply concepts to real-world applications and lifelike situations. Answers for selected problems are located in Appendix D.

5. *Critical Thinking Problems.* The ability to think may be the most important skill that you can develop while in college. To develop this all-important skill, we have included a special group of problems that are called "Critical Thinking Problems." Take a moment and look at a typical Critical Thinking Problem on page 216. Although problems 19 and 20 require just calculations, you must summarize or apply your findings in two or three short sentences in order to answer question 21.

6. *Personal Math Problems.* "Personal Math Problems" are located in most chapters and involve a life situation where you must use math concepts to reach a decision. Look at the Personal Math Problem on page 386. Notice in this problem that questions 36 and 37 require calculations, but question 38 requires that you make a decision based on the facts and summarize your findings in two or three short sentences.

Instant Replay Summary

To help you review the material in each chapter, a concise outline called "Instant Replay" summarizes important concepts in each chapter. There is also an example or completed problem that illustrates each concept.

Mastery Quiz

Each Mastery Quiz contains 20 problems that can be used to test your comprehension of all the chapter concepts. Answers for Mastery Quiz problems are located in Appendix D.

Cumulative Review

At the end of Chapters 4, 8, 12, and 15, you can test your comprehension of chapter material by completing the cumulative review. Each cumulative review contains 20 problems and answers are located in Appendix D.

A FINAL WORD

A text should always be evaluated by the students and instructors who use it. I will welcome and sincerely appreciate your comments and suggestions and will acknowledge your assistance in the next edition of *Business Mathematics*.

Bob Hughes
Dallas County Community College District
12800 Abrams Road
Dallas, TX 75243

ACKNOWLEDGMENTS

I wish to express a great deal of appreciation to the following individuals who have helped improve and refine this text and instructional package. For the generous giving of their time and their thoughtful and useful comments and suggestions, I am indebted to the following reviewers:

Robert Bonnewell
  Danville Community College
John A. Brown
  Savannah Technical Institute
Anthony Brunswick
  Delaware Technical & Community College
Calvin Holt
  Paul D. Camp Community College
John Mastriani
  El Paso Community College
Catherine Merriken
  Pearl River Community College
Charles Trester
  Northeast Wisconsin Technical College
Catherine Vollstedt
  Northcentral Technical College
Nancy Wallace
  Chattahoochee Technical Institute
Zonell Webster-Miller
  Rose State College

I also wish to thank John Balek, Morton College, and Charles Trester, Northeast Wisconsin Technical College, for helping ensure the accuracy of the text material problem solutions. For sharing their ideas I wish to thank Wilene Landfair, Gwen Hester, Wanda Matheson, and Becky Jones, all business math instructors at Richland College.

Finally, many talented editorial, production, and marketing professionals at Richard D. Irwin have contributed to this edition of *Business Mathematics.* I am especially grateful to Dick Hercher, Gail Korosa, Denise Santor-Mitzit, Heidi Baughman, and Brian Kibby for their inspiration, patience, and friendship.

CONTENTS

Number Values, Addition, and Subtraction

In this chapter, we look at the basics that are necessary for working any mathematical problem. First we examine the number system used in business today. Next we discuss the rules for rounding numbers. Then our focus shifts to two operations—addition and subtraction—that are the basis for most business application problems.

MATH TODAY

BIG NUMBERS FOR BLOCKBUSTER VIDEO

Wow! What a difference. Those four words describe the 3,258 Blockbuster Video superstores in the United States, Great Britain, Canada, Chile, Austria, Mexico, and Japan. Each store is big, well-lit, and carries a comprehensive selection of between 5,000 and 6,000 recent motion picture titles. That's two to three times the number of titles that most independents—sometimes called mom-and-pop video stores—carry. Incidentally, the firm doesn't carry X-rated movies because sex videos go against Blockbuster's emphasis on family entertainment.

Blockbuster Video is the world's largest leading video retailer. In 1992, the last year for which complete financial results were available, the firm had sales revenues of over $1.2 billion. Net profits for 1992 were $142 million, which represents a 52 percent increase over 1991 profits. And at this writing, the first six months of 1993 were looking extremely promising with both revenues and profits expected to continue to improve over 1992 figures. Not bad for a company that had just over 200 stores back in 1987.

Although revenue growth and overall profits are expected to slow down in the video-rental industry between now and the year 2000, Blockbuster is taking steps to ensure that its own sales revenues and profits continue to increase for at least the next 10 years. With a corporate goal of 5,000 stores by 1995, Blockbuster continues to open new retail outlets in both the United States and abroad. Blockbuster has also diversified by purchasing over 200 Sound Warehouse stores and acquiring a 21 percent interest in a children's recreational fitness center called Discovery Zone. Finally, the firm plans to open a chain of virtual reality amusement parks throughout the United States.

Source: For more information, see Peter Katel, "New Kid on the Block, Buster," *Newsweek,* January 11, 1993, p. 48; *Moody's Handbook of Common Stocks* (New York, NY: Moody's Investors Service, Fall 1993); Gary Hoover, *Hoover's Handbook of American Business 1993* (Austin, TX: The Reference Press, 1992), p. 159; and Gail DeGeorge, "They Don't Call It Blockbuster for Nothing," *Business Week,* October 19, 1992, pp. 113–14.

UNIT 1   WHOLE NUMBERS

LEARNING OBJECTIVES

After completing this unit you will be able to:

1. identify a specified place value within the decimal number system.
2. convert written whole numbers to numeral form.
3. change whole numbers in numeral form to written form.

3,258 Blockbuster Video superstores.
A selection of 5,000 to 6,000 motion picture titles.
Sales revenues of over $1.2 billion.
Net profits of $142 million.
Corporate goal of 5,000 stores by 1995.

All of the above numbers can be used to describe Blockbuster Video. These same numbers help management determine if the firm is meeting its goals and operating efficiently. And these numbers help lenders, suppliers, stockholders, and government agencies evaluate the firm. Finally, the above numbers underscore the importance of our number system in today's business world.

Let's begin by *examining* that number system. Our **number system** is based on ten individual digits: 0, 1, 2, 3, 4, 5, 6, 7, 8, and 9. When digits are used to form a number, each digit stands for a specified value. For example, each digit in the number 5,683,427,891 represents a specific value as illustrated in Example 1.

EXAMPLE 1 Place Values for the Number 5,683,427,891

Billions Group		Millions Group			Thousands Group			Units or Ones Group		
5	,	6	8	3	,	4	2	7	,	8	9	1

5 — Billions
6 — Hundred millions
8 — Ten millions
3 — Millions
4 — Hundred thousands
2 — Ten thousands
7 — Thousands
8 — Hundreds
9 — Tens
1 — Units or ones

In this number, there are 5 billions, 6 hundred millions, 8 ten millions, 3 millions, 4 hundred thousands, 2 ten thousands, 7 thousands, 8 hundreds, 9 tens, and 1 unit. Let's look at another example.

EXAMPLE 2 Place Values for the Number 7,654,932

In the number 7,654,932, the value in the

units place is 2 (equal to 2).
tens place is 3 (equal to 30).
hundreds place is 9 (equal to 900).
thousands place is 4 (equal to 4,000).
ten thousands place is 5 (equal to 50,000).
hundred thousands place is 6 (equal to 600,000).
millions place is 7 (equal to $7,000,000).

Each group of three digits is separated by a comma. The commas help us read the number or write it in words. You read each group of three digits followed by the name of that specific group of numbers. For example, notice where the commas are placed in the above number 7,654,932. This same number is read aloud and written as "seven million, six hundred fifty-four thousand, nine hundred thirty-two." Also, note that a hyphen is used to separate compound numbers like fifty-four or thirty-two when expressed in written form. Finally, note that you do not use the word *and* when you read or write these numbers. The word *and* represents the decimal point and will be discussed in Unit 2.

Identify the Digit or Place Value in the Following Problems

1. In the number 32,649,108

    a. the digit in the hundreds place is _____ *1* _____.

   b. the digit in the thousands place is _____.

   c. the digit in the millions place is _____.

   d. the digit in the units place is _____.

   e. the digit in the ten millions place is _____.

2. In the number 2,456,138

   a. the digit in the _____ *units* _____ place is 8.

   b. the digit in the _____ place is 6.

   c. the digit in the _____ place is 5.

   d. the digit in the _____ place is 2.

   e. the digit in the _____ place is 3.

Write the Following Whole Numbers in Numeral Form

3. a. One hundred ninety-three   ___193___

   b. Seven thousand, four hundred fifteen   _____

   c. Twelve thousand, two hundred forty-six   _____

   d. Three million, two hundred thousand   _____

   e. Two hundred sixty-nine thousand, seven hundred eighty-four   _____

   f. Five hundred two   _____

   g. Seventy thousand, four hundred thirty-three   _____

Write the Following Whole Numbers in Words

4. a. 147 _____ *one hundred forty-seven* _____

_____

   b. 1,678 _____

_____

   c. 14,569 _____

_____

   d. 723,659 _____

_____

   e. 65,432 _____

_____

   f. 3,459,700 _____

_____

   g. 444 _____

_____

UNIT 2	DECIMAL NUMBERS

LEARNING OBJECTIVES

After completing this unit, you will be able to:

1. identify decimal place values to the right of the decimal.
2. convert written numbers containing decimals to numeral form.
3. change numbers containing decimals in numeral form to written form.

The **decimal point** separates the whole number part of a number from the decimal part. Like whole numbers, decimal numbers also have assigned place values. Take, for example, the decimal number 0.562418.

EXAMPLE 1   Place Values for the Decimal Number 0.562418

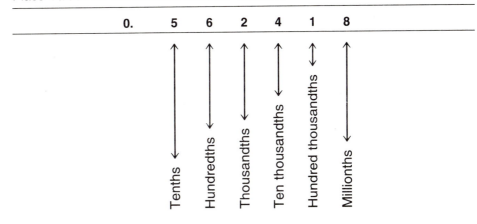

In this number, there are 5 tenths, 6 hundredths, 2 thousandths, 4 ten thousandths, 1 hundred thousandth, and 8 millionths. Notice that each of these place values ends with the letters "th(s)" to indicate that it is a decimal number that is less than one—not to be confused with a whole number.

When a number contains both whole numbers and decimal numbers, the whole number part is to the left of the decimal point, and the decimal part is to the right of the decimal point.

EXAMPLE 2   Place Values for the Number 1,246.985

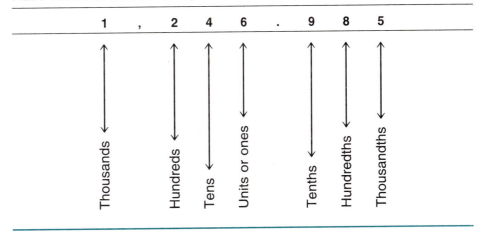

When reading or writing numbers that contain decimals, extra care must be taken to make sure that other people who may hear or see the number understand that it contains a decimal value.

EXAMPLE 3   How to Read or Write Decimal Numbers

1. The number 0.4 is read or written as "four ten*ths*."
2. The number 0.35 is read or written as "thirty-five hundred*ths*."
3. The number 0.276 is read or written as "two hundred seventy-six thousand*ths*."

When the number contains a whole number and a decimal value, read the whole number first and then the decimal value. The word *and* is used to separate the whole number from the decimal value.

EXAMPLE 4   How to Read or Write Numbers That Contain Both Whole Numbers and Decimal Numbers

1. The number 23.6 is read or written as "twenty-three *and* six ten*ths*."
2. The number 125.68 is read or written as "one hundred twenty-five *and* sixty-eight hundred*ths*."

If the number represents dollars and cents, the words *dollars and* are inserted after the whole number and before the decimal value. The word *cents* is placed after the decimal value.

EXAMPLE 5   How to Read or Write Numbers That Represent Dollars and Cents

1. The number $129.87 is read or written as "one hundred twenty-nine *dollars and* eighty-seven *cents*."
2. The number $2,312.15 is read or written as "two thousand, three hundred twelve *dollars and* fifteen *cents*."

Identify the Digit or Place Value in the Following Problems

1. In the number 35.129

    a. the digit in the thousandths place is _____ 9 _____.

    b. the digit in the units place is _____.

    c. the digit in the tens place is _____.

    d. the digit in the hundredths place is _____.

2. In the number 2,478.1956

    a. the digit in the _____ ten thousandths _____ place is 6.

    b. the digit in the _____ place is 1.

    c. the digit in the _____ place is 4.

    d. the digit in the _____ place is 5.

Convert the Following Written Numbers to Numeral Form

3. a. thirty-nine hundredths   _0.39_

   b. four hundred sixty-five thousandths   _____

   c. one hundred forty-three and two tenths   _____

   d. sixty-five dollars and nineteen cents   _____

   e. forty-six thousandths   _____

   f. thirty-nine dollars and sixty-two cents   _____

   g. three hundred fifty-five dollars and twenty-one cents   _____

   h. twelve thousand, two hundred seventy dollars and ninety-nine cents   _____

Convert the Following Numbers to Written Form

4. a. 0.38   _thirty-eight hundredths_

   b. 25.4

   c. $532.49

   d. $23,670.31

   e. $120,450.20

   f. 0.356

   g. $19.11

   h. $4,567,000

UNIT 3   ROUNDING NUMBERS

LEARNING OBJECTIVE   *After completing this unit, you will be able to:*

1. round numbers to a specified place value.

When we say a corporation—like WalMart—employs 180,000 employees, we are using an approximation or estimate of the actual number of employees. It may be that the actual number is 182,340, but it is easier to read the rounded number. It is also easier to remember the rounded number.

Whole numbers can be rounded to the nearest unit, ten, hundred, thousand, and beyond. In business, it is common to round decimal numbers to hundredths—or two decimal places—because the rounded answer represents dollars and cents. It is also possible to round numbers to other decimal places.

When rounding numbers, follow these steps:

STEP [1]   Identify the place value to which you will round the number.

STEP [2]   If the digit to the right of the identified place value (step 1)
         a. is 5 or greater, increase the identified digit by 1.
         b. is between 1 and 4, do not change the identified digit.

STEP [3]   Change all digits to the right of the rounded place value to zeros.

EXAMPLE 1   Round 4,738 to the Nearest Ten

STEP [1]   Identify the digit in tens place.

$$4, 7 \; ③ \; 8$$

STEP [2]   Because the digit to the right of the 3 in the tens place is greater than 5, increase the 3 by 1. The 3 becomes a 4.

$$4, 7 \; \overset{+1}{\underset{4}{3}} \; 8$$

STEP [3]   Change all digits to the right of the digit in the tens place to zeros.

$$4,738 \text{ is rounded to } 4,740$$
$$40$$

EXAMPLE 2   Round 567.84392 to the Nearest Hundredth

STEP [1]   Identify the digit in the hundredths place.

$$5 \; 6 \; 7 \; . 8 \; ④ \; 3 \; 9 \; 2$$

STEP [2]   Because the digit to the right of the 4 in the hundredths place is less than 5, do not change the 4.

$$5 \; 6 \; 7 \; . 8 \; 4 \; 3 \; 9 \; 2$$

STEP [3]   Change all digits to the right of the digit in the hundredths place to zeros.

$$567.84392 = 567.84000 \text{ or } 567.84$$
$$000$$

| EXAMPLE 3 | Round 24.95984 to the Nearest Thousandth |

STEP 1 Identify the digit in the thousandths place.

2 4 . 9 5 ⑨ 8 4

STEP 2 Because the digit to the right of the 9 in the thousandths place is greater than 5, increase the 9 by 1.

2 4 . 9 5 $\overset{+1}{\underset{10}{9}}$ 8 4

STEP 3 Because the 9 in the thousandths place is increased by 1 and becomes a 10, there is a chain reaction. The 10 in the thousandths place becomes a zero when the 1 is carried over to the 5 in the hundredths position. The 5 in the hundredths place becomes a 6.

2 4 . 9 $\overset{+1}{\underset{6}{5}}$ $\overset{+1}{\underset{\underset{0}{10}}{9}}$ 8 4

STEP 4 Change all digits to the right of the digit in the thousandths place to zeros.

24.95984 = 24.96000 or 24.960

Round Each Number as Indicated

1. 563 to the nearest ten —————— *560*
2. 67,345 to the nearest thousand ——————
3. 2,745 to the nearest hundred ——————
4. 884,689 to the nearest ten thousand ——————
5. 802 to the nearest ten ——————
6. 28,650 to the nearest hundred ——————
7. 456,794 to the nearest hundred thousand ——————
8. 114.567 to the nearest tenth ——————
9. 45,678 to the nearest thousand ——————
10. 0.542 to the nearest unit ——————
11. 0.44897 to the nearest tenth ——————
12. 17.802 to the nearest unit ——————
13. 1,776.54 to the nearest unit ——————
14. 23,450.985 to the nearest hundredth ——————
15. 567,594.995 to the nearest hundredth ——————

Round Each Dollar Amount as Indicated

	Nearest Dollar Amount	Nearest Hundredth (2 decimal places)
16. $ 478.296	———	———
17. $1,355.623	———	———

	Nearest Dollar Amount	Nearest Hundredth (2 decimal places)
18. $ 58.957	_____	_____
19. $6,780.115	_____	_____
20. $ 9.394	_____	_____

Convert the Following Written Numbers to Numeral Form and Round Each Number as Indicated

21. Round seventeen to the nearest ten   _____ *20* _____

22. Round one thousand, four hundred thirty to the
nearest thousand   _____

23. Round sixty-two dollars and forty-nine cents to the
nearest dollar   _____

24. Round seven hundred ninety-four dollars and
ninety-three cents to the nearest ten   _____

25. Round one hundred thirty-three thousand,
two hundred twelve dollars and ten cents to the
nearest thousand   _____

Round Each Number as Indicated and Then Write Your Answer in Written Form

26. 430 to the nearest 100 _____ *four hundred* _____

_____

27. $2,456.985 to the nearest hundred _____

_____

28. $49.50 to the nearest dollar _____

_____

29. 0.4522 to the nearest thousandth _____

_____

30. 3,496 to the nearest ten _____

_____

UNIT 4	ADDING BASIC NUMBER COMBINATIONS

LEARNING OBJECTIVE   *After completing this unit, you will be able to:*

   1.   add basic number combinations.

Addition is the process of combining two or more numbers to obtain a total. Correct answers in addition are no accident. This fact is even more important because addition is often the most frequently used mathematical operation in business.

The numbers that are being added are called **addends.** The answer is called the **sum** or *total.*

EXAMPLE 1   Add 5 + 4

```
   5     addend
 + 4     addend
 ───
   9     sum or total
```

Often people get a "feeling" that an answer is wrong. Of course you can work the problem the same way a second time, but most people usually make the same mistake. Since the addends can be added in any order, a much better procedure is to add again, in the reverse direction. In Example 2, the answer is calculated by adding "down." To check the problem, the same numbers are added again, but in the opposite direction.

EXAMPLE 2   Add and Check 5 + 4

STEP [1]   Add downward.

```
   5
 + 4
 ───
   9
```

STEP [2]   *Check* by adding upward.

```
   9
 + 5
   4
```

HINT: It is important to memorize addition facts—that is, combinations of numbers from 0 through 9. After you have mastered these combinations, you will be able to increase your accuracy and speed on more advanced problems.

Complete the Following Addition Problems

1.
a	b	c	d	e	f	g	h	i	j
0	0	0	0	0	0	0	0	0	0
+ 0	+ 1	+ 2	+ 3	+ 4	+ 5	+ 6	+ 7	+ 8	+ 9
0									

2.
a	b	c	d	e	f	g	h	i	j
1	1	1	1	1	1	1	1	1	1
+ 0	+ 1	+ 2	+ 3	+ 4	+ 5	+ 6	+ 7	+ 8	+ 9
1									

3.

a	b	c	d	e	f	g	h	i	j
2	2	2	2	2	2	2	2	2	2
+ 0	+ 1	+ 2	+ 3	+ 4	+ 5	+ 6	+ 7	+ 8	+ 9

4.

a	b	c	d	e	f	g	h	i	j
3	3	3	3	3	3	3	3	3	3
+ 0	+ 1	+ 2	+ 3	+ 4	+ 5	+ 6	+ 7	+ 8	+ 9

5.

a	b	c	d	e	f	g	h	i	j
4	4	4	4	4	4	4	4	4	4
+ 0	+ 1	+ 2	+ 3	+ 4	+ 5	+ 6	+ 7	+ 8	+ 9

6.

a	b	c	d	e	f	g	h	i	j
5	5	5	5	5	5	5	5	5	5
+ 0	+ 1	+ 2	+ 3	+ 4	+ 5	+ 6	+ 7	+ 8	+ 9

7.

a	b	c	d	e	f	g	h	i	j
6	6	6	6	6	6	6	6	6	6
+ 0	+ 1	+ 2	+ 3	+ 4	+ 5	+ 6	+ 7	+ 8	+ 9

8.

a	b	c	d	e	f	g	h	i	j
7	7	7	7	7	7	7	7	7	7
+ 0	+ 1	+ 2	+ 3	+ 4	+ 5	+ 6	+ 7	+ 8	+ 9

9.

	a	b	c	d	e	f	g	h	i	j
	8	8	8	8	8	8	8	8	8	8
	+ 0	+ 1	+ 2	+ 3	+ 4	+ 5	+ 6	+ 7	+ 8	+ 9

10.

	a	b	c	d	e	f	g	h	i	j
	9	9	9	9	9	9	9	9	9	9
	+ 0	+ 1	+ 2	+ 3	+ 4	+ 5	+ 6	+ 7	+ 8	+ 9

Add and Check the Following Addition Problems

Note there is a line below the problem and a line above the problem for your answers.

11.

	a	b	c	d	e	f	g	h	i	j
	3									
	0	0	1	1	1	2	2	3	3	4
	+ 3	+ 7	+ 1	+ 5	+ 9	+ 4	+ 9	+ 4	+ 9	+ 4
	3									

UNIT 5	ADDING LARGER NUMBERS

LEARNING OBJECTIVES   *After completing this unit, you will be able to:*

1. add whole numbers that contain two, three, four, or more digits.
2. estimate correct answers for addition problems.
3. complete word or statement problems.

    The single digit number combinations presented in Unit 4 provide the basis for adding numbers with more digits, but there is one additional step. When one column of numbers totals more than 10, you write only the *last* digit of the total under that column. You must "carry" the remaining digits to the next column of numbers.

EXAMPLE 1   Add 456 + 380 + 678

```
  21
 456
+ 380
+ 678
─────
1,514
```

STEP 1   Add the numbers in the units column: 6 + 0 + 8 = 14. Write the 4 in the units column, and carry the 1 to the tens column.

STEP 2   Add the numbers in the tens column, including the 1 carried from the units column: 1 + 5 + 8 + 7 = 21. Write the 1 in the tens column, and carry the 2 to the hundreds column.

STEP   3   Add the numbers in the hundreds column, including the 2 carried from the tens column: $2 + 4 + 3 + 6 = 15$.

In this example, the final answer is 1,514. You may find it helpful to write the carried digit above the column to which it is carried, as shown in the example.

ESTIMATING CORRECT ANSWERS

To improve accuracy, you may want to mentally estimate the correct answer to a math problem before actually solving the problem. Notice in the next example, each number has been rounded to the nearest hundred before completing the addition problem.

EXAMPLE 2   Estimate the correct answer for the following problem: $222 + 589 + 375$

$222 = 200$
$589 = 600$
$375 = 400$

STEP   1   Round each number to the nearest hundred.

Estimate

```
   200
 + 600
 + 400
 ------
 1,200
```

STEP   2   Mentally add the rounded numbers.

Actual

```
   222
 + 589
 + 375
 ------
 1,186
```

STEP   3   Compare the estimate with the actual answer.

When the estimate for the above example is compared with the actual answer, there is only a small difference between the two numbers. A large difference would have indicated a possible error. In the case of a large difference, you should work the problem again to ensure the actual answer is correct.

Complete the Following Addition Problems

1.	a	b	c	d	e	f	g	h
	17	34	65	45	87	80	37	26
	+ 59	+ 46	+ 15	+ 40	+ 90	+ 30	+ 52	+ 39
	76							

2.
a	b	c	d	e	f	g	h
161	524	246	971	380	545	287	341
+ 34	+ 820	+ 111	+ 856	+ 452	+ 678	+ 108	+ 284
195							

3.
a	b	c	d	e	f
2,345	6,762	1,006	238	5,670	20,456
+ 2,670	+ 1,567	+ 340	+ 4,562	+ 1,234	+ 34,560
		+ 4,610	+ 43	+ 4,567	
		+ 35	+ 246		

Arrange Each of the Following Problems in a Column and Then Add

HINT: When working addition problems, be sure to align each number in the units column with other numbers in the units column, each number in the tens column with other numbers in the tens column, and so on.

4. 23 + 45 + 67 + 83 + 55 _____ 273

5. 45 + 135 + 256 _____

6. 1,278 + 3,456 _____

7. 4,570 + 1,345 + 28 _____

8. 123,500 + 34,600 _____

9. 2,341 + 344 _____

10. 29 + 123 + 8,903 _____

HINT: The following problems are word, or statement, problems. If you have had trouble working this kind of problem in the past, try following these steps:

STEP 1 Before working the problem, read *and* reread the problem.

STEP 2 Analyze the problem by asking yourself the following questions:
a. What must I find to solve the problem?
b. What steps are necessary to solve this problem?
c. Can I check my results?

STEP 3 Before working the problem, estimate what the answer should be.

STEP 4 Check your work by comparing your answer with your estimated answer.

11. On a recent business trip to New York City, Ann Cook spent $174 for food, $368 for hotels, and $420 for airfare. How much did Ann spend on this trip?

12. Mike Johnson, owner of Johnson's Automotive, has four employees. Two employees earn $425 a week each. The third employee earns $375 a week. The fourth employee earns $287. What is the total salary expense for Johnson Automotive?

13. During the first week of November, Jim Boyer sold merchandise valued at the following amounts:

Monday $1,278
Tuesday   965
Wednesday   546
Thursday   876
Friday   2,110

What were Jim's total sales for the week?

14. In November, All-Star Appliances sold 115 washing machines, 46 electric dryers, 60 gas dryers, 34 air conditioners, 24 microwave ovens, and six trash compactors. How many appliances did All-Star sell during the month?

15. For the problem below (1) determine the total number of hours that each employee worked during the week; (2) find the total number of hours worked by all employees on each day; and (3) cross check the grand total number of hours for each day of the week with the grand total number of hours for all employees.

Employee	Monday	Tuesday	Wednesday	Thursday	Friday	Totals
Appleby	8	8	9	6	9	
Carson	10	8	8	10	9	
Jackson	5	10	7	9	7	
Totals						

16. During the month of October, Mike Gonzales paid the following business expenses:

Telephone $ 45
Electricity 190
Water 32
Gas 84
Rent 375

What is the total amount of business expenses for the month of October?

17. During January, Jane Mathias earned $3,200. In February, she earned $2,100. And in March, she earned $2,400. What is the amount of her total earnings for the three-month period?

UNIT 6	ADDING NUMBERS THAT CONTAIN DECIMALS

LEARNING OBJECTIVE   *After completing this unit, you will be able to:*

1.   add numbers that contain decimals.

Adding numbers with decimal parts is just like adding whole numbers. To make sure that you are adding digits with the same place value, you must make sure that when the addends are stacked vertically, all decimal points are in a straight line. When you add a whole number to numbers that contain decimals, you may find it helpful to insert a decimal point after the whole number. You might also want to insert zeros after the decimal point to help align the columns.

EXAMPLE 1   Add the Following Numbers: 4.52 + 3.821 + 6

$$
\begin{array}{r}
4.520 \\
+3.821 \\
+6.000 \\
\hline
14.341
\end{array}
$$

4.52 ⎫
3.821 ⎬ becomes
6 ⎭

STEP 1   Insert a decimal point, and add zeros after the whole number 6.

STEP 2   Align the decimal points.

STEP 3   Add the numbers.

HINT: When you have completed your addition, take a moment to look at your answer. Often, you can discover incorrect decimal placement while estimating what the correct answer should be.

Complete the Following Addition Problems

1. a	b	c	d	e	f
23.54	67.40	234.61	474.10	2,378.92	4,567.89
+ 1.14	+20.40	+ 45.99	+123.62	+ 554.29	+1,234.78
24.68					

2. a	b	c	d	e	f
2.34	226.37	1.256	23.8	5,560.22	23.56
+ 12.4	+ 19.67	+ 20.5	+ 0.075	+ 34.95	+ 450.2
14.74		+ 11.45	+ 2.44	+ 4.50	+ 0.5

In the Space Provided, (a) Rewrite the Following Problems in Vertical Form and (b) Determine the Answer for Each Problem

3. Add 0.35 + 4.785 + 12.3541 + 0.5671

0.35
+ 4.785
+ 12.3541
+ 0.5671
18.0562

4. Add 56 + 2.43 + 4.1123 + 5.67 + 3

5. Add 0.1 + 156.4 + 1,234 + 4.5 + 6.003

6. Add $1,245.30 + 34.95 + 10.99

7. Add $34.56 + 167.80 + 3.29 + 19.99

8. During the month of November, Charles Lopez wrote checks in the following amounts:

$23.76 $56.70 $345.67 $429.90 $45.00 $6.78

What is the total dollar amount of checks that Lopez wrote?

9. During the first week of March, Sally Bell worked 7.5, 8, 6.75, 8, and 7.5 hours per day. What is the total number of hours Sally worked?

10. On February 1, Bob Appleby had $14,567.30 in his savings account. During the month he made two deposits of $245.60 and 567.89. What is Bob's saving account balance at the end of February?

11. Global Electronics has divided the United States into four sales territories. The following sales amounts were reported for the month of August.

North $45,678.32
East   56,782.10
South   72,345.90
West   89,432.50

What were Global's total sales during the month of August?

12. On Monday, Handy Hardware Store received checks for $19.00, $24.50, 56.75, 100.00, $243.70, $54.67, and $23.40 from customers. What is the total of the checks that were received?

13. Prime Properties had rental income of $15,678.40, $12,455.00, $16,432.20, $19,865.00, and $15,435.67 for the last five months. What is the company's total rental income for this five-month period?

14. For the problem below, (1) calculate the daily sales totals for all departments combined; (2) find the weekly sales totals for each department; and (3) cross check the grand totals for daily sales with the grand total for weekly departmental sales.

Department	Monday	Tuesday	Wednesday	Thursday	Friday	Totals
Fashion	$ 985.20	$ 630.10	$ 542.32	$ 892.36	$ 1,035.20	
Sportswear	$1,020.30	805.40	613.20	914.00	720.30	
Shoes	$ 310.35	94.30	135.60	224.35	426.80	
Totals						

UNIT 7   SUBTRACTING WHOLE NUMBERS

LEARNING OBJECTIVES   *After completing this unit, you will be able to:*

1. subtract whole numbers.

Subtraction is the process of finding the difference between two numbers. In a subtraction problem, the top number is called the **minuend.** The bottom number is called the **subtrahend.** The answer is called the **difference.**

EXAMPLE 1   Subtract 5 from 8

```
  8      minuend
- 5      subtrahend
  3      difference
```

In this problem, the subtrahend (5) is subtracted from the minuend (8) to obtain the difference (3).

THE BORROWING PROCESS

To complete many subtraction problems, it is necessary to borrow. For example, when subtracting 57 from 95, you need to borrow from the tens column.

EXAMPLE 2 Subtract 57 from 95

$$\begin{array}{r} \overset{8\,1}{\cancel{9}5} \\ -\ 57 \\ \hline 38 \end{array}$$

STEP 1 Borrow 1 from the 9 in the tens column. When you do so, cross out the 9 and write in an 8.

STEP 2 Combine the borrowed digit from the tens column with the 5 in the units column. Now, subtract 7 from 15.

STEP 3 Subtract the numbers in the tens column.

CHECKING SUBTRACTION PROBLEMS

Answers that are close just aren't good enough. Like in addition, you can always rework the problem a second time, but most people will make the same mistake. Because addition is the opposite (or reverse) of subtraction, you can check a subtraction problem by adding the subtrahend to the difference.

EXAMPLE 3 Subtract 578 from 764

	Subtraction	*Check*	
minuend	764	186	difference
subtrahend	− 578	+ 578	subtrahend
difference	186	764	minuend

If the check answer doesn't agree with the top number (minuend) of the original problem, you have made a mistake and need to work the problem again.

Complete the Following Subtraction Problems

1.

a	b	c	d	e	f	g	h	i	j
1	1	2	2	2	3	3	3	3	4
− 0	− 1	− 0	− 1	− 2	− 0	− 1	− 2	− 3	− 0
1									

2.

a	b	c	d	e	f	g	h	i	j
4	4	4	4	5	5	5	5	5	5
− 1	− 2	− 3	− 4	− 0	− 1	− 2	− 3	− 4	− 5
3									

3.

a	b	c	d	e	f	g	h	i	j
6	6	6	6	6	6	6	7	7	7
− 0	− 1	− 2	− 3	− 4	− 5	− 6	− 0	− 1	− 2

4.

a	b	c	d	e	f	g	h	i	j
7	7	7	7	7	8	8	8	8	8
− 3	− 4	− 5	− 6	− 7	− 0	− 1	− 2	− 3	− 4

5.

a	b	c	d	e	f	g	h	i	j
8	8	8	8	9	9	9	9	9	9
− 5	− 6	− 7	− 8	− 0	− 1	− 2	− 3	− 4	− 5

6.

a	b	c	d	e	f	g	h	i	j
9	9	9	9	27	57	42	70	136	99
− 6	− 7	− 8	− 9	− 15	− 34	− 11	− 29	− 48	− 50

7.

a	b	c	d	e	f	g
32	47	62	91	125	154	152
− 17	− 36	− 54	− 28	− 92	− 78	− 104

8.

a	b	c	d	e	f	g
572	711	897	1,300	3,452	4,565	7,777
− 393	− 429	− 503	− 456	− 1,210	− 2,963	− 6,525

Complete the Following Subtraction Problems; Check Each Problem in the Space Provided

9. **a**
	Check
23,451	*22,257*
− 1,194	*+ 1,194*
22,257	*23,451*

b   Check

85,427
−12,352

c   Check

12,700
−   561

d   Check

24,562
− 4,005

10. Last week, Winnie Chung purchased a coffee maker that was on sale. The original selling price was $34. The reduced sale price was $21. How much did she save on this purchase?

$34
−21
$13

11. On May 1, Theresa Lambert had a $1,879 balance in her checking account. On Saturday, she wrote a check to The Home Depot for $187. What is her new balance?

12. On Friday, Martha Anderson spent $56 at the grocery store, $38 at the drug store, and $25 at the hardware store. Assuming that Martha had $225 when she started this shopping trip, how much money does she have now?

13. At the end of the week, Ross Wholesale Supply had gross sales of $42,356. The firm also had sales returns that totaled $3,900. What is the firm's sales total for the week? **Hint: Subtract sales returns from gross sales.**

14. All-American Consumer Electronics sells a compact disc player for $249. It costs All-American $167 for the same product. How much profit before expenses does All-American make on this product?

15. On August 1, Juan Creel owed $167 to Champion Department Store. On August 15, Mr. Creel purchased merchandise valued at $154. Then, on August 18, he returned merchandise valued at $78. On August 22, he purchased additional merchandise valued at $220. On August 28, he made a payment of $120. What is his current balance at the end of August?

Critical Thinking Problem

Wendy's International is a US corporation that specializes in providing quality food at fast-food prices. In the third quarter of 1993, sales revenues were $338,890,000. For the same period in 1992, sales revenues were $317,282,000. Net income for 1993 was $24,484,000. During the same period for 1992, net income was $19,910,000. For the third quarter, a share of Wendy's stock earned $0.24 during 1993 compared with 0.20 per share in 1992.

16. How much did sales revenues increase in 1993 when compared with sales revenues for 1992?

17. How much did net income increase in 1993 when compared with net income for 1992?

18. If you were a manager for Wendy's, how could this financial information help you manage this firm in the future? _____

_____

_____

_____

19. If you owned stock in Wendy's, how could you use the above financial information to evaluate your investment? _____

_____

_____

_____

UNIT 8   SUBTRACTING NUMBERS THAT CONTAIN DECIMALS

LEARNING OBJECTIVES

After completing this unit, you will be able to:

1. subtract numbers that contain decimals.

With the exception of having to make sure decimal points are aligned, subtracting numbers with decimals is just like subtracting whole numbers. In Example 1, the original numbers are rewritten so that all decimal points are in a straight line. Then the numbers are subtracted.

EXAMPLE 1   Subtract 2.98 from 14

$$\left.\begin{array}{r}14\\-2.98\\\hline ?\end{array}\right\}\text{ becomes }\left\{\begin{array}{r}14.00\\-2.98\\\hline 11.02\end{array}\right.$$

A whole number (in this case 14) is assumed to have a decimal point to its right. Also, if necessary, zeros can be added to the top number (to the right of the decimal point) before subtracting the bottom number from the top (in this case 2.98 from 14.00). Finally, the decimal point in the answer is placed directly below the decimal points in the problem.

Complete the Following Subtraction Problems

1.
a	b	c	d	e	f
0.52	0.66	1.92	29.72	56.78	57.94
− 0.39	− 0.48	− 0.57	− 3.40	− 29.89	− 31.14
0.13					

2.
a	b	c	d	e	f
$5.99	$34.50	$234.75	$2,356.92	22	0.35
− 2.64	− 7.77	− 44.99	− 399.50	− 0.45	− 0.19
$3.35					

In the Space Provided, (a) Rewrite the Following Problems in Vertical Form and (b) Determine the Answer for Each Problem

3. Subtract 0.345 from 2.21

$$\begin{array}{r}2.21\\-0.345\\\hline 1.865\end{array}$$

4. Subtract 29.81 from 456.7

5. Subtract 111.90 from 2,340.085

6. Subtract $29.95 from $100

7. Subtract $4,999 from $10,567.20

8. Subtract $20,500.20 from $76,750

9. Tom Young uses his automobile for both business and personal use. During the month of December, he drove a total of 2,456.8 miles. Mileage for personal use was 1,087.5. How many miles did Tom drive on business?

10. Sarah Weinstein purchased a new car for $14,359.69. She made a down payment of $3,500. How much does she owe on the car?

11. Blackstone Electronics installs a security system for $499. The system costs the company $275.29. How much profit does Blackstone make on this item?

12. The price of a new oak desk at Miami Office Supply is 675.95. The price of the same type of desk at Barton Office Supply is $619.99. How much would you save if you purchased the desk at Barton Office Supply?

13. Denise Martin earns $483.50 each week. Her employer deducts the following amounts from her paycheck.

Federal withholding $65.45
FICA (Social Security)   36.32
Voluntary savings plan   50.00

What is the amount of Ms. Martin's take-home pay?

14. On a recent loan application, The Campside Camping Center listed assets and property valued at $1,065,439.27. It also listed debts that totaled $456,702.42. What is the difference between the firm's assets and its debts?

15. Last year Mike Gonzales paid quarterly tax payments that totaled $10,650.40 to the federal government. At the end of the year, his accountant told him his actual tax liability is $9,356.00. What is the amount of the refund check that Mike will receive?

16. Peggy Hess has a charge account with a local department store. During the month of September, she made purchases that totaled 289.65, and she also made a payment of $325.50. At the beginning of September, her balance was $176.50. What is her balance at the end of September?

17. One year ago, Business Press, Inc., borrowed a total of $89,420.58 to finance a new warehouse. According to the loan agreement, Business Press must make quarterly payments of $2,350.75 to repay the loan. During the past year, the firm made four quarterly payments. What is the current balance for this loan?

Personal Math Problem

Joan Towns has decided to purchase a new color television. Normally, a local discount store sells a 25-inch Magnavox TV for $599, but as part of a special promotion, the store has reduced the price by $100 for purchases made during the next two months. If Joan purchases the television from the local retailer, she must pay $40 for sales tax. An out-of-town mail-order retailer advertises the same television for $450. This retailer does charge $57.95 for shipping, delivery, and insurance costs, but no sales tax.

18. What is the total purchase cost if Joan buys the television from the local discounter?

19. What is the total purchase cost if Joan buys the television from the out-of-town mail-order retailer?

20. While the out-of-town mail-order retailer does offer the same television at a cheaper price, describe the other factors that should be considered before making a decision to purchase the television from either retailer. _____

_____

_____

_____

_____

_____

INSTANT REPLAY

UNIT	IMPORTANT POINTS TO REMEMBER	EXAMPLES
UNIT 1 Whole Numbers	Each digit in a number stands for a specified value. Place values are units, tens, hundreds, thousands, ten thousands, hundred thousands, millions, and so on.	7,529 = seven thousand, five hundred twenty-nine
UNIT 2 Decimal Numbers	Place values to the right of the decimal point are tenths, hundredths, thousandths, ten thousandths, millionths, and so on.	5.827 = five *and* eight hundred twenty-seven thousandths
UNIT 3 Rounding Numbers	Identify the place value to which you will round the number. If the digit to the *right* of the identified place value is 5 or greater, increase the identified place value by 1. If the digit to the right of the identified place value is between 1 and 4, do not change the identified place value. Change all digits to the right of the rounded place value to zeros.	67 rounded to the nearest ten = 70 $157.432 rounded to the nearest hundredth = $157.43
UNIT 4 Adding Basic Number Combinations	The numbers being added are called addends, and the answer is called the sum or total.	5 addend +3 addend 8 sum or total
UNIT 5 Adding Larger Numbers	When adding a column of numbers that total more than 10, carry the first digit of that column total to the next column of numbers.	1 37 addend +56 addend 93 sum or total

continued from page 28

UNIT	IMPORTANT POINTS TO REMEMBER	EXAMPLES
UNIT 6 Adding Numbers That Contain Decimals	When adding numbers that contain decimal values, make sure all decimal points are in a straight line. Place the decimal point in the final answer directly below the decimal points in the problem.	$4.52 +2.40 $6.92
UNIT 7 Subtracting Whole Numbers	In a subtraction problem, the top number is called the minuend, the bottom number is called the subtrahend, and the answer is called the difference.	125 minuend – 76 subtrahend 49 difference
UNIT 8 Subtracting Numbers That Contain Decimals	For subtraction problems that contain decimals, the decimal points in the problem must be aligned in a straight line. Place the decimal point in the final answer directly below the decimal points in the problem.	$48.53 – 6.21 $42.32

NOTES

CHAPTER 1	MASTERY QUIZ

DIRECTIONS   *If necessary, round each answer to two decimal places or hundredths. (Each answer counts 5 points.)*

In the number 2,459.176

1. the digit in the hundreds place is _____.

2. the digit in the units place is _____.

3. the digit in the tenths place is _____.

4. the digit in the hundredths place is _____.

Convert the following numbers to written form.

5. 0.45 _____.

6. 29.6 _____.

7. $56.70 _____.

8. $56,789.04 _____

_____.

In the space provided, (a) solve the problem and (b) round all decimal answers to two decimal places (hundredths).

9. $\begin{array}{r} 56 \\ +49 \\ +21 \\ \hline \end{array}$ 10. $\begin{array}{r} 2.473 \\ +93.98 \\ +11.6 \\ \hline \end{array}$ 11. $\begin{array}{r} 356.6712 \\ -3.46 \\ \hline \end{array}$ 12. $\begin{array}{r} \$4,231.90 \\ -1,349.35 \\ \hline \end{array}$

In the space provided: (1) convert the following written numbers to numeral form, (2) solve each problem, and (3) round all answers to two decimal places (hundredths).

13. Four hundred sixty-nine and twelve hundredths plus fifty-nine and two tenths

14. Three thousand, five hundred sixty plus eleven and forty-six hundredths

15. On Monday, Mollie Jones purchased a new calculator. The purchase price was $59.95 plus $4.80 sales tax. What was the total cost for the calculator?

16. During the first week of October, Mack Cooper sold the following amounts:

Monday	$1,456.78
Tuesday	OFF
Wednesday	892.56
Thursday	2,082.40
Friday	1,542.67
Saturday	998.28

What were Mack's total sales for the week?

17. During the first week of April, Jean Sullivan worked 6.5, 7.25, 8.75, 8, and 5.5 hours. What is the total number of hours Jean worked?

18. On October 1, Sight & Sound Music Store purchased 2,200 compact discs. At the end of October, 1,135 compact discs remained unsold. How many compact discs did the store sell during the month of October?

19. The manager for Video Concepts has decided to reduce the price on a VCR to $199.95. If the VCR originally cost $269.50, what is the dollar amount of the reduction?

20. On June 24, Carol Baxter had a $678.10 checking account balance. The next day she wrote a check for $26.59 to Saks Fifth Avenue. She also made a bank deposit for $275.00. What is Carol's new checking account balance?

CHAPTER 2

Multiplication and Division

CHAPTER PREVIEW

In Chapter 1, we discussed addition and subtraction. Here we look at two more mathematical functions—multiplication and division. For both multiplication and division, we examine basic number combinations, larger number combinations, problems that contain decimals, and common shortcuts.

MATH TODAY

WHICH INVESTMENT IS RIGHT FOR YOU?

Assume that it is January 1992 and you have saved $25,000 to fund an investment program. Also assume that you are considering the following two investment options:

▌ *Option 1.* You could invest $25,000 in a certificate of deposit (CD) for two years and earn a guaranteed 5.5 percent return each year.
▌ *Option 2.* You could invest $25,000 in PepsiCo, Inc.—the company famous for Pepsi-Cola, Mountain Dew, and other soft drink and snack food products. A share of PepsiCo stock sells for $33 a share, and the firm pays a $0.46 per-share dividend each year to stockholders.

Before you read on, answer the following questions. (1) Based on the above information, would you invest $25,000 in a certificate of deposit or purchase shares of stock in PepsiCo, Inc.? (2) What other factors should you consider before making this investment decision?

Two years later, the certificate of deposit has increased in value and is now worth $27,826. To determine the interest for the first year, the investor must multiply the original investment amount ($25,000) by the interest rate (5.5 percent). The interest is then added to the original investment amount, and the process is repeated for the second year. Because CDs are insured by the federal government, there is virtually no risk involved in this investment.

Two years later, the PepsiCo stock investment is now worth $30,782. To determine the number of shares you can purchase, you must first divide the original investment amount ($25,000) by the original stock price ($33). Also, you must use multiplication, addition, and subtraction to determine the total for dividends paid during the two-year period, the commissions paid to buy and sell your stock, and the dollar value of the investment at the end of two years. Because this investment involves purchasing stock in a corporation, you should know that there are no guarantees that PepsiCo will continue to pay dividends or that a share of stock will increase in value.

Source: For more information, see *The Wall Street Journal,* December 14, 1993, p. C5; *Hoover's Handbook of American Business 1994* (Austin, TX.: The Reference Press, Inc., 1993), pp. 868–9; *Moody's Handbook of Common Stocks* (New York, NY: Moody's Investors Service, Fall 1993); *Moody's Handbook of Dividend Achievers.* (New York, NY: Moody's Investors Service, 1992).

UNIT 9   MULTIPLYING BASIC NUMBER COMBINATIONS

LEARNING OBJECTIVES   *After completing this unit, you will be able to:*

1. multiply basic number combinations.
2. check multiplication problems.

There are two factors that are obvious about the investment options presented in the opening case. First, the investment in PepsiCo, Inc. returned almost $3,000 more than the certificate of deposit. But while the increased return is substantial, the stock investment in PepsiCo is not without risk; although the stock's value increased, the price for a share of *any* stock can also decline—a factor that good investors never forget.

A second factor is also obvious. It is impossible for anyone to calculate the return on an investment without the ability to add, subtract, multiply, and divide. In this chapter, we begin our discussion with multiplication—a math function used almost as much as addition by people in business.

Multiplication is really a shortcut method of addition. This idea is more easily understood if you look at an example.

EXAMPLE 1   Multiply 15 × 5

Multiplication	Addition
15	15
× 5	15
75	15
	15
	15
Answer ⟶	75

Multiplication can do the job more quickly, more easily, and usually more accurately than repeated addition of the same number.

In a multiplication problem, the top number is called the **multiplicand.** The bottom number is called the **multiplier.** The multiplicand and multiplier are sometimes called *factors.* The answer is called the **product.**

EXAMPLE 2   Multiply 9 × 7

9	multiplicand
× 7	multiplier
63	product (answer)

In this problem, the multiplicand (9) was multiplied by the multiplier (7). The product (63) is the final answer in a multiplication problem.

CHECKING MULTIPLICATION PROBLEMS

To check a multiplication problem, reverse the top and bottom number in a multiplication problem.

EXAMPLE 3   Multiply 3 × 4

$$
\begin{array}{ccc}
3 & & 4 \\
\times\ 4 & & \times\ 3 \\
\hline
12 & \longleftrightarrow & 12
\end{array}
$$

When you reverse the top number and the bottom number, the answer is the same. If the two answers do not agree, you have made a mistake in the calculation and need to work the problem again.

Complete the Following Multiplication Problems

(Note: when you finish this exercise, you will have constructed your own multiplication table that will enable you to work more difficult problems.)

1.
a	b	c	d	e	f	g	h	i
1	2	3	4	5	6	7	8	9
×0	×0	×0	×0	×0	×0	×0	×0	×0
0								

2.
a	b	c	d	e	f	g	h	i
1	2	3	4	5	6	7	8	9
×1	×1	×1	×1	×1	×1	×1	×1	×1
1								

3.
a	b	c	d	e	f	g	h	i
1	2	3	4	5	6	7	8	9
×2	×2	×2	×2	×2	×2	×2	×2	×2

4.
a	b	c	d	e	f	g	h	i
1	2	3	4	5	6	7	8	9
×3	×3	×3	×3	×3	×3	×3	×3	×3

5.
a	b	c	d	e	f	g	h	i
1	2	3	4	5	6	7	8	9
×4	×4	×4	×4	×4	×4	×4	×4	×4

6.
a	b	c	d	e	f	g	h	i
1	2	3	4	5	6	7	8	9
×5	×5	×5	×5	×5	×5	×5	×5	×5

7.
	a	b	c	d	e	f	g	h	i
	1	2	3	4	5	6	7	8	9
	× 6	× 6	× 6	× 6	× 6	× 6	× 6	× 6	× 6

8.
	a	b	c	d	e	f	g	h	i
	1	2	3	4	5	6	7	8	9
	× 7	× 7	× 7	× 7	× 7	× 7	× 7	× 7	× 7

9.
	a	b	c	d	e	f	g	h	i
	1	2	3	4	5	6	7	8	9
	× 8	× 8	× 8	× 8	× 8	× 8	× 8	× 8	× 8

10.
	a	b	c	d	e	f	g	h	i
	1	2	3	4	5	6	7	8	9
	× 9	× 9	× 9	× 9	× 9	× 9	× 9	× 9	× 9

Complete the Following Multiplication Problems; Check Each Problem in the Space Provided

11.
a		Check	b		Check	c		Check	d		Check
2		3	6			3			8		
× 3		× 2	× 2			× 3			× 3		
6		6									

12.
a		Check	b		Check	c		Check	d		Check
5		4	9			3			4		
× 4		× 5	× 4			× 5			× 6		
20		20									

UNIT 10 | MULTIPLYING LARGER NUMBER COMBINATIONS

LEARNING OBJECTIVE *After completing this unit, you will be able to:*

1. multiply whole numbers that contain two, three, or more digits.

The single-digit number combinations presented in Unit 9 provide the basis for multiplying larger numbers, but there is one additional step. When each number in a multiplication problem has more than one digit, you need to multiply each digit in the top number by each digit in the bottom number. Then you add each **partial product** to obtain the final product. An example will show this principle more clearly.

EXAMPLE 1   Multiply 639 × 27

```
  639     STEP  1    Vertically line up the multiplicand and multiplier so
× 27                 that their digits begin in the units or ones column.
```

```
  639     STEP  2    Multiply 639 by the 7 in the multiplier to obtain the
× 27                 first partial product.
4473
```

```
  639     STEP  3    Multiply 639 by the 2 in the multiplier to obtain the
× 27                 second partial product. Because the 2 in the
4473                 multiplier is in the tens position, place the last digit
1278                 of the second partial product in the tens column.
```

```
  639     STEP  4    Add the two partial products.
× 27
4473
1278
17253
```

When partial products are aligned as illustrated in the example, tens will be added to tens, hundreds to hundreds, thousands to thousands, and so on. With a comma inserted between the 7 and the 2, our final answer is 17,253.

Complete the Following Multiplication Problems

	a	b	c	d	e	f	g
1.	27 × 2 54	36 × 3	45 × 8	19 × 4	26 × 7	62 × 9	43 × 5

2.

	a	b	c	d	e	f	g
	31	46	58	63	28	58	41
	× 23	× 41	× 52	× 29	× 18	× 32	× 33

3.

	a	b	c	d	e	f	g
	124	230	345	489	711	456	520
	× 47	× 79	× 42	× 51	× 28	× 112	× 104

4.

	a	b	c	d
	1,234	45,670	34,561	256,701
	× 56	× 124	× 562	× 1,125

5. Jack Bailey just developed a new advertising campaign for Clayworks Pottery Company. As part of the campaign, he purchased six full-page ads in a well-known hobby magazine. The cost for each page of advertising was $825. What was the total cost of advertising in this magazine?

$$
\begin{array}{r}
\$\ 825 \\
\times\ \ \ \ \ 6 \\
\hline
\$4,950
\end{array}
$$

6. John Evans is a clothing salesperson for Crow & Sons, a local retail merchant. On Monday, he sold nine shirts priced at $22, two suits priced at $299, and one sport coat priced at $149. What were John's total sales for the day?

7. Harwood Manufacturing signed a maintenance agreement with Gonzales & Martin Office Supply. The terms of the agreement are that the office-supply firm will provide maintenance on each of Harwood's typewriters for an annual cost of $69 per typewriter. If Harwood has 18 typewriters, what is the total cost of maintenance?

8. Carol d'Amour purchased 140 shares of stock in General Dynamics. She paid $32 for each share of stock. If the brokerage firm charged $81 for processing the transaction, what was the total amount that Carol owed the brokerage firm?

9. Nelson Floral Supply purchased a new refrigerated display unit on credit. The company paid $500 down and agreed to make 18 monthly installments of $189 each. At the end of 18 payments, how much did the new display unit cost the floral supply company?

10. Determine the total for the invoice below.

SOLD TO: Bartlett's Home Furnishings 727 LBJ Freeway Dallas, TX 75233			Date: 2-15-XX
Quantity	Description	Unit Cost	Total Cost
7	Floor lamps	$119	
10	Chrome table lamps	$ 82	
12	Ginger jar lamps	$ 49	
		Total	

UNIT 11   MULTIPLYING NUMBERS THAT CONTAIN DECIMALS

LEARNING OBJECTIVE   After completing this unit, you will be able to:

1. multiply numbers that contain decimals.

Finding the correct answer to a multiplication problem containing decimals is like working a multiplication problem with whole numbers, with one exception: You have to "place" the decimal point in the answer. To correctly place the decimal point in the answer, follow these three steps:

STEP 1   Multiply the top number (multiplicand) and the bottom number (multiplier) just like you would whole numbers without decimals.

STEP   2   Count the total number of decimal places in the top number and bottom number.

STEP   3   Starting with the last digit on the right in the answer (product), count out the number of places based on the total from step 2. Then insert the decimal point. There should be as many decimal places in the answer as there are in the top and bottom numbers combined.

The following example illustrates the multiplication process when numbers contain decimals.

EXAMPLE 1   Multiply 6.44 × 2.51

$$
\begin{array}{r}
6.44 \\
\times 2.51 \\
\hline
644 \\
3220 \\
1288 \\
\hline
16.1644
\end{array}
$$

STEP   1   Multiply the numbers.

STEP   2   Count the number of decimal places in the top number (two) and bottom number (two). There are a total of four decimal places.

STEP   3   Starting with the last digit on the right in the product, count out the number of places based on the total from step 2. Then insert the decimal point.

In this example, the answer may be left with four decimal places (ten thousandths) or it may be rounded off.

To correctly place the decimal point in some multiplication problems, it may be necessary to add zeros in the final answer.

EXAMPLE 2   Multiply 0.4672 × 0.034

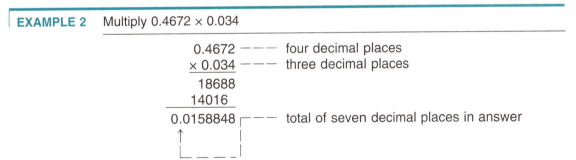

$$
\begin{array}{r}
0.4672 \\
\times 0.034 \\
\hline
18688 \\
14016 \\
\hline
0.0158848
\end{array}
$$

0.4672 — — — four decimal places
× 0.034 — — — three decimal places
0.0158848 — — total of seven decimal places in answer

Because the top and bottom numbers have seven decimal places, the final answer must also have seven decimal places. Therefore, it is necessary to add one zero to the left of the 1 before placing the decimal point.

Complete the Following Multiplication Problems

1.

	a	b	c	d	e	f
	23.4	4.56	0.345	345	2,371	1,200
	× 7	× 5	× 6	× 1.2	× 2.3	× 4.5
	163.8					

2.

	a	b	c	d	e	f
	1.456	3.561	23.40	2.367	0.45	29.1432
	× 20	× 0.42	× 7.39	× .115	× .225	× 0.7

3. Rusty Quinn is paid $6.27 an hour. If he worked 35 hours last week, how much did he earn?

$6.27
× 35
$219.45

4. At the end of the year, Ajax Lawn Supply had an inventory of 17 lawn mowers. Each mower was valued at $168.50. What is the total value of inventory for this particular item?

5. Olga Zotman's boss sent her to the print shop to make photocopies of some legal documents. There were 169 pages to be copied. The print shop charges 5 cents a copy. Calculate how much it will cost Olga to copy these documents.

6. Able Management Services ordered three new computer systems. Each system costs $1,025.49. What is the total cost?

7. At the Office Depot, Everwrite ballpoint pens sell for $13.50 a dozen. If you purchase six dozen pens, what is the total cost?

8. The law firm of Jones & Reece just purchased the following items: three secretarial desks at $310.20 each; two executive desks at $789.00 each; six side chairs at $45.90 each; and eight filing cabinets at $199.50 each. What is the total amount that the firm spent on office furniture?

9. Mike and Tom Jackson just installed a lawn sprinkler system that required 544.6 feet of plastic pipe that cost $0.24 a foot. They also used 42 sprinkler heads that cost $1.04 each, six valves that cost $9.98 each, and one control panel that cost $49.50. How much did they spend on materials?

Critical Thinking Problem

Jane and Bob Brown are planning to open a children's clothing store in Kansas City, Kansas. They are considering two locations. The first location is a space that has 1,750 square feet in a large shopping mall. The mall charges $29.15 a year for each square foot of space. The second location is a space that also has 1,750 square feet, but is in a strip shopping center. The second location costs $20.45 a year for each square foot of space.

10. What is the cost of the location in the shopping mall?

11. What is the cost of the location in the strip shopping center?

12. If you were the Browns, which location would you choose for a children's clothing store? Why? _____

_____

_____

_____

UNIT 12	MULTIPLICATION SHORTCUTS

LEARNING OBJECTIVES   *After completing this unit, you will be able to*

1. multiply numbers using three shortcuts.
2. estimate correct answers for multiplication problems.

If you can use shortcuts to increase speed and to improve accuracy, why not use them? Three different multiplication shortcuts can be used to increase speed and improve accuracy.

MULTIPLYING BY 10, 100, OR 1,000

When you multiply a decimal number by 10, 100, or 1,000, move the decimal point in the top number (multiplicand) as many places to the right as there are zeros in the bottom number (multiplier).

EXAMPLE 1   Multiply 987.1 × 10

$$
\begin{array}{r}
987.1 \\
\times\quad 10 \\
\hline
0000 \\
9871 \\
\hline
9871.0
\end{array}
$$

Shortcut Method

$$987.1 \times 10 = 987.1 = 9,871$$

If the top number is a whole number without decimals, simply add the number of zeros in the bottom number (multiplier) to the top number. If you are multiplying by 10, add one zero to the top number. If you are multiplying by 100, add two zeros to the top number. And if you are multiplying by 1,000, add three zeros to the top number.

EXAMPLE 2   Multiply 456 × 100

$$
\begin{array}{r}
456 \\
\times 100 \\
\hline
000 \\
000 \\
456 \\
\hline
45600
\end{array}
$$

Shortcut Method

$$456 \times 100 = 45600 = 45,600$$

MULTIPLYING BY A MULTIPLE OF 10, 100, OR 1,000

You can use a variation of the shortcut just illustrated to help with a multiplication problem where the bottom number (multiplier) is a multiple of 10, 100, or 1,000. Examples of such multiples are 30 (3 × 10 = 30), 400 (4 × 100 = 400), and 7,000 (7 × 1,000).

EXAMPLE 3   Multiply 45 × 30

STEP 1   Multiply 45 by 3.

$$\begin{array}{r} 45 \\ \times\ 3 \\ \hline 135 \end{array}$$

STEP 2   Multiply the step 1 answer by 10.

$$\begin{array}{r} 135 \\ \times\ 10 \\ \hline 1{,}350 \end{array}$$

MULTIPLYING NUMBERS THAT END IN ZEROS

If the top number, bottom number, or both contain end zeros, disregard the end zeros and multiply the remaining numbers. Then, add the number of end zeros in both the top number and bottom number to the answer.

EXAMPLE 4   Multiply 840 × 400

STEP 1   Disregard the zeros and multiply the remaining numbers.

$$\begin{array}{r} 84 \\ \times\ 4 \\ \hline 336 \end{array}$$

STEP 2   Add three zeros (the combined number of zeros from 840 and 400) to the answer.

$$336\underline{000} = 336{,}000$$

ESTIMATING CORRECT ANSWERS

Before working multiplication problems, you should take a few seconds to estimate the correct answer. As illustrated in the next example, the multiplication shortcut procedures presented in this unit can help.

EXAMPLE 5   Estimate the Correct Answer for the Following Problem: 311 × 95

---

311 = 300
95 = 100

STEP 1   Round each number to simpler numbers that you can multiply easily.

Estimate

$$\begin{array}{r} 300 \\ \times\ 100 \\ \hline 30{,}000 \end{array}$$

STEP 2   Multiply the rounded numbers.

Actual

$$
\begin{array}{r}
311 \\
\times \quad 95 \\
\hline
1555 \\
2799 \\
\hline
29{,}545
\end{array}
$$

STEP $\boxed{3}$ Compare the estimate with the actual answer.

When the estimate for the above example is compared with the actual answer, there is only a small difference between the two numbers. A large difference would have indicated a possible error. In the case of a large difference, you should work the problem again to ensure the actual answer is correct.

Using the Shortcuts Presented in This Chapter, Complete the Following Multiplication Problems

	a	**b**	**c**	**d**	**e**
1.	3.4 × 10 = 34	256 × 10	1,241 × 10	78.91 × 100	456 × 100
2.	32,974 × 1,000	125,672 × 1,000	110,792 × 100	278 × 1,000	45 × 100
3.	45 × 20	3,456 × 30	4,562 × 40	13,525 × 70	25,630 × 80
4.	50 × 30	4,500 × 400	5,650 × 200	1,450 × 2,000	3,370 × 3,000

For Multiplication Problems 5 and 6, Estimate the Correct Answer, Determine the Actual Answer, and Compare the Estimate with the Actual Answer.

	a	Estimate	**b**	Estimate	**c**	Estimate
5.	480 × 12		1,030 × 23		85 × 79	

6.

	a	Estimate		b	Estimate		c	Estimate
	365			2,340			66	
	× 37			× 226			× 53	

7. If Debra Martin's new Chevrolet gets 29.7 miles per gallon, how far can she travel on 10 gallons of gas?

8. Joe and Martha Hernandez purchased 100 shares of General Motors stock for $51.25 a share. If the brokerage firm charged a $43 commission, what is the total amount they must pay the brokerage firm?

9. Sandra Peterson purchased 24 pen and pencil sets. If each set costs $20, what is the total cost for all 24 sets?

10. Malcolm Nations earns $7.25 an hour as chief cashier for an auto repair shop. If he works 40 hours a week, what is his gross pay before deductions?

11. All-Star Office Supply just purchased 120 calculators for resale to retail customers. If each calculator costs $40, what is the total amount that All-Star paid for the calculators?

12. The Houston Independent School District purchased 2,000 student desks for use in a new high school. If each desk costs $60, what is the total cost for student desks?

Personal Math Problem

During a seven-day California vacation, Jane Kelley spent $154 for food, $483 for hotels, and $235 for gifts and miscellaneous expenses. She also rented a Ford Mustang convertible from Western Rent-a-Car. Western charges $42 a day plus $0.29 a mile.

13. If Jane drove the car 500 miles during the seven-day period, how much did the rental car cost?

14. What is the total cost for Jane's seven-day vacation?

UNIT 13	DIVIDING BASIC NUMBER COMBINATIONS

LEARNING OBJECTIVES   *After completing this unit, you will be able to:*

1. divide numbers by single-digit divisors.
2. check division problems.

Last week, Joan Murphy worked 40 hours and made eight sales calls. To determine how long the average sales call took, she divided the total number of hours worked by the number of sales calls. By using division, she determined that she spent an average of five hours on each sales call.

This calculation is just one practical use for division. There are many personal and business applications that justify a brief review of the division process.

Division is the process of finding how many times one number is contained in another number. In a division problem, the number being divided is called the **dividend**. The **divisor** is the number that does the dividing. The answer is called the **quotient**.

EXAMPLE 1   Divide 40 by 8

divisor $8\overline{)40}$ dividend

STEP 1 Write the dividend (40) under the division bracket and the divisor (8) to the left of the bracket.

$$\begin{array}{r} 5 \\ 8\overline{)40} \\ \underline{40} \end{array}$$ quotient (answer)

STEP 2 Divide the dividend by the divisor.

In this example, 40 is the dividend or the number being divided, 8 is the divisor or the number by which you divide, and 5 is the quotient or the answer.

Division problems can be written three different ways.

$$8\overline{)40}^{\,5}$$

or

$$40 \div 8 = 5$$

or

$$\frac{40}{8} = 5$$

The answer is the same regardless of the method used to present the problem. If a problem does not work out evenly, there will be a remainder. A **remainder** is a number that is left over and that cannot be divided by the divisor.

CHECKING DIVISION PROBLEMS

To check a division problem, multiply the answer (quotient) by the divisor *and* add any remainder.

EXAMPLE 2   Divide 12 by 3 and then Check the Answer by Multiplication

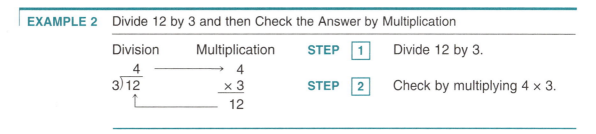

Division	Multiplication		

STEP 1   Divide 12 by 3.

STEP 2   Check by multiplying 4 × 3.

The answer to the multiplication problem should be the same number as the dividend in the original division problem. Notice in the next example that the remainder (3) is added to the product of 9 × 7.

EXAMPLE 3   Divide 66 by 7 and then Check the Answer by Multiplication

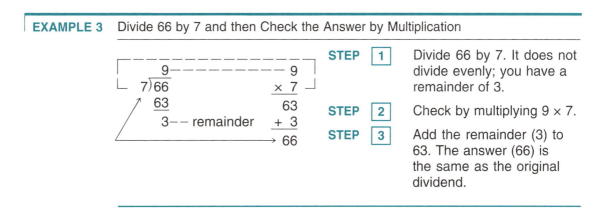

STEP 1   Divide 66 by 7. It does not divide evenly; you have a remainder of 3.

STEP 2   Check by multiplying 9 × 7.

STEP 3   Add the remainder (3) to 63. The answer (66) is the same as the original dividend.

Complete the Following Division Problems

(Note: When you finish this exercise, you will be ready to work division problems with larger numbers.)

1.	a	b	c	d	e	f	g	h	i
	$0\overline{)1}$	$0\overline{)2}$	$0\overline{)3}$	$0\overline{)4}$	$0\overline{)5}$	$0\overline{)6}$	$0\overline{)7}$	$0\overline{)8}$	$0\overline{)9}$

(1a answer: 0)

2.	a	b	c	d	e	f	g	h	i
	$1\overline{)1}$	$1\overline{)4}$	$1\overline{)2}$	$1\overline{)5}$	$1\overline{)6}$	$1\overline{)8}$	$1\overline{)9}$	$1\overline{)3}$	$1\overline{)7}$

3.	a	b	c	d	e	f	g	h	i
	$2\overline{)2}$	$2\overline{)14}$	$2\overline{)12}$	$2\overline{)16}$	$2\overline{)10}$	$2\overline{)8}$	$2\overline{)6}$	$2\overline{)4}$	$2\overline{)18}$

4.	a	b	c	d	e	f	g	h	i
	$3\overline{)3}$	$3\overline{)18}$	$3\overline{)6}$	$3\overline{)21}$	$3\overline{)9}$	$3\overline{)24}$	$3\overline{)15}$	$3\overline{)12}$	$3\overline{)27}$

5.	a	b	c	d	e	f	g	h	i
	$4\overline{)4}$	$4\overline{)12}$	$4\overline{)16}$	$4\overline{)8}$	$4\overline{)24}$	$4\overline{)36}$	$4\overline{)20}$	$4\overline{)32}$	$4\overline{)28}$

6.	a	b	c	d	e	f	g	h	i
	$5\overline{)5}$	$5\overline{)30}$	$5\overline{)10}$	$5\overline{)35}$	$5\overline{)40}$	$5\overline{)15}$	$5\overline{)45}$	$5\overline{)20}$	$5\overline{)25}$

7.	a	b	c	d	e	f	g	h	i
	$6\overline{)6}$	$6\overline{)36}$	$6\overline{)42}$	$6\overline{)12}$	$6\overline{)48}$	$6\overline{)18}$	$6\overline{)54}$	$6\overline{)24}$	$6\overline{)30}$

8.	a	b	c	d	e	f	g	h	i
	$7\overline{)7}$	$7\overline{)42}$	$7\overline{)49}$	$7\overline{)14}$	$7\overline{)56}$	$7\overline{)21}$	$7\overline{)63}$	$7\overline{)28}$	$7\overline{)35}$

9.	a	b	c	d	e	f	g	h	i
	$8\overline{)8}$	$8\overline{)48}$	$8\overline{)64}$	$8\overline{)16}$	$8\overline{)24}$	$8\overline{)56}$	$8\overline{)32}$	$8\overline{)72}$	$8\overline{)40}$

Complete the Following Division Problems; Check Each Problem in the Space Provided

10. **a**   **b**   **c**   **d**

| Check | | Check | | Check | | Check |

$$\frac{3}{9)\overline{27}}$$ Check $$\begin{array}{r} 3 \\ \times\ 9 \\ \hline 27 \end{array}$$

$$9)\overline{45}$$ Check

$$9)\overline{54}$$ Check

$$9)\overline{81}$$ Check

11. **a**   **b**   **c**   **d**

Check   Check   Check   Check

$$9)\overline{36}$$

$$9)\overline{63}$$

$$9)\overline{72}$$

$$9)\overline{18}$$

| **UNIT 14** | **DIVIDING LARGER NUMBER COMBINATIONS** |

LEARNING OBJECTIVE   *After completing this unit, you will be able to:*

1. divide whole numbers that contain two, three, or more digits.

The rules and number combinations presented in Unit 13 can be used to work a problem that contains larger numbers.

EXAMPLE 1   Divide 5,842 by 23

$$\begin{array}{r} 254 \\ 23)\overline{5842} \\ \underline{46} \quad\text{------- step 1} \\ 124 \quad\text{------- step 2} \\ \underline{115} \quad\text{------- step 3} \\ 92 \quad\text{------- step 4} \\ \underline{92} \quad\text{------- step 5} \end{array}$$

In this example, the following steps are required to find the answer:

STEP **1**   Estimate how many times 23 will go into 58. In this example, 23 goes into 58 *two* times with a remainder of 12.

STEP **2**   Bring down the 4 in the dividend, and place it beside the remainder from step 1.

STEP **3**   Estimate how many times 23 will go into 124. Twenty-three goes into 124 *five* times with a remainder of 9.

STEP **4**   Bring down the 2 in the dividend, and place it beside the remainder from step 3.

STEP **5**   Estimate how many times 23 will go into 92. Twenty-three goes into 92 *four* times with no remainder.

When you have a remainder, place "R" and the number remaining beside the answer.

EXAMPLE 2   Divide 24,216 by 781

$$
\begin{array}{r}
31\ \text{R5} \\
781\overline{)24216} \\
2343 \\
\hline
786 \\
781 \\
\hline
5
\end{array}
$$

Complete the Following Division Problems

Show any remainders by placing an "R" and the number left over beside your answer.

1. **a**

$$
\begin{array}{r}
78 \\
14\overline{)1{,}092} \\
98 \\
\hline
112 \\
112 \\
\hline
\end{array}
$$

b $26\overline{)1{,}352}$

c $44\overline{)3{,}828}$

d $61\overline{)5{,}307}$

e $70\overline{)3{,}150}$

2. **a**

$$
\begin{array}{r}
23\ \text{R1} \\
15\overline{)346} \\
30 \\
\hline
46 \\
45 \\
\hline
1
\end{array}
$$

b $71\overline{)2{,}558}$

c $82\overline{)8{,}367}$

d $42\overline{)2{,}229}$

e $50\overline{)2{,}904}$

3. **a** $104\overline{)3{,}744}$

b $210\overline{)5{,}040}$

c $215\overline{)8{,}385}$

d $302\overline{)16{,}912}$

e $412\overline{)15{,}244}$

4.

	a	b	c	d	e
	191)3,822	147)4,120	111)4,445	221)5,530	784)21,954

5. Jane Malone is responsible for planning the annual Christmas party for the employees at Southwestern Plumbing. The caterer gave her a bid of $1,352 for food and beverage service. If there are 104 employees, what will each employee's meal cost?

6. During the first week of July, Mike Hong made four sales that totaled $36,748. What was the average dollar amount for each sale? **HINT: An average is found by dividing the total by the number of individual sales transactions.**

```
      $9,187
  4) $36,748
     36
      7
      4
     34
     32
      28
      28
```

7. Hadduck Printing Company is considering the purchase of a new printing press to replace one that is 15 years old. The new press will cost $34,584, including finance charges. If the firm pays for the press over 24 months, what will be the amount of each monthly installment?

8. Bill Harris, Jane Siegel, and Mark Sidney formed a three-way partnership two years ago. During the first two years, the firm did not earn any profit. During the third year, the firm showed a profit of $47,610. According to the original partnership agreement, each partner is to receive an equal share of all profits. What is the amount of each partner's share?

9. During the last 12 months, Four Seasons Nursing Center had the following electric bills:

January	$435	July	$330
February	245	August	410
March	200	September	280
April	170	October	185
May	215	November	175
June	265	December	330

What is the average of this year's electric bills?

Personal Math Problem

Financial planners suggest that an individual should establish an emergency fund that is equal to at least three months' living expenses before beginning an investment program. With this fact in mind, Jill Santana has saved $9,000. Now she wants to establish an emergency fund and use the remainder of the $9,000 to purchase stock in the Wrigley Corporation—the firm that manufactures and distributes chewing gum.

10. If Jill's living expenses total $1,450 a month, how much money does she need to set aside for an emergency fund?

11. If a share of Wrigley stock cost $38 a share and the brokerage firm charges $90 commission to purchase the stock, how many shares can she purchase with the money remaining after she establishes her emergency fund?

12. Jill Santana chose to establish an emergency fund and purchase stock in the Wrigley Corporation. Would you have done the same thing? Why or why not? _____

_____

_____

UNIT 15	DIVIDING NUMBERS THAT CONTAIN DECIMALS

LEARNING OBJECTIVE   *After completing this unit, you will be able to:*

1. divide numbers that contain decimals.

Often you will have to work division problems that contain decimal points. When the dividend (the number being divided) is a decimal number, simply divide as usual. But you must be sure to insert a decimal point in the answer.

EXAMPLE 1   Divide 243.30 by 15

```
                    16.22——answer  (quotient)   STEP  1    Divide as usual.
divisor ——15)243.30——dividend
             15                                  STEP  2    Insert a decimal
             93                                              point in the
             90                                              answer directly
              3 3                                            above the
              3 0                                            decimal point in
               30                                            the dividend.
               30
```

When the divisor contains a decimal point, it is necessary to change the divisor to a whole number without a decimal point. To make the divisor a whole number, follow these two steps: (a) move the decimal point in the divisor as many places to the right as needed to make the divisor a whole number, and (b) move the decimal point in the dividend the same number of places to the right as there were decimal places in the divisor.

56   CHAPTER TWO

EXAMPLE 2   Divide 1,075.25 by 4.25

```
                      2 53    answer        STEP  1    Move the decimal point
 divisor  4.25 )1,075.25  dividend                     in the divisor two
                                                       places to the right to
                                                       make the whole number
           850                                         425.
           225 2
           212 5                           STEP  2    Move the decimal point
            12 75                                      in the dividend two
            12 75                                      places to the right to
                                                       make the whole number
                                                       107525.

                                           STEP  3    Divide as usual.

                                           STEP  4    Place the decimal point
                                                       in the answer above the
                                                       new decimal location in
                                                       the dividend.
```

If a division problem does not work out exactly, it is possible to increase the number of decimal places in the answer if you add additional zeros in the dividend. In the following problem, three zeros have been added to the original dividend 836.

EXAMPLE 3   Divide 836 by 32

```
              26.125    or 26.13 rounded off to two places
     32)836.000
        64
        196
        192
          4 0
          3 2
            80
            64
            160
            160
```

The answer for this example is 26.125. This answer may be left with three decimal places (thousandths), or it may be rounded off.

EXAMPLE 4   Divide 108 by 9.76

$$
\begin{array}{r}
11.0655 \\
9.76\overline{)108.000000} \\
\end{array}
$$

                                          976
                                         1040
                                          976
                                         6400
                                         5856
                                          5440
                                          4880
                                           5600
                                           4880
                                            720

To increase the accuracy of the answer in the above problem, we added six zeros to the right of the whole number 8 in the dividend. But regardless of the number of zeros added to the above problem, there will always be a remainder. In this situation, you must decide how many decimal places you want in your final answer. Once this decision is made, you can use the rules for rounding off (presented in Unit 3) to determine your final answer. In the above problem, the answer is

11.1 if rounded to tenths
11.07 if rounded to hundredths
11.066 if rounded to thousandths

HINT: You may want to review the material in Unit 3 on rounding numbers before completing this exercise.

Complete the Following Division Problems

Round off each answer to hundredths (two decimal places).

1.   **a**   **b**   **c**   **d**

$$
\begin{array}{r}
12.25 \\
34\overline{)416.50} \\
\end{array}
$$
              $26\overline{)551.20}$ $45\overline{)217.8}$ $56\overline{)691.04}$

             34
             76
             68
              85
              68
              170
              170

2.   **a**   **b**   **c**   **d**

$12.5\overline{)452.50}$ $10.65\overline{)26.625}$ $78.1\overline{)184.316}$ $22.4\overline{)45.61}$

3.   **a**   **b**   **c**   **d**

$1.45\overline{)56.25}$ $3.45\overline{)204}$ $124.5\overline{)256}$ $245.12\overline{)600}$

4. Scott Peterson earns $568 per week as a freight handler for Allied Steel. If he works 40 hours per week, how much does he make an hour?

5. Sandra Stewart is the accountant for Mills Paper Company. When reviewing the firm's monthly bank statement, she noticed an activity fee of $24.50 for processing the firm's checks. During the month, the firm wrote 98 checks. What does the bank charge to process each check?

6. On a recent four-day business trip to Chicago, Karen Southern spent $875.42. On average, what did each day in Chicago cost Karen? **HINT: An average is found by dividing the trip's total expense by the number of days.**

7. Reynolds Production Company paid the following phone charges during a five-month period:

January	$234.50
February	190.45
March	345.62
April	298.60
May	400.56

What was the firm's average dollar amount for phone service each month? **HINT: An average is found by dividing the total by the number of months.**

8. Martha Edwards is the office manager for Pro Real Estate. It is time to order ribbons for three printers that are used with the computers in the office. Company A advertises 12 ribbons for $40.45, and Company B offers 24 ribbons for $74.40. Which company offers the lowest price on this particular item?

9. J.D.'s Boottown and Western Wear pays $2,450 a month for a retail location in a strip shopping center. The store occupies 1,750 square feet. How much does a square foot cost in this location?

10. Jackson Delivery Service has just purchased a new Ford delivery van. The first week the van was driven 1,245.6 miles. At the end of the first week, the van had used 67.3 gallons of diesel fuel. How many miles per gallon did the van get during the first week?

11. During one Saturday, $8,958.50 was collected at the Surfside Amusement Park. If individual tickets cost $9.50 each, how many tickets were sold?

12. There are 109 vacation homes in The Barker Creek Recreational Development on Lake Michigan. Each owner is required to join the homeowner's association. This year the homeowner's association has approved a major project to build a retaining wall on the waterfront. The total cost for this project is $144,819.58. According to the association's by-laws, each homeowner must pay an equal share for all construction projects. How much should each homeowner pay?

UNIT 16	DIVISION SHORTCUTS

LEARNING OBJECTIVES   *After completing this unit, you will be able to:*

1. divide numbers using two shortcut methods.
2. estimate correct answers for division problems.

The following two shortcuts can be used to increase your speed and improve your accuracy when working division problems.

DIVIDING BY 10, 100, OR 1,000

When you divide a decimal number by 10, move the decimal point in the dividend one place to the left.

EXAMPLE 1   Divide 512.4 by 10

$$
\begin{array}{r}
51.24 \\
10\overline{)512.40} \\
\underline{50} \\
12 \\
\underline{10} \\
24 \\
\underline{20} \\
40 \\
\underline{40} \\
\end{array}
$$

Shortcut Method

$512.4 \div 10 = 51\underset{\smile}{2.4} = 51.24$

If you are dividing by 100, move the decimal point two places to the left. If you are dividing by 1,000, move the decimal point three places to the left.

This shortcut will also work when the dividend is a whole number. In this type of problem, it is assumed that the decimal point follows the digit in the units or ones column.

EXAMPLE 2   Divide 4,321 by 100

$$
\begin{array}{r}
43.21 \\
100\overline{)4,321.00} \\
\underline{4\ 00} \\
321 \\
\underline{300} \\
210 \\
\underline{200} \\
100 \\
\underline{100} \\
\end{array}
$$

Shortcut Method

$4,321 \div 100 = 43\underset{\smile}{21} = 43.21$

DIVIDING BY A MULTIPLE OF 10, 100, OR 1,000

You can use a variation of the previous shortcut to work a division problem where the divisor is a *multiple* of 10, 100, or 1,000. Examples of such multiples are 50 (10 × 5), 300 (100 × 3 = 300), and 8,000 (1,000 × 8 = 8,000).

EXAMPLE 3   Divide 2,736 by 300

$$
\begin{array}{r}
9.12 \\
300\overline{)2,736.00} \\
2,700 \\
\hline
360 \\
300 \\
\hline
600 \\
600 \\
\hline
\end{array}
$$

Shortcut Method

STEP 1 $2{,}736 \div 100 = 2{,}736 = 27.36$

STEP 2
$$
\begin{array}{r}
9.12 \\
3\overline{)27.36} \\
27 \\
\hline
3 \\
3 \\
\hline
6 \\
6 \\
\hline
\end{array}
$$

In this example, the following steps are required to find the answer:

STEP 1   Move the decimal point in the dividend as many places to the left as there are zeros in the divisor.

STEP 2   Divide the new dividend by the remaining digit.

ESTIMATING CORRECT ANSWERS

Before working division problems, you should take a few seconds to estimate the correct answer. As illustrated in the next example, the division shortcut procedures presented in this unit can help.

EXAMPLE 4   Estimate the Correct Answer for the Following Problem: 5,970 ÷ 495

5,970 = 6,000
495 =   500

STEP 1   Round each number to simpler numbers that you can divide easily.

Estimate
$$
\begin{array}{r}
12 \\
500\overline{)6,000} \\
500 \\
\hline
1000 \\
1000 \\
\hline
\end{array}
$$

STEP 2   Divide the rounded numbers.

STEP 3   Compare the estimate with the actual number.

Actual
$$
\begin{array}{r}
12.06 \\
495\overline{)5,970.00} \\
495 \\
\hline
1020 \\
990 \\
\hline
3000 \\
2970 \\
\hline
30 \\
\end{array}
$$

When the estimate for the above example is compared with the actual answer, there is only a small difference between the two numbers. A large difference would have indicated a possible error. In the case of a large difference, you should work the problem again to ensure that the actual answer is correct.

Using the Shortcuts Presented in This Unit, Complete the Following Division Problems

1.
a	b	c	d	e

$$\begin{array}{r}51.14\\10\overline{)511.40}\end{array}$$ $10\overline{)67.42}$ $100\overline{)345.2}$ $100\overline{)456.25}$ $1{,}000\overline{)2{,}345.1}$

2.
a	b	c	d	e

$$\begin{array}{r}4.56720\\1{,}000\overline{)4{,}567.20}\end{array}$$ $10\overline{)5{,}430}$ $10\overline{)345}$ $100\overline{)112}$ $100\overline{)5{,}600}$

3.
a	b	c	d	e

$1{,}000\overline{)23{,}400}$ $1{,}000\overline{)34{,}500}$ $30\overline{)660}$ $40\overline{)1280}$ $50\overline{)650}$

4.
a	b	c	d	e

$20\overline{)34.2}$ $30\overline{)57.0}$ $200\overline{)456.2}$ $700\overline{)1{,}610}$ $2{,}000\overline{)91{,}600}$

5. $1{,}246.56 \div 10 =$ 6. $23.45 \div 100 =$

7. $4.5621 \div 10 =$ 8. $2{,}678.2 \div 1{,}000 =$

For Division Problems 9 and 10, Estimate the Correct Answer, Determine the Actual Answer, and Compare the Estimate with the Actual Answer

9.   **a**   **b**

               Estimate   Estimate

$$\begin{array}{r}9\\11\overline{)99}\\\underline{99}\end{array}\qquad\begin{array}{r}10\\10\overline{)100}\end{array}\qquad 129\overline{)3870}$$

10. **a** **b**

Estimate Estimate

108)1810 2303)50600

11. Jane Petroni purchased 10 calculator ribbons for $31.50. What was the cost of each calculator ribbon?

$$\frac{\$3.15}{10)\$31.50}$$

12. During a 10-day period, Joe Highlands drove 720 miles while delivering new appliances for Wichita Appliance Sales and Repair. What is the average number of miles that he drove each day?

13. All-Star Office Supply advertised 1,000 file folders for $149.50. What is the cost of one file folder?

14. Detroit Paper Products, Inc., sells 30 small cardboard boxes for $23.70. What is the cost of one box?

15. Cross Town Cab of New York purchased 200 tires to replace worn tires on company cabs. The company received an invoice for $15,630. How much did each tire cost?

INSTANT REPLAY

UNIT	IMPORTANT POINTS TO REMEMBER	EXAMPLES
UNIT 9 Multiplying Basic Number Combinations	In a multiplication problem, the top number is called the multiplicand. The bottom number is called the multiplier. The answer is called the product.	$\begin{array}{r} 5 \\ \times\ 3 \\ \hline 15 \end{array}$ multiplicand multiplier product
UNIT 10 Multiplying Larger Number Combinations	With larger multiplication problems, multiply each digit in the top number by each digit in the bottom number. Then add each partial product to obtain the final product.	$\begin{array}{r} 24 \\ \times\ 27 \\ \hline 168 \\ 48 \\ \hline 648 \end{array}$ partial products
UNIT 11 Multiplying Numbers That Contain Decimals	To correctly place the decimal point follow these steps: (a) multiply the top number and bottom number just like whole numbers; (b) count the total number of decimal places in the top and bottom numbers; (c) starting with the last digit on the right side of the product, count out the number of decimal places (equal to the total in step b); (d) then insert the decimal point.	$\begin{array}{r} 2.34 \\ \times\ 0.12 \\ \hline 468 \\ 234 \\ \hline 0.2808 \end{array}$ two decimals two decimals four decimals
UNIT 12 Multiplication Shortcuts	When you multiply a number by 10, 100, or 1,000, move the decimal point in the top number as many places to the right as there are zeros in the bottom number. A variation of this shortcut can be used when the bottom number is a *multiple* of 10, 100, or 1,000 or when the top number, bottom number, or both contain end zeros.	$875.43 \times 100 = 875.43 =$ 87,543
UNIT 13 Dividing Basic Number Combinations	Division is the process of finding how many times one number is contained in another number. In a division problem, the number being divided is called the dividend. The divisor is the number that does the dividing. The answer is called the quotient.	$\text{divisor } 5\overline{)30}$ quotient dividend

continued from page 65

UNIT	IMPORTANT POINTS TO REMEMBER	EXAMPLES
UNIT 14 Dividing Larger Number Combinations	The rules presented in Unit 13 (Dividing Basic Number Combinations) can be used to work problems that contain larger numbers.	$\begin{array}{r} 124 \\ 35\overline{)4340} \\ \underline{35} \\ 84 \\ \underline{70} \\ 140 \\ \underline{140} \end{array}$
UNIT 15 Dividing Numbers That Contain Decimals	If the divisor does not contain a decimal place, the decimal point in the answer is placed directly above the decimal point in the dividend. When the divisor contains a decimal point, move the decimal point in the divisor to the extreme right. Then, move the decimal point in the dividend the same number of places to the right. Place the answer's decimal point directly above the *new* decimal point in the dividend.	$\begin{array}{r} 2.5 \\ 4.2\overline{)10.50} \\ \underline{8\,4} \\ 2\,10 \\ \underline{2\,10} \end{array}$
UNIT 16 Division Shortcuts	When you divide a decimal number by 10, 100, 1,000, and so on, move the decimal point in the dividend as many places to the left as there are zeros in the divisor. A variation of this shortcut can be used to work a division problem in which the divisor is a *multiple* of 10, 100, or 1,000.	$542.3 \div 10 = 542.3 = 54.23$

| CHAPTER 2 | MASTERY QUIZ |

DIRECTIONS   *Round each answer to two decimal places (or hundredths). (Each answer counts 5 points).*

1.
$$126 \\ \times\ 38$$

2.
$$171 \\ \times\ 89$$

3.
$$6.72 \\ \times\ 25$$

4.
$$24.55 \\ \times\ 0.36$$

5.
$$\$2{,}345.60 \\ \times\quad 1.23$$

6.
$$2{,}345.98 \\ \times\qquad 10$$

7.
$$24{,}562 \\ \times\qquad 100$$

8. $16\overline{)448}$

9. $125\overline{)16{,}500}$

10. $23\overline{)155.25}$

11. $14.7\overline{)345.45}$

12. $100\overline{)4{,}710.2}$

13. $400\overline{)3{,}680}$

14. Tavia Ali earns $485 a week. How much will she earn in a 52-week year?

15. The US government allows taxpayers to deduct $0.275 cents for each mile traveled in a car on business. If Mike Kontas drove 4,676.5 miles on business during the year, what is his tax deduction for business mileage?

16. Betty Peoples earns $6.45 an hour as a clerk for a local insurance firm. If she works 37.5 hours a week, what is her gross pay?

17. The management of Stratton Creek Apartments needs to purchase 225 yards of carpet to replace worn carpet in two apartments. If the carpet costs $7.98 a yard, what is the total cost of 225 yards?

18. There are four equal partners in the Apple Construction Company. Last year, the company earned $289,400. If the profits are divided equally among the four partners, how much will each partner receive?

19. For the month of January, the electric bill for Tri-County Appliance Store was $325.40. What was the firm's average cost per day for electricity?

20. On January 15, Janice Sneed bought 200 shares of Kellogg stock. The total cost for the stock including commission was $11,832. If the brokerage firm charged $232 commission to buy the stock, how much did each share of stock cost?

Fractions

CHAPTER PREVIEW

Now that we have reviewed the basic math functions—addition, subtraction, multiplication, and division—we are ready to move on to fractions. We begin by defining different types of fractions; then we discuss the procedures that are required to add, subtract, multiply, and divide numbers with fractions.

MATH TODAY

ARE FRACTIONS STILL USED IN BUSINESS?

Today, some people may wonder if fractions are still used in the business world. The answer is a definite *yes.* Consider the three business applications described below.

▌ *Payroll Calculations.* Many employees are paid an hourly wage for the first 40 hours worked in any week. They are then paid $1\frac{1}{2}$ times their hourly wage for time worked in excess of 40 hours. Mary Jensen works for Wendy's International—the third largest quick-service hamburger chain in the United States. She earns $6.80 an hour as an assistant shift manager. Last week she worked $43\frac{1}{2}$ hours and earned $307.70.

▌ *Stock Quotations.* All of the major stock exchanges report stock prices in dollars and fractional equivalents of $\frac{1}{8}$. For example, *The Wall Street*

Journal recently reported that the price for a share of stock for Black & Decker, a global manufacturer and marketer of power tools and household appliances, was $20\frac{1}{8}$, or $20.125.

▪ *Manufacturing Applications.* Today, most of the products that you purchase have gone through standardization—a process that establishes uniform, exacting specifications for manufactured goods. And while you may not realize the importance of standardization, consider the following example: Polaroid Corporation, the firm that is known for the instant camera and film, also produces videotapes for VCRs. The tape in each videotape cassette is exactly $\frac{1}{2}$ of an inch wide. If the tape were just a little wider—say $\frac{3}{4}$ of an inch wide—the videotape would not work in the millions of VCRs that have been sold since their invention.

Source: For more information, see *The Wall Street Journal*, January 10, 1994, p. C3; *Moody's Handbook of Common Stock* (New York, NY: Moody's Investors Service, Fall 1993); and Gary Hoover, *Hoover's Handbook of American Business 1994*, (The Reference Press, Inc., 1993), pp. 1116–7, 884–5.

UNIT 17	DIFFERENT TYPES OF FRACTIONS

LEARNING OBJECTIVES

After completing this unit, you will be able to:

1. define different types of fractions.
2. convert improper fractions to mixed numbers.
3. convert mixed numbers to improper fractions.

The payroll calculations, stock quotations, and the manufacturing applications described in the opening case are just a few of the business applications that involve the use of fractions. There are many others. You must be able to work with fractions to be successful in today's business world.

Let's begin with a definition. A **fraction** is a number that is used to describe part of a whole. For example, the fraction $\frac{1}{4}$ describes the one shaded part of a circle that has been divided into four parts.

EXAMPLE 1   The Fraction One-Fourth ($\frac{1}{4}$)

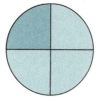

numerator ¼
denominator

The top number in a fraction is called the **numerator,** and the bottom number is called the **denominator.** Fractions can be classified as proper fractions, improper fractions, or mixed numbers. A **proper fraction** is a fraction whose numerator (top number) is smaller than its denominator (bottom number). A proper fraction is always less than one.

EXAMPLE 2   Proper Fractions

$$\frac{1}{2} \qquad \frac{4}{5} \qquad \frac{5}{6} \qquad \frac{7}{9} \qquad \frac{11}{13}$$

An **improper fraction** is a fraction whose numerator (top number) is equal to *or* larger than its denominator (bottom number).

EXAMPLE 3   Improper Fractions

$$\frac{2}{2} \qquad \frac{4}{3} \qquad \frac{6}{3} \qquad \frac{8}{6} \qquad \frac{15}{11}$$

A **mixed number** is a number that contains a whole number and a proper fraction.

EXAMPLE 4   Mixed Numbers

$$1\frac{2}{3} \qquad 5\frac{3}{5} \qquad 10\frac{4}{5} \qquad 21\frac{8}{11} \qquad 36\frac{12}{25}$$

In order to solve a business application, it is often necessary to convert an improper fraction to a mixed number or a mixed number to an improper fraction.

CONVERTING IMPROPER FRACTIONS TO MIXED NUMBERS

To change an improper fraction to a mixed number, first divide the numerator (top number) by the denominator (bottom number). Then place the remainder over the old denominator.

EXAMPLE 5   Convert $\frac{11}{4}$ to a Mixed Number

$$\frac{11}{4} = 4\overline{)11} \begin{array}{c} 2 \\ \\ \underline{8} \\ 3 \end{array} = 2\frac{3}{4}$$

STEP [1]   Divide the numerator (11) by the denominator (4).

STEP [2]   Place the remainder (3) over the old denominator.

CONVERTING MIXED NUMBERS TO IMPROPER FRACTIONS

To change a mixed number to an improper fraction, first multiply the whole number by the denominator (bottom number). Then add the numerator (top number) of the *original* fraction. The total will be the numerator of the *new* fraction. Place the new numerator over the original denominator.

EXAMPLE 6 Convert $5\frac{2}{3}$ to an Improper Fraction

$5\frac{2}{3} = \frac{(5 \times 3) + 2}{3} = \frac{17}{3}$

STEP [1] Multiply the whole number (5) by the denominator (3).

STEP [2] Add the original numerator (2) to the total obtained in Step 1 (15) to determine the new numerator.

STEP [3] Place the new numerator (17) over the original denominator (3).

Answer the Following Questions

1. In the fraction $\frac{4}{5}$, the number 4 is the _____ *numerator.* _____.

2. In the fraction $\frac{6}{7}$, the number 7 is the _____.

3. In the fraction, $\frac{19}{11}$, the number 11 is the _____.

4. In the fraction, $\frac{2}{17}$, the number 2 is the _____.

5. The fraction $2\frac{1}{4}$ is an example of a(n) _____.

6. The fraction $\frac{5}{8}$ is an example of a(n) _____.

7. The fraction $\frac{6}{5}$ is an example of a(n) _____.

8. The fraction $\frac{7}{8}$ is an example of a(n) _____.

9. The fraction $\frac{4}{2}$ is an example of a(n) _____.

10. The fraction $3\frac{5}{21}$ is an example of a(n) _____.

Convert the Following Improper Fractions to Mixed Numbers

11. a $\frac{21}{4} =$ _____ $5\frac{1}{4}$ _____   b $\frac{15}{2} =$ _____

12. a $\frac{14}{3} =$ _____   b $\frac{36}{5} =$ _____

13. a $\frac{27}{2} =$ _____   b $\frac{41}{6} =$ _____

14.   **a**   **b**

$\frac{73}{9} =$ _____ $\frac{61}{7} =$ _____

15.   **a**   **b**

$\frac{107}{15} =$ _____ $\frac{99}{10} =$ _____

Convert the Following Mixed Numbers to Improper Fractions

16.   **a**   **b**

$5\frac{1}{4} =$ _____ $\frac{21}{4}$ _____ $6\frac{1}{8} =$ _____

17.   **a**   **b**

$2\frac{5}{8} =$ _____ $6\frac{3}{5} =$ _____

18.   **a**   **b**

$7\frac{5}{12} =$ _____ $8\frac{1}{6} =$ _____

19.   **a**   **b**

$3\frac{2}{3} =$ _____ $10\frac{7}{9} =$ _____

20.   **a**   **b**

$12\frac{1}{7} =$ _____ $14\frac{2}{5} =$ _____

UNIT 18	REDUCING AND RAISING FRACTIONS

LEARNING OBJECTIVES   *After completing this unit, you will be able to:*

1. reduce fractions to their lowest terms.
2. raise fractions to higher terms.

REDUCING FRACTIONS TO LOWEST TERMS

When an answer to a problem contains a fraction, the fraction should be reduced to its lowest terms. Reducing a fraction does not change the value of a fraction. For example, the fraction $\frac{2}{4}$ reduced to lowest terms is $\frac{1}{2}$, yet the fraction $\frac{2}{4}$ and the fraction $\frac{1}{2}$ have the same numerical value. It is always easier to work with reduced fractions because a reduced fraction is made up of smaller numbers.

To reduce a fraction to its lowest terms, you must first find a common divisor—that is, a number that will divide into both the numerator and the denominator with no remainder. When reducing fractions, most people begin the search for a common divisor with the prime numbers. A **prime number** is a number that is *only* divisible by itself and the number 1. (The first seven prime numbers are 2, 3, 5, 7, 11, 13, or 17.) It should be pointed out that *not all numbers are prime numbers.* For example the number 8 is not a prime number because it can be divided by both 2 and 4 (not by just 1 and 8).

In Example 1, the number 5 is the first prime number that will divide evenly into both the numerator (5) and the denominator (20) with no remainder.

EXAMPLE 1   Reduce the Fraction $\frac{5}{20}$

$$\frac{5}{20} = \frac{5 \div 5}{20 \div 5} = \frac{1}{4}$$

When no number other than 1 will divide evenly into both the numerator and the denominator, a fraction has been reduced to its lowest terms.

With fractions that contain larger numbers, it is often necessary to check and make sure that a reduced fraction cannot be reduced even further.

EXAMPLE 2   Reduce the Fraction $\frac{20}{48}$

STEP **1**   By observation, determine the common divisor is 2.

STEP **2**   Divide the numerator (20) by the common divisor (2).

$\frac{20}{48} = \frac{20 \div 2}{48 \div 2} = \frac{10}{24}$ STEP **3**   Divide the denominator (48) by the common divisor (2).

STEP **4**   Check to see if the fraction $\frac{10}{24}$ can be reduced.

STEP **5**   By observation, determine the common divisor is 2.

$\frac{10}{24} = \frac{10 \div 2}{24 \div 2} = \frac{5}{12}$ STEP **6**   Divide the numerator (10) by the common divisor (2).

STEP $\boxed{7}$ Divide the denominator (24) by the common divisor (2).

STEP $\boxed{8}$ Check to see if the fraction $\frac{5}{12}$ can be reduced.

Of course, the fraction $\frac{20}{48}$ could have been reduced by dividing by 4 instead of 2. Choosing 4 would have eliminated the last four steps.

RAISING FRACTIONS TO HIGHER TERMS

It is sometimes necessary to raise some fractions to higher terms to complete addition and subtraction problems. To raise a fraction to a higher term, multiply both the numerator (top) and denominator (bottom) by the same number.

EXAMPLE 3   Raise the Fraction $\frac{3}{4}$ to $\frac{6}{8}$

$\frac{3}{4} = \frac{3 \times 2}{4 \times 2} = \frac{6}{8}$

STEP $\boxed{1}$ Multiply the numerator (3) by 2.

STEP $\boxed{2}$ Multiply the denominator (4) by 2.

Both $\frac{3}{4}$ and $\frac{6}{8}$ are equivalent in numerical value. When you already know the new denominator but need to know the new *numerator,* first divide the larger denominator by the smaller one. Then multiply that answer by the numerator in the original fraction.

EXAMPLE 4   Raise the Fraction $\frac{3}{8}$ to $\frac{12}{32}$

$\frac{3}{8} = \frac{?}{32} = \frac{12}{32}$

STEP $\boxed{1}$ Divide the larger denominator (32) by the smaller one (8).

STEP $\boxed{2}$ Multiply the answer from step 1 (4) by the numerator of the original fraction (3).

Note both $\frac{3}{8}$ and $\frac{12}{32}$ are equivalent in numerical value.

Reduce the Following Fractions to Their Lowest Terms

1.   **a**   **b**   **c**

$\frac{2}{6} = \underline{\ \ \frac{1}{3}\ \ }$ $\frac{5}{10} = \underline{\hspace{2cm}}$ $\frac{2}{8} = \underline{\hspace{2cm}}$

2.   **a**   **b**   **c**

$\frac{6}{8} = \underline{\hspace{2cm}}$ $\frac{10}{20} = \underline{\hspace{2cm}}$ $\frac{16}{36} = \underline{\hspace{2cm}}$

3.   **a**   **b**   **c**

$\frac{18}{36} = $ _____ $\frac{12}{60} = $ _____ $\frac{8}{24} = $ _____

4.   **a**   **b**   **c**

$\frac{21}{42} = $ _____ $\frac{15}{70} = $ _____ $\frac{54}{72} = $ _____

5.   **a**   **b**   **c**

$\frac{20}{42} = $ _____ $\frac{6}{66} = $ _____ $\frac{28}{86} = $ _____

6.   **a**   **b**   **c**

$\frac{54}{126} = $ _____ $\frac{84}{112} = $ _____ $\frac{70}{130} = $ _____

7.   **a**   **b**   **c**

$\frac{32}{128} = $ _____ $\frac{5}{70} = $ _____ $\frac{42}{63} = $ _____

Raise the Following Fractions to Higher Terms

8.   **a**   **b**   **c**

$\frac{2}{3} = \frac{8}{12}$ $\frac{4}{5} = \frac{}{25}$ $\frac{5}{6} = \frac{}{42}$

9.   **a**   **b**   **c**

$\frac{3}{7} = \frac{}{21}$ $\frac{2}{9} = \frac{}{27}$ $\frac{1}{4} = \frac{}{28}$

10.   **a**   **b**   **c**

$\frac{1}{6} = \frac{}{54}$ $\frac{2}{5} = \frac{}{15}$ $\frac{2}{7} = \frac{}{35}$

11.	a	b	c
	$\frac{5}{14} = \frac{}{42}$	$\frac{9}{17} = \frac{}{51}$	$\frac{4}{34} = \frac{}{136}$

12.	a	b	c
	$\frac{23}{36} = \frac{}{108}$	$\frac{11}{13} = \frac{}{65}$	$\frac{34}{35} = \frac{}{210}$

13.	a	b	c
	$\frac{45}{55} = \frac{}{275}$	$\frac{3}{32} = \frac{}{160}$	$\frac{14}{19} = \frac{}{152}$

14.	a	b	c
	$\frac{3}{10} = \frac{}{400}$	$\frac{5}{31} = \frac{}{124}$	$\frac{10}{21} = \frac{}{126}$

UNIT 19　ADDING AND SUBTRACTING PROPER FRACTIONS WITH THE SAME DENOMINATORS

LEARNING OBJECTIVES　*After completing this unit, you will be able to:*

1. add fractions that have the same denominators.
2. subtract fractions that have the same denominators.

ADDING PROPER FRACTIONS WITH THE SAME DENOMINATORS

When you add fractions that have the same denominators, add the numerators (top numbers) and place the total over the original denominator. REMEMBER: Reduce the fraction in your answer to its lowest terms.

EXAMPLE 1　Add $\frac{1}{8} + \frac{3}{8}$

$$\begin{array}{l} \frac{1}{8} \\ +\frac{3}{8} \\ \hline \frac{4}{8} = \frac{1}{2} \end{array}$$

STEP 1　Add the numerators.

STEP 2　Place the total over the original denominator.

STEP 3　If necessary, reduce the fraction in the answer to its lowest terms.

Sometimes the total of added fractions is an improper fraction. If it is, convert the answer to a mixed number as explained in Unit 17.

EXAMPLE 2   Add $\frac{3}{5} + \frac{1}{5} + \frac{4}{5}$

$\frac{3}{5}$
$+\frac{1}{5}$
$+\frac{4}{5}$
$\frac{8}{5} = 1\frac{3}{5}$

STEP $\boxed{1}$ Add the numerators.

STEP $\boxed{2}$ Place the total over the original denominator.

STEP $\boxed{3}$ Convert the answer to a mixed number.

STEP $\boxed{4}$ If necessary, reduce the fraction in the answer to its lowest terms.

SUBTRACTING PROPER FRACTIONS WITH THE SAME DENOMINATORS

To subtract proper fractions with the same denominators, subtract the smaller numerator from the larger numerator. Place the answer over the original denominator. Like all fraction problems, the answer should be reduced to its lowest terms.

EXAMPLE 3   Subtract $\frac{1}{8}$ from $\frac{7}{8}$

$\frac{7}{8}$
$-\frac{1}{8}$
$\frac{6}{8} = \frac{3}{4}$

STEP $\boxed{1}$ Subtract the smaller numerator from the larger numerator.

STEP $\boxed{2}$ Place the answer over the original denominator.

STEP $\boxed{3}$ If necessary, reduce the fraction in the answer to its lowest terms.

HINT: Before working the following problems, you may want to review the material presented in Unit 17 on converting improper fractions to mixed numbers and the material in Unit 18 on reducing fractions to lowest terms.

Add the Following Fractions

Be sure to (a) convert improper fractions to mixed numbers and (b) reduce the fraction in each answer to its lowest terms.

1.

a

$\frac{1}{3}$
$+\frac{2}{3}$
$\frac{3}{3} = 1$

b

$\frac{1}{5}$
$+\frac{2}{5}$

c

$\frac{4}{7}$
$+\frac{2}{7}$

2.

a

$\frac{4}{9}$
$+\frac{3}{9}$

b

$\frac{2}{3}$
$+\frac{1}{3}$

c

$\frac{3}{4}$
$+\frac{3}{4}$

3.   **a**

$$\begin{array}{r}\frac{1}{15}\\+\frac{8}{15}\\\hline\end{array}$$

   b

$$\begin{array}{r}\frac{3}{20}\\+\frac{11}{20}\\\hline\end{array}$$

   c

$$\begin{array}{r}\frac{7}{8}\\+\frac{5}{8}\\\hline\end{array}$$

4.   **a**

$$\begin{array}{r}\frac{1}{4}\\\frac{3}{4}\\+\frac{3}{4}\\\hline\end{array}$$

   b

$$\begin{array}{r}\frac{5}{6}\\\frac{1}{6}\\+\frac{5}{6}\\\hline\end{array}$$

   c

$$\begin{array}{r}\frac{3}{17}\\\frac{6}{17}\\+\frac{15}{17}\\\hline\end{array}$$

Complete the Following Subtraction Problems That Contain Fractions

Be sure to reduce each answer to its lowest terms.

5.   **a**

$$\begin{array}{r}\frac{7}{9}\\-\frac{5}{9}\\\hline\frac{2}{9}\end{array}$$

   b

$$\begin{array}{r}\frac{4}{5}\\-\frac{3}{5}\\\hline\end{array}$$

   c

$$\begin{array}{r}\frac{6}{7}\\-\frac{3}{7}\\\hline\end{array}$$

6.   **a**

$$\begin{array}{r}\frac{2}{3}\\-\frac{1}{3}\\\hline\end{array}$$

   b

$$\begin{array}{r}\frac{3}{4}\\-\frac{1}{4}\\\hline\end{array}$$

   c

$$\begin{array}{r}\frac{5}{6}\\-\frac{1}{6}\\\hline\end{array}$$

7.   **a**

$$\begin{array}{r}\frac{5}{12}\\-\frac{1}{12}\\\hline\end{array}$$

   b

$$\begin{array}{r}\frac{11}{13}\\-\frac{9}{13}\\\hline\end{array}$$

   c

$$\begin{array}{r}\frac{9}{14}\\-\frac{5}{14}\\\hline\end{array}$$

8. On Monday, $\frac{1}{8}$ of the fried pies produced by the Better Taste Pie Company were found to be defective. What fraction of the fried pies were acceptable?

$$1 - \frac{1}{8} = \frac{8}{8} - \frac{1}{8} = \frac{7}{8}$$

9. During the first week of September, $\frac{1}{10}$ of Martin Manufacturing's employees contributed to a special fund-raising drive for the local hospital. During the second week of September, an additional $\frac{3}{10}$ of the firm's employees contributed. To date, what fraction of the employees have contributed?

10. On Monday, a local hardware store sold $\frac{2}{5}$ of a special group of merchandise that had been marked down. On Tuesday, the store sold an additional $\frac{1}{5}$ of the merchandise. How much of the merchandise has been sold after two days?

11. Given the facts in problem 10, how much of the merchandise remains unsold?

12. Max Jones, Matthew Garcia, and Betty Schwartz formed a three-way partnership. According to the agreement, Max owns $\frac{5}{9}$ of the business, and Matthew owns $\frac{2}{9}$ of the business. What portion of the business does Betty own?

UNIT 20	FINDING THE LOWEST COMMON DENOMINATOR

LEARNING OBJECTIVE   *After completing this unit, you will be able to:*

1.   find the lowest common denominator for fraction problems.

The last unit explained how to add and subtract fractions with the same denominator. Sometimes you will need to add or subtract fractions with different denominators. When that happens, all denominators in a problem must be converted to a common denominator. It is best to find the smallest number that is evenly divisible by each denominator in the problem. This number is called the **lowest common denominator** (LCD). There are three ways to find a common denominator.

1.   Finding the LCD by observation: Sometimes it is possible to determine the LCD by just looking at the denominators in the problem.

EXAMPLE 1   Find the LCD for $\frac{1}{4}$, $\frac{1}{3}$, and $\frac{1}{6}$

$$\frac{1}{4} = \frac{}{12}$$

$$\frac{1}{3} = \frac{}{12}$$

$$\frac{1}{6} = \frac{}{12}$$

The number 12 is divisible by all of the denominators in the original problem and is therefore the LCD. When searching for an LCD, many people begin the search by testing multiples of the largest denominator against the other denominator(s) in the problem. The smallest number divisible by all denominators is the LCD.

2. Finding the LCD by using prime numbers: As discussed in Unit 18, a *prime number* is a number that is divisible only by itself and the number one. Examples of the first seven prime numbers are 2, 3, 5, 7, 11, 13, and 17. Example 2 shows how the prime number method works.

EXAMPLE 2   Find the LCD for $\frac{1}{4}$, $\frac{1}{10}$, $\frac{1}{16}$, and $\frac{1}{25}$

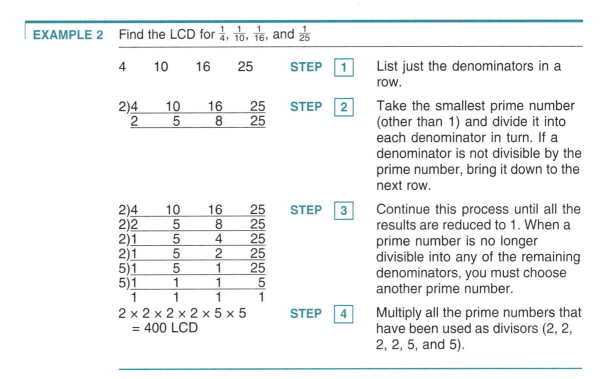

| 4 | 10 | 16 | 25 | **STEP** 1 | List just the denominators in a row. |

STEP 2 — Take the smallest prime number (other than 1) and divide it into each denominator in turn. If a denominator is not divisible by the prime number, bring it down to the next row.

STEP 3 — Continue this process until all the results are reduced to 1. When a prime number is no longer divisible into any of the remaining denominators, you must choose another prime number.

$2 \times 2 \times 2 \times 2 \times 5 \times 5$
$= 400$ LCD

STEP 4 — Multiply all the prime numbers that have been used as divisors (2, 2, 2, 2, 5, and 5).

---

3. Finding a common denominator by multiplying all the denominators: It is possible to find a common denominator by multiplying all the denominators in a problem.

EXAMPLE 3   Find a Common Denominator for $\frac{1}{6}$, $\frac{1}{8}$, and $\frac{1}{9}$

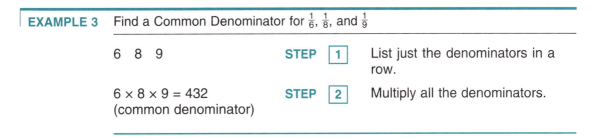

6 8 9   **STEP** 1   List just the denominators in a row.

$6 \times 8 \times 9 = 432$
(common denominator)   **STEP** 2   Multiply all the denominators.

This method will always produce a common denominator, *but* it may not be the *lowest* common denominator. Had you used either observation (method 1) or prime numbers (method 2), you would have found that the lowest common denominator for the preceding example was 72 and not 432. As you will see in the next two units, it is much easier to complete addition and subtraction problems if you use the LCD and not just any common denominator.

Find the Lowest Common Denominator (LCD) for the Following Fractions

1.   **a**   **b**

   Fractions $\frac{1}{8}$ $\frac{1}{3}$ Fractions $\frac{1}{3}$ $\frac{1}{4}$
   LCD _____ *24* _____   LCD _____

2.   **a**   **b**

   Fractions $\frac{1}{4}$ $\frac{1}{6}$ Fractions $\frac{1}{10}$ $\frac{1}{5}$
   LCD _____   LCD _____

3.   **a**   **b**

   Fractions $\frac{1}{6}$ $\frac{3}{5}$ Fractions $\frac{1}{8}$ $\frac{1}{6}$
   LCD _____   LCD _____

4.   **a**   **b**

   Fractions $\frac{1}{2}$ $\frac{1}{4}$ $\frac{1}{8}$ Fractions $\frac{1}{5}$ $\frac{1}{2}$ $\frac{5}{6}$
   LCD _____   LCD _____

5.   **a**   **b**

   Fractions $\frac{1}{6}$ $\frac{1}{8}$ $\frac{1}{12}$ Fractions $\frac{1}{9}$ $\frac{1}{4}$ $\frac{1}{18}$
   LCD _____   LCD _____

6.   **a**   **b**

   Fractions $\frac{1}{7}$ $\frac{1}{11}$ $\frac{1}{154}$ Fractions $\frac{1}{5}$ $\frac{1}{15}$ $\frac{1}{30}$
   LCD _____   LCD _____

7.   **a**   **b**

Fractions $\frac{1}{2}$ $\frac{1}{18}$ $\frac{1}{72}$ Fractions $\frac{1}{3}$ $\frac{1}{9}$ $\frac{1}{27}$

LCD _____   LCD _____

8.   **a**   **b**

Fractions $\frac{1}{5}$ $\frac{1}{2}$ $\frac{1}{80}$ Fractions $\frac{1}{7}$ $\frac{1}{21}$ $\frac{1}{105}$

LCD _____   LCD _____

9.   **a**   **b**

Fractions $\frac{1}{4}$ $\frac{1}{3}$ $\frac{1}{2}$ $\frac{1}{5}$ Fractions $\frac{1}{3}$ $\frac{1}{2}$ $\frac{1}{8}$ $\frac{1}{12}$

LCD _____   LCD _____

10.   **a**   **b**

Fractions $\frac{1}{10}$ $\frac{1}{2}$ $\frac{1}{5}$ $\frac{1}{8}$ Fractions $\frac{1}{20}$ $\frac{1}{15}$ $\frac{1}{6}$ $\frac{1}{8}$

LCD _____   LCD _____

UNIT 21	ADDING FRACTIONS AND MIXED NUMBERS

LEARNING OBJECTIVES   *After completing this unit, you will be able to:*

1. add fractions with different denominators.
2. add mixed numbers.

ADDING PROPER FRACTIONS WITH DIFFERENT DENOMINATORS

To add proper fractions with different denominators, you must (a) find the lowest common denominator; (b) raise all fractions to equivalent values with the new lowest common denominator; (c) add the new numerators, and place the total over the common denominator; and (d) if necessary, convert the improper fraction to a mixed number and reduce to lowest terms.

EXAMPLE 1 Add $\frac{1}{3} + \frac{5}{8} + \frac{5}{12}$

The LCD is 24.

STEP $\boxed{1}$ Find the lowest common denominator (LCD).

$\frac{1}{3} = \frac{8}{24}$

$\frac{5}{8} = \frac{15}{24}$

$\frac{5}{12} = \frac{10}{24}$

STEP $\boxed{2}$ Raise each fraction so that each denominator will be 24.

$$\frac{1}{3} = \frac{8}{24}$$
$$+\frac{5}{8} = \frac{15}{24}$$
$$+\frac{5}{12} = \frac{10}{24}$$
$$\frac{33}{24}$$

STEP $\boxed{3}$ Add the new numerators, and place the total over the common denominator.

$\frac{33}{24} = 1\frac{9}{24} = 1\frac{3}{8}$

STEP $\boxed{4}$ If necessary, convert the improper fraction to a mixed number. Reduce to lowest terms.

ADDING MIXED NUMBERS

To add mixed numbers, first add the whole numbers, and then add the fractions, finding the lowest common denominator if necessary. Finally, combine the whole number total and the fraction total.

EXAMPLE 2 Add $4\frac{5}{8} + 6 + 3\frac{1}{2}$

$$4\frac{5}{8}$$
$$+6$$
$$3\frac{1}{2}$$
$$\overline{13}$$

STEP $\boxed{1}$ Add the whole numbers.

$$4\frac{5}{8} \qquad \frac{5}{8}$$
$$+6 \quad = \quad +$$
$$+3\frac{1}{2} \qquad +\frac{4}{8}$$
$$\qquad \frac{9}{8} = 1\frac{1}{8}$$

STEP $\boxed{2}$ Find the LCD, and add the fractions. If the answer is an improper fraction, convert it to a mixed number.

$$13$$
$$+ \ 1\frac{1}{8}$$
$$\overline{14\frac{1}{8}}$$

STEP $\boxed{3}$ Combine the totals from steps 1 and 2.

Hint: Before completing this unit, you may want to review the following units:

1. Unit 17: Different Types of Fractions
2. Unit 18: Reducing and Raising Fractions
3. Unit 20: Finding the Lowest Common Denominator

Add the Following Fractions and Mixed Numbers

Be sure to (a) find the lowest common denominator, (b) convert improper fractions to mixed numbers, and (c) reduce the fraction in each answer to its lowest term.

1.   **a**   **b**

$$\frac{1}{4} \qquad\qquad \frac{3}{12} \qquad\qquad\qquad \frac{3}{5}$$
$$+\frac{2}{3} \qquad\qquad +\frac{8}{12} \qquad\qquad\qquad +\frac{1}{6}$$
$$\qquad\qquad\qquad \frac{11}{12}$$

2.   **a**   **b**

$$\frac{3}{4} \qquad\qquad\qquad\qquad\qquad\qquad \frac{1}{12}$$
$$+\frac{5}{8} \qquad\qquad\qquad\qquad\qquad\qquad +\frac{1}{3}$$

3.   **a**   **b**

$$\frac{1}{10} \qquad\qquad\qquad\qquad\qquad\qquad \frac{2}{5}$$
$$+\frac{3}{4} \qquad\qquad\qquad\qquad\qquad\qquad +\frac{1}{8}$$

4.   **a**   **b**

$$\frac{1}{4} \qquad\qquad\qquad\qquad\qquad\qquad \frac{7}{12}$$
$$+\frac{1}{3} \qquad\qquad\qquad\qquad\qquad\qquad +\frac{5}{9}$$
$$+\frac{1}{8} \qquad\qquad\qquad\qquad\qquad\qquad +\frac{3}{4}$$

5.   **a**   **b**

$$\frac{3}{8} \qquad\qquad\qquad\qquad\qquad\qquad \frac{5}{6}$$
$$+\frac{2}{10} \qquad\qquad\qquad\qquad\qquad\qquad +\frac{4}{5}$$
$$+\frac{1}{5} \qquad\qquad\qquad\qquad\qquad\qquad +\frac{3}{4}$$

6.   **a**   **b**

$$4\frac{3}{4} \qquad\qquad\qquad\qquad\qquad\qquad 3\frac{1}{2}$$
$$+2\frac{5}{7} \qquad\qquad\qquad\qquad\qquad\qquad +2\frac{5}{12}$$

7. a

$$10\frac{1}{5}$$
$$+12\frac{3}{4}$$

b

$$8\frac{3}{10}$$
$$+18\frac{2}{5}$$

8. a

$$41\frac{4}{5}$$
$$+\ 8\frac{9}{11}$$

b

$$10\frac{7}{8}$$
$$+\ 2\frac{4}{9}$$

9. a

$$4\frac{3}{4}$$
$$+5\frac{3}{11}$$
$$+6\frac{3}{22}$$

b

$$14\frac{5}{7}$$
$$+\ 6\frac{2}{3}$$
$$+\ 5\frac{1}{3}$$

10. At the close of trading on Monday, the price for a share of General Motors stock was $62\frac{1}{4}$. During trading on Tuesday, the stock increased in value by $\frac{7}{8}$ of a dollar. What was a share of General Motors worth at the end of the day on Tuesday?

$$62\frac{1}{4} = 62\frac{2}{8}$$
$$+\ \ \frac{7}{8} \ \ \ \ +\ \frac{7}{8}$$
$$62\frac{9}{8} = 63\frac{1}{8}$$

11. Mary Lopez worked the following hours: Monday, $8\frac{1}{4}$ hours; Tuesday, off; Wednesday $9\frac{3}{4}$ hours; Thursday, $6\frac{1}{4}$ hours; Friday, 8 hours; and Saturday, $9\frac{1}{2}$ hours. How many hours did she work during this week?

12. The delivery truck for Pronto Delivery Service used $4\frac{1}{2}$ gallons of gas on Monday. On Tuesday, it took $11\frac{3}{10}$ gallons of gas to fill the tank. On Wednesday, it took $17\frac{3}{4}$ gallons of gas to fill the tank. How many gallons of gasoline did Pronto Delivery purchase during the first three days of the week?

13. On Friday, Marty's Barbeque sold the following items: $42\frac{3}{8}$ pounds of beef, $24\frac{3}{4}$ pounds of ham, and $8\frac{1}{3}$ pounds of ribs. How many pounds of meat did the restaurant sell?

Personal Math Problem

For the past three months, Sandra and Bob Hernandez have been on a diet. Sandra lost the following amounts: June—$3\frac{3}{4}$ pounds; July—$4\frac{1}{8}$ pounds; and August—$3\frac{3}{10}$ pounds. Bob lost the following amounts: June—$1\frac{7}{10}$ pounds; July—$3\frac{1}{3}$ pounds; and August—$2\frac{4}{5}$ pounds.

14. How many pounds did Sandra lose during the three-month period?

15. How many pounds did Bob lose during the three-month period?

16. Based on your calculations, what should this couple do if their goal was to lose a combined total of 20 pounds? _____

_____

UNIT 22	SUBTRACTING FRACTIONS AND MIXED NUMBERS

LEARNING OBJECTIVES   *After completing this unit, you will be able to:*

1. subtract fractions with different denominators.
2. subtract mixed numbers.

SUBTRACTING PROPER FRACTIONS WITH DIFFERENT DENOMINATORS

To subtract proper fractions with different denominators, you must (a) find the lowest common denominator, (b) raise all fractions to equivalent values with the new lowest common denominator, (c) subtract the smaller numerator from the larger numerator, and (d) place the answer over the common denominator. Remember to reduce the answer to lowest terms if necessary.

EXAMPLE 1   Subtract $\frac{1}{4}$ from $\frac{6}{15}$

The LCD is 60.

$\frac{6}{15}$ $\frac{24}{60}$
$-\frac{1}{4}$ $-\frac{15}{60}$

STEP 1   Find the lowest common denominator (LCD).

STEP 2   Raise each fraction so that each denominator is the LCD (60).

$\frac{6}{15} = \frac{24}{60}$
$-\frac{1}{4}$ $-\frac{15}{60}$
          $\frac{9}{60}$

STEP 3   Subtract the smaller numerator from the larger numerator, and place the answer over the common denominator.

$\frac{9}{60} = \frac{3}{20}$

STEP 4   If necessary, reduce the answer to lowest terms.

SUBTRACTING MIXED NUMBERS

Subtracting mixed numbers is similar to subtracting proper fractions, with one exception: It may be necessary to borrow from a whole number if the numerator in the bottom number is larger than the numerator in the top number. Borrowing involves taking from the whole number and adding it—written as an improper fraction—to the fraction part of a mixed number.

EXAMPLE 2   Subtract $3\frac{5}{6}$ from $5\frac{1}{8}$

The LCD is 24.

$5\frac{1}{8} \rightarrow 5\frac{3}{24}$
$-3\frac{5}{6} \rightarrow -3\frac{20}{24}$

STEP 1   Find the lowest common denominator (LCD).

STEP 2   Raise each fraction so that each denominator is the LCD (24).

$5\frac{1}{8} \rightarrow 5\frac{3}{24}$
$-3\frac{5}{6} \rightarrow -3\frac{20}{24}$

STEP 3   If the bottom fraction ($\frac{20}{24}$) cannot be subtracted from the top fraction ($\frac{3}{24}$), borrow 1 from the whole number adjacent to the top fraction (5). Convert the borrowed 1 to a fraction with the LCD as its denominator ($1 = \frac{24}{24}$), and add it to the original numerator ($\frac{3}{24}$).

$4\frac{27}{24}$
$-3\frac{20}{24}$
$1\frac{7}{24}$

STEP 4   Subtract the whole numbers, and then subtract the numerators of the fractions.

$1\frac{7}{24}$ cannot be reduced

STEP 5   Check to see if the fraction can be reduced.

Hint: Before completing this unit, you may want to review the following units:

1. Unit 7: The Borrowing Process
2. Unit 18: Reducing and Raising Fractions
3. Unit 20: Finding the Lowest Common Denominator

Complete the Following Subtraction Problems

Be sure to (a) find the lowest common denominator and (b) reduce the fraction in each answer to its lowest term.

1. **a** $\dfrac{3}{5}$ $-\dfrac{1}{4}$ $\dfrac{12}{20}$ $\dfrac{5}{20}$ $\dfrac{7}{20}$ **b** $\dfrac{3}{4}$ $-\dfrac{1}{2}$

2. **a** $\dfrac{7}{8}$ $-\dfrac{2}{9}$ **b** $\dfrac{7}{10}$ $-\dfrac{2}{5}$

3. **a** $\dfrac{5}{6}$ $-\dfrac{1}{3}$ **b** $\dfrac{7}{8}$ $-\dfrac{1}{3}$

4. **a** $\dfrac{8}{9}$ $-\dfrac{1}{5}$ **b** $\dfrac{7}{9}$ $-\dfrac{1}{3}$

5. **a** $7\dfrac{5}{8}$ $-3\dfrac{1}{4}$ **b** $10\dfrac{5}{16}$ $-2\dfrac{1}{4}$

6. **a** $12\dfrac{1}{4}$ $-5\dfrac{3}{8}$ **b** $34\dfrac{2}{3}$ $-6\dfrac{8}{9}$

7. **a**

$$14\frac{1}{2}$$
$$-\ 6\frac{1}{8}$$

    b

$$20$$
$$-11\frac{2}{7}$$

8. **a**

$$118\frac{5}{6}$$
$$-\ 90\frac{10}{11}$$

    b

$$430\frac{5}{24}$$
$$-\ 124\frac{11}{12}$$

9. Last week, the 120 workers at Cameron Machine Works were asked to vote on a new union contract. If $\frac{3}{8}$ of the workers voted for the union, what fraction of the workers voted against the union?

10. For Denver-based Mountain Terrain Skiwear, Inc., $\frac{1}{3}$ of total sales revenues received by the firm are used to purchase merchandise for resale. $\frac{1}{2}$ of total sales revenues are used to pay operating expenses. Given this information, what fraction of total sales revenues remains for profits?

Hint: Sales Revenues – Cost of Merchandise Sold – Operating Expenses = Profit or Loss.

11. At the beginning of the month, Nathaniel Brown had accumulated $38\frac{3}{8}$ days of sick leave. Because he was sick with the flu, he missed $5\frac{1}{2}$ workdays during the month. How much sick leave does Mr. Brown still have?

12. At the end of trading on Thursday, a share of stock in IBM was selling for $\$104\frac{3}{8}$. At the end of trading on Friday, the share of stock was selling for $\$97\frac{3}{4}$. How much did the stock drop in value?

13. Great Lakes Development and Land Company purchased 91 acres of land on the north side of town. The company planned to use $68\frac{5}{6}$ acres of the land for a shopping center. The remainder of the property will be given to the city for a park. How much land will be given to the city?

Critical Thinking Problem

Nancy Vega is trying to decide which of two investments she should make. The first investment is a certificate of deposit at a local credit union that pays $3\frac{1}{2}$ percent interest annually. The second investment is a three-way partnership in a small retail clothing store. The original owner of the store has retained a $\frac{3}{5}$ ownership interest in the store. A second owner has a $\frac{1}{4}$ ownership interest in the store.

14. If Ms. Vega invests in the business, what fraction of the business will she own?

15. If you were Ms. Vega, which investment would you choose? Why? _____

_____

_____

_____

_____

UNIT 23	MULTIPLYING FRACTIONS AND MIXED NUMBERS

LEARNING OBJECTIVES   *After completing this unit, you will be able to:*

1. multiply fractions.
2. multiply mixed numbers.

MULTIPLYING PROPER FRACTIONS

Compared with adding and subtracting fractions, multiplying fractions is relatively simple because you do not need to find a common denominator. Multiplying fractions involves three steps: (a) multiplying the numerators of the fractions, (b) multiplying the denominators of the fractions, and (c) reducing the answer to its lowest terms.

EXAMPLE 1   Multiply $\frac{3}{4} \times \frac{2}{5}$

$\frac{3}{4} \times \frac{2}{5} = \frac{6}{?}$	STEP 1	Multiply the numerators of the fractions.
$\frac{3}{4} \times \frac{2}{5} = \frac{6}{20}$	STEP 2	Multiply the denominators of the fractions.
$\frac{6}{20} = \frac{3}{10}$	STEP 3	Reduce the answer to its lowest terms.

THE MULTIPLICATION SHORTCUT

It is possible to simplify the multiplication process if the numerator of one fraction and the denominator of another fraction can be divided evenly. Simply reduce the particular numerator and denominator and pencil in the answer you obtain by dividing them by the same number. Once this process is completed, multiply as usual. This shortcut does not change the answer, but, in most cases, it does eliminate the need to reduce your answer to lowest terms.

EXAMPLE 2   Multiply $\frac{2}{3} \times \frac{1}{5} \times \frac{3}{4}$

$\frac{\cancel{2}}{\cancel{3}_1} \times \frac{1}{5} \times \frac{\cancel{3}^1}{4}$	STEP 1	Cross out the numerator 3 and the denominator 3 because both are divisible by 3.
$\frac{\cancel{2}^1}{\cancel{3}_1} \times \frac{1}{5} \times \frac{\cancel{3}^1}{\cancel{4}_2}$	STEP 2	Cross out the numerator 2 and change the denominator 4 to 2 because both are divisible by 2.
$\frac{\cancel{2}^1}{\cancel{3}_1} \times \frac{1}{5} \times \frac{\cancel{3}^1}{\cancel{4}_2} = \frac{1}{10}$	STEP 3	Multiply as usual.

MULTIPLYING MIXED NUMBERS

RULE: To complete a multiplication problem that contains mixed numbers, it is necessary to convert all mixed numbers to improper fractions.

Once mixed numbers have been converted to improper fractions, multiply as usual, using the shortcut described above if possible. If the answer is an improper fraction, convert it to a mixed number.

EXAMPLE 3   Multiply $2\frac{2}{9} \times 4\frac{1}{5} \times 2$

$2\frac{2}{9} \times 4\frac{1}{5} \times 2$ $\frac{20}{9} \times \frac{21}{5} \times \frac{2}{1}$	STEP 1	Convert the mixed numbers to improper fractions.
$\frac{\cancel{20}^4}{\cancel{9}_3} \times \frac{\cancel{21}^7}{\cancel{5}_1} \times \frac{2}{1}$	STEP 2	Use the multiplication shortcut.
$\frac{\cancel{20}^4}{\cancel{9}_3} \times \frac{\cancel{21}^7}{\cancel{5}_1} \times \frac{2}{1} = \frac{56}{3}$	STEP 3	Multiply as usual.
$\frac{56}{3} = 18\frac{2}{3}$	STEP 4	If necessary, convert the improper fraction to a mixed number and reduce the fraction to lowest terms.

Before multiplying, any whole number in a multiplication problem must be converted to a fraction by placing the whole number over the number 1. Note that in the above example, the whole number 2 has been converted to the improper fraction $\frac{2}{1}$.

Complete the Following Multiplication Problems
Be sure to convert any mixed number to an improper fraction before multiplying and reduce any fractions to lowest terms.

1. **a** **b**

$\frac{2}{3} \times \frac{3}{4} = \frac{1}{2}$ $\frac{1}{5} \times \frac{5}{6} =$

2. **a** **b**

$\frac{5}{7} \times \frac{2}{5} =$ $\frac{1}{8} \times \frac{4}{5} =$

3. **a** **b**

$\frac{8}{9} \times \frac{3}{11} =$ $\frac{3}{20} \times \frac{5}{21} =$

4. **a** **b**

$\frac{2}{3} \times \frac{1}{2} \times \frac{3}{4} =$ $\frac{5}{6} \times \frac{3}{8} \times \frac{7}{9} =$

5. **a** **b**

$\frac{1}{10} \times \frac{5}{6} \times \frac{7}{8} =$ $\frac{3}{10} \times \frac{2}{9} \times \frac{1}{11} =$

6. **a** **b**

$1\frac{5}{8} \times 3\frac{1}{2} =$ $10\frac{2}{3} \times 4\frac{1}{2} =$

7.

a

$2\frac{3}{7} \times 1\frac{4}{5} =$

b

$8\frac{5}{6} \times 5\frac{3}{8} =$

8.

a

$12\frac{1}{4} \times 5\frac{1}{2} =$

b

$16\frac{1}{4} \times 4\frac{1}{3} =$

9.

a

$7\frac{5}{11} \times 2\frac{1}{2} =$

b

$24\frac{1}{4} \times 2\frac{2}{3} \times 4 =$

10. According to financial planners, your home mortgage payment should not exceed $\frac{1}{4}$ of your take-home pay. If your monthly take-home pay is $2,200, what is the maximum amount that you can afford to pay for a home mortgage payment?

$$\frac{1}{\underset{1}{\cancel{4}}} \times \frac{\overset{\$550}{\cancel{\$2,200}}}{1} = \$550$$

11. Jack Jackson, Martha Black, and Sam Tishler formed a three-way partnership. According to the partnership agreement Jack Jackson owns $\frac{1}{2}$ of the business; Martha Black owns $\frac{1}{3}$ of the business; and Sam Tishler owns $\frac{1}{6}$ of the business. If the firm's profits for the year are $54,000, how much does each partner receive?

Jack   **Martha**   **Sam**

12. Expenses for the Southwest Art Gallery total $42,600. According to the firm's accountant, $\frac{5}{8}$ of the firm's expenses are classified as marketing expenses. What is the dollar amount of the marketing expenses?

13. During the month of March, Becky Garrot called on an average of $3\frac{4}{5}$ customers a day. If she worked 20 days during the month, how many customers did she call on?

14. Each of the five employees at the Iowa Speed Skate Company produces $11\frac{1}{2}$ pairs of skates an hour. Assuming that all five employees work eight hours a day, how many pairs of skates are produced in a five-day week?

15. On a recent business trip, Matt Johnson averaged $14\frac{1}{4}$ miles per gallon. When he stopped for gas, it took $4\frac{2}{3}$ gallons of gas to fill the car's tank. How many miles did he drive on this tank of gas?

Personal Math Problems

When Martin Martinez received his $353 paycheck, he was sure that some-one in the payroll office had made a mistake. By his calculations, he had worked $8\frac{1}{2}$ hours on Monday, 8 hours on Tuesday, $9\frac{1}{4}$ hours on Wednesday, $10\frac{1}{2}$ hours on Thursday, and $10\frac{1}{4}$ hours on Friday.

16. At $8 an hour with time-and-a-half for all hours in excess of 40 hours, how much should Mr. Martinez have received?

17. What is the dollar difference between what Mr. Martinez should have been paid and the amount he was paid?

18. If you were Mr. Martinez, what would you do in this situation? _____

_____

_____

_____

_____

UNIT 24	DIVIDING FRACTIONS AND MIXED NUMBERS

LEARNING OBJECTIVES   *After completing this unit, you will be able to:*

1. divide fractions.
2. divide mixed numbers.
3. convert fractions to decimals.

DIVIDING PROPER FRACTIONS

RULE: When dividing proper fractions, it is necessary to invert the divisor.

Dividing proper fractions is similar to multiplying fractions. The difference is that first you must invert the divisor—that is, turn it upside down so that the numerator becomes the denominator and vice versa. Then you multiply the fractions. Do not use the multiplication shortcut described in Unit 23 until *after* you invert the divisor. If your answer is an improper fraction, convert it to a mixed number. Then, as always, check to see if the fraction can be reduced.

EXAMPLE 1   Divide $\frac{3}{16}$ by $\frac{3}{8}$

$\frac{3}{16} \div \frac{3}{8}$ $\frac{3}{16} \times \frac{8}{3}$	**STEP**	1	Invert (turn upside down) the divisor and change the division sign to a multiplication sign.
$\frac{\overset{1}{3}}{\underset{2}{16}} \times \frac{\overset{1}{8}}{\underset{1}{3}}$	**STEP**	2	Use the multiplication shortcut.
$\frac{\overset{1}{3}}{\underset{2}{16}} \times \frac{\overset{1}{8}}{\underset{1}{3}} = \frac{1}{2}$	**STEP**	3	Multiply the fractions.
$\frac{1}{2}$ cannot be reduced	**STEP**	4	Reduce to lowest terms, if necessary.

DIVIDING MIXED NUMBERS

To work a division problem with mixed numbers, it is necessary to convert all mixed numbers to improper fractions. Once you have converted all numbers to improper fractions, follow the same procedures for dividing proper fractions.

EXAMPLE 2   Divide $6\frac{3}{4}$ by $2\frac{2}{5}$

$6\frac{3}{4} \div 2\frac{2}{5}$ $\frac{27}{4} \times \frac{12}{5}$	**STEP**	1	Convert all mixed numbers to improper fractions.
$\frac{27}{4} \div \frac{12}{5}$ $\frac{27}{4} \times \frac{5}{12}$	**STEP**	2	Invert (turn upside down) the divisor. Change the division sign to a multiplication sign.
$\frac{\overset{9}{27}}{4} \times \frac{5}{\underset{4}{12}}$	**STEP**	3	Use the multiplication shortcut.

$$\frac{\overset{9}{\cancel{27}}}{4} \times \frac{5}{\underset{4}{\cancel{12}}} = \frac{45}{16}$$ **STEP** 4   Multiply the fractions.

$$\frac{45}{16} = 2\frac{13}{16}$$ **STEP** 5   Convert the improper fraction to a mixed number and reduce if necessary.

If either number in the division problem is a whole number, you can form an improper fraction by putting the whole number over 1. That is, $16 = \frac{16}{1}$.

EXAMPLE 3   Divide 16 by $3\frac{3}{5}$

$$16 \div 3\frac{3}{5}$$
$$\frac{16}{1} \div \frac{18}{5}$$ **STEP** 1   Convert all mixed numbers to improper fractions.

$$\frac{16}{1} \div \frac{18}{5}$$
$$\frac{16}{1} \times \frac{5}{18}$$ **STEP** 2   Invert (turn upside down) the divisor. Change the division sign to a multiplication sign.

$$\frac{\overset{8}{\cancel{16}}}{1} \times \frac{5}{\underset{9}{\cancel{18}}}$$ **STEP** 3   Use the multiplication shortcut.

$$\frac{\overset{8}{\cancel{16}}}{1} \times \frac{5}{\underset{9}{\cancel{18}}} = \frac{40}{9}$$ **STEP** 4   Multiply the fractions.

$$\frac{40}{9} = 4\frac{4}{9}$$ **STEP** 5   Convert the improper fraction to a mixed number and reduce if necessary.

CONVERTING FRACTIONS TO DECIMALS

Sometimes, it is easier or necessary to convert a fraction to its decimal equivalent. To do so, divide the numerator (top number) by the denominator (bottom number).

EXAMPLE 4   Convert $\frac{1}{8}$ to its Decimal Equivalent

$$\begin{array}{r} 0.125 \\ 8\overline{)1.000} \\ \underline{8} \\ 20 \\ \underline{16} \\ 40 \\ \underline{40} \end{array}$$

To improve the accuracy of your work, you may want to carry out your answers to at least thousandths (three decimal places) when converting fractions to decimals. Notice in Example 5, that five zeros were added to the numerator 2 to increase the accuracy of the answer.

EXAMPLE 5   Convert $\frac{2}{3}$ to Its Decimal Equivalent

$$
\begin{array}{r}
.66666 \\
3\overline{)2.00000} \\
\underline{18} \\
20 \\
\underline{18} \\
20 \\
\underline{18} \\
20 \\
\underline{18} \\
20 \\
\underline{18} \\
20 \\
\underline{18} \\
2 \\
\end{array}
$$

Even with this many zeros, the problem still does not work out evenly without a remainder. On occasion, you may choose to round off decimal answers to this type of problem. In Example 5, 0.66666 could be rounded to the following decimal values:

0.7 (tenths)
0.67 (hundredths)
0.667 (thousandths)
0.6667 (ten thousandths)

HINT: When converting fractions to decimals, most people use a calculator. The steps required to complete this type of division problem—and to complete addition, subtraction, and multiplication problems—with the help of a calculator are presented in Appendix A between Chapters 3 and 4.

Complete the Following Division Problems

Be sure to (a) convert any mixed numbers to improper fractions before dividing, (b) convert the final answer to a mixed number if necessary, and (c) reduce the fraction part of the final answer if possible.

1.   **a**   **b**

$\frac{1}{3} \div \frac{1}{4} = 1\frac{1}{3}$ $\frac{4}{5} \div \frac{2}{3} =$

$\frac{1}{3} \times \frac{4}{1} = \frac{4}{3} = 1\frac{1}{3}$

2.   **a**   **b**

$\frac{5}{8} \div \frac{3}{16} =$ $\frac{4}{9} \div \frac{1}{11} =$

3.   **a**

$\frac{7}{16} \div \frac{5}{6} =$

b

$\frac{9}{14} \div \frac{2}{7} =$

4.   **a**

$\frac{8}{15} \div \frac{3}{48} =$

b

$\frac{7}{8} \div \frac{1}{3} =$

5.   **a**

$2\frac{1}{4} \div 1\frac{1}{2} =$

b

$3\frac{1}{8} \div 1\frac{1}{16} =$

6.   **a**

$4\frac{5}{6} \div 2\frac{1}{3} =$

b

$5\frac{1}{6} \div 1\frac{3}{4} =$

7.   **a**

$12 \div 4\frac{3}{4} =$

b

$8 \div 2\frac{1}{3} =$

8.   **a**

$20 \div 2\frac{5}{6} =$

b

$8\frac{1}{4} \div 3\frac{3}{10} =$

9.   **a**

$130 \div \frac{4}{5} =$

b

$155 \div \frac{5}{6} =$

NOTE: For problems 10 through 14, round all answers to three decimal places.

10. Convert the fraction $\frac{1}{5}$ to its decimal equivalent.

11. Convert the fraction $\frac{2}{3}$ to its decimal equivalent.

12. Convert the fraction $\frac{5}{9}$ to its decimal equivalent.

13. Convert the fraction $\frac{1}{15}$ to its decimal equivalent.

14. Convert the fraction $\frac{22}{45}$ to its decimal equivalent.

15. Joe Barnes does installation work for National Auto Sound. On Monday, he was paid $26 for a job that took $3\frac{1}{4}$ hours to complete. How much does Joe earn per hour?

$$\frac{\$26}{1} \div \frac{13}{4} = \frac{\overset{2}{\$\cancel{26}}}{1} \times \frac{4}{\underset{1}{\cancel{13}}} = \$8$$

16. Midwest Land Development Company purchased a tract of land that contains 120 acres. The company wants to develop the property into a subdivision for homes. If each lot is $\frac{3}{4}$ of an acre, how many lots will there be in this project?

17. On Monday, Tom Boles filled his company truck with gasoline. During the day, he drove 162 miles. At the end of the day, it took $8\frac{1}{10}$ gallons to refill the truck's tank. How many miles per gallon did the truck get on Monday?

18. Westpoint Textiles purchases bolts of terry cloth that are 120 feet long. In order to make dish towels, the cloth is cut into $1\frac{1}{2}$ foot lengths. How many towels can be cut from one bolt of cloth?

19. Joan Littlefield purchases $12\frac{1}{2}$ acres of land for $26,100. How much did she pay for each acre of land?

20. In problem 19, what is the total purchase price for the property if Ms. Littlefield must pay $1,305 commission to the real estate agent?

UNIT	IMPORTANT POINTS TO REMEMBER	EXAMPLES
UNIT 17 Different Types of Fractions	A proper fraction is a number in which the top number is smaller than the bottom number. An improper fraction is a number in which the top number is equal to or larger than the bottom number. A mixed number contains both a whole number and a proper fraction. You can convert an improper fraction to a mixed number and a mixed number to an improper fraction.	$\frac{17}{8} = 2\frac{1}{8}$ $3\frac{3}{4} = \frac{15}{4}$
UNIT 18 Reducing and Raising Fractions	When an answer for a problem contains a fraction, the fraction part of the answer should be reduced to its lowest terms. To raise a fraction to a higher term, multiply both the numerator (top number) and the denominator (bottom number) by the same number.	$\frac{3}{12} = \frac{1}{4}$ $\frac{2}{3} = \frac{16}{24}$
UNIT 19 Adding and Subtracting Proper Fractions with the Same Denominators	When you add fractions that have the same denominators, add the numerators and place the total over the original denominator. To subtract proper fractions with the same denominators, subtract the smaller numerator from the larger and place the total over the original denominator. In both cases, check and see if your answer can be reduced.	$\frac{5}{8} + \frac{2}{8} = \frac{7}{8}$ $\frac{5}{6} - \frac{1}{6} = \frac{4}{6} = \frac{2}{3}$
UNIT 20 Finding the Lowest Common Denominator	The lowest common denominator (LCD) is the smallest number that is evenly divisible by each denominator in the problem. You can find the LCD by observation or by using prime numbers. You can also find a common denominator by multiplying all the denominators.	For the fractions $\frac{1}{4}, \frac{1}{3}, \frac{1}{8}, \frac{1}{12}$ the common denominator is 24.

continued from page 101 UNIT	IMPORTANT POINTS TO REMEMBER	EXAMPLES
UNIT 21 Adding Fractions and Mixed Numbers	To add proper fractions with different denominators, (a) find the LCD; (b) raise all fractions to their equivalent values with the LCD; and (c) add the new numerators, placing the total over the LCD. To add mixed numbers, (a) add the whole numbers, (b) add the fractions, and (c) combine the totals.	$\frac{3}{5} = \frac{12}{20}$ $+\frac{1}{4} = +\frac{5}{20}$ $\frac{17}{20}$ $2\frac{3}{8} = 2\frac{3}{8}$ $+1\frac{3}{4}\quad +1\frac{6}{8}$ $3\frac{9}{8} = 4\frac{1}{8}$
UNIT 22 Subtracting Fractions and Mixed Numbers	To subtract proper fractions with different denominators, (a) find the LCD, (b) raise each fraction to its equivalent value with the LCD, and (c) subtract the smaller numerator from the larger, placing the answer over the LCD. To subtract mixed numbers, it may be necessary to borrow from a whole number if the numerator in the bottom number is larger than the numerator in the top number.	$\frac{4}{5} = \frac{12}{15}$ $-\frac{1}{3} = -\frac{5}{15}$ $\frac{7}{15}$ $3\frac{1}{4} = \overset{2\ 5}{3\frac{1}{4}} = 2\frac{5}{4}$ $-1\frac{1}{2}\quad -1\frac{2}{4}\quad -1\frac{2}{4}$ $1\frac{3}{4}$
UNIT 23 Multiplying Fractions and Mixed Numbers	To multiply fractions, (a) multiply the numerators, (b) multiply the denominators, and (c) reduce the answer to its lowest terms. To complete a problem that contains mixed numbers, convert all mixed numbers to improper fractions before multiplying.	$\frac{1}{4} \times \frac{5}{6} = \frac{5}{24}$ $1\frac{2}{3} \times 3\frac{1}{9} =$ $\frac{5}{3} \times \frac{28}{9} = \frac{140}{27} = 5\frac{5}{27}$
UNIT 24 Dividing Fractions and Mixed Numbers	To divide proper fractions, (a) invert the divisor, (b) multiply the fractions, and (c) reduce the answer if necessary. To work a division problem with mixed numbers, convert all mixed numbers to improper fractions before dividing. It is possible to find the decimal equivalent for any proper fraction by dividing the top number by the bottom number.	$\frac{5}{8} \div \frac{1}{2} =$ $\frac{5}{8} \times \frac{\overset{1}{2}}{1} = \frac{5}{4} = 1\frac{1}{4}$ $2\frac{1}{10} \div 1\frac{1}{2} =$ $\frac{\overset{7}{21}}{10} \times \frac{\overset{1}{2}}{\underset{1}{3}} = \frac{7}{5} = 1\frac{2}{5}$ $\frac{3}{4} = 4\overline{)3.00}^{0.75}$ $\underline{28}$ 20 $\underline{20}$

CHAPTER 3   MASTERY QUIZ

DIRECTIONS   *Complete the following problems. Be sure to reduce all answers. (Each answer counts 5 points).*

1. In the fraction $\frac{5}{16}$, the number 16 is the _____.

2. In the fraction $\frac{3}{11}$, the number 3 is the _____.

3. The fraction $\frac{1}{3}$ is an example of a(n) _____ fraction.

4. The fraction $1\frac{5}{6}$ is an example of a(n) _____ number.

5. The fraction $\frac{10}{4}$ is an example of a(n) _____ fraction.

6. Reduce $\frac{24}{64}$ to lowest terms _____.

7. Raise $\frac{7}{15}$ to $\frac{?}{90}$ _____.

8. Find the LCD for $\frac{1}{3}$, $\frac{1}{9}$, $\frac{1}{18}$, and $\frac{1}{27}$. _____

9. Add $\frac{1}{5} + \frac{3}{8}$ = _____.

10. Add $2\frac{1}{4} + 3\frac{1}{8} + 5\frac{1}{2}$ = _____.

11. Subtract $\frac{7}{15} - \frac{1}{3}$ = _____.

12. Subtract $6\frac{1}{12} - 5\frac{1}{10}$ = _____.

13. Multiply $\frac{3}{5} \times \frac{1}{9} \times \frac{5}{12}$ = _____.

14. Multiply $2\frac{1}{17} \times 3\frac{1}{2}$ = _____.

15. Divide $\frac{7}{12} \div \frac{1}{2}$ = _____.

16. Divide $10\frac{3}{4} \div 2\frac{1}{2}$ = _____.

17. Convert $\frac{4}{9}$ to its decimal equivalent. _____.

18. Juan Garcia worked 9 hours on Monday, $8\frac{1}{2}$ hours on Tuesday, 7 hours on Wednesday, $10\frac{3}{4}$ hours on Thursday, and 6 hours on Friday. How many hours did he work during this one-week period?

19. Jones Meat Packing, Inc., sells sliced ham for $3 a pound. If the average restaurant orders $22\frac{2}{3}$ pounds of ham a week, what is the cost of the average order?

20. Last week, Mildred Prescott earned $435. If she worked $36\frac{1}{4}$ hours, how much does she earn each hour?

The Hand-Held Calculator

Many people rely on the hand-held calculator to complete the routine mathematical calculations required in daily life. With current technology, you can buy a small calculator that will add, subtract, multiply, and divide for less than $20.

Although there are many different brands and models to choose from, the material in this appendix provides basic instructions applicable to all hand-held calculators. Most hand-held calculators have the following keys:

All Clear Key	Clears all numbers in the calculator including the memory register.
Clear Key	Clears all numbers entered in the calculator except for numbers in memory.
CE Key	Clears an incorrect entry out of the keyboard before the +, −, ×, or ÷ key is depressed.
.	Decimal point key used to enter a decimal point in a number.
+ Key	Depressed after each number in an addition problem.
− Key	Depressed after the minuend (top number) is entered in a subtraction problem.
× Key	Depressed after each multiplicand when working a multiplication problem.
÷ Key	Depressed after the dividend is entered in a division problem.
= Key	Completes a mathematical problem.
% Key	Used to find the portion when a base is multiplied by a rate. Using this key eliminates the step of converting a percent to a decimal before multiplying.
+/− Key	Changes a positive number to a negative number or changes a negative number to a positive number.
M+ Key	Registers a positive number in the memory part of the calculator.
M− Key	Registers a negative number in the memory part of the calculator.
MRC Key	Sometimes referred to as the memory recall key—totals all numbers that have been stored in the memory part of the calculator; also clears the memory part of the calculator.

As in larger models of electronic calculators, the hand-held calculator will correctly place the decimal in the answer *if* the decimal is entered along with the numbers in a mathematical problem. Most hand-held calculators are equipped with a "floating" decimal. This means your answer will be decimally correct if you entered the decimal point along with the numbers in the problem.

ADDITION PROBLEMS

When working addition problems on a hand-held calculator, enter the first number on the keyboard, depress the + key, enter the next number on the keyboard, depress the + key, and continue this process until all numbers have been entered. When the last number has been entered into the machine and the = key depressed, the answer will be displayed on the calculator.

EXAMPLE 1   Add 67 + 145 + 389

Depress clear key

Enter 67   Depress   +
Enter 145   Depress   +
Enter 389   Depress   =

Read the answer   601

When decimals appear in a problem, set them along with the numbers. Then your calculator will place the decimal in the answer.

EXAMPLE 2   Add 2.45 + 3.67

Depress clear key

Enter 2.45   Depress   +
Enter 3.67   Depress   =

Read the answer   6.12

Addition Tryout Problems

NOTE: For all problems in this appendix that do not work out evenly, round each answer to two decimal places or hundredths.

1. Add 46 + 32   *78*

2. Add 257 + 453   _____

3. Add 5,782 + 4,789   _____

4. Add 3.57 + 5.76 + 1.12   _____

5. Add $21.47 + 4.56 + 123.40   _____

6. Add 2.4 + 457.29 + 5.63   _____

7. Add $2,340 + 45.75 + 2.23   _____

Note: Answers to odd numbered tryout problems are included in the back of the text in Appendix D.

SUBTRACTION PROBLEMS

To complete a subtraction problem on a hand-held calculator, enter the first number on the keyboard, depress the – key, enter the second number on the keyboard, depress the = key. The answer will be displayed on the calculator.

EXAMPLE 3 Subtract 76.50 – 32.45

Depress clear key

Enter 76.50   Depress   –
Enter 32.45   Depress   =

Read the answer   44.05

Subtraction Tryout Problems

8. Subtract 789 – 312   _477_

9. Subtract 1,245 – 489   _____

10. Subtract 12,348 – 2,224   _____

11. Subtract 145.86 – 65.42   _____

12. Subtract 3,210.4 – 1,178.9   _____

13. Subtract $456.11 – 106.82   _____

14. Subtract $6,700.26 – 3,556.90   _____

MULTIPLICATION PROBLEMS

To complete a multiplication problem on a hand-held calculator, enter the first number on the keyboard, depress the × key, enter the second number on the keyboard, depress the = key. The answer will be displayed on the calculator.

EXAMPLE 4 Multiply 32.48 × 12.3

Depress clear key

Enter 32.48   Depress   ×
Enter 12.3   Depress   =

Read the answer 399.504 or 399.50 (rounded to 2 places)

Multiplication Tryout Problems

15. Multiply 24 × 12   *288*

16. Multiply 256 × 940   _____

17. Multiply 3,456 × 821   _____

18. Multiply 32.5 × 7.82   _____

19. Multiply 311.6 × 34.68   _____

20. Multiply $19.95 × 32   _____

21. Multiply $129.99 × 1.25   _____

DIVISION PROBLEMS

To complete a division problem on a hand-held calculator, enter the first number on the keyboard, depress the ÷ key, enter the second number on the keyboard, depress the = key. The answer will be displayed on the calculator.

EXAMPLE 5   Divide 2,769.9 ÷ 140

Depress clear key

Enter 2,769.9   Depress   ÷
Enter   140   Depress   =

Read the answer 19.785 or 19.79 (rounded to 2 places)

Division Tryout Problems

22. Divide 248 ÷ 110   *2.25*

23. Divide 6,700 ÷ 125   _____

24. Divide 23.41 ÷ 1.78   _____

25. Divide 120.68 ÷ 56.40   _____

26. Divide $15.20 ÷ 16   _____

27. Divide $167.80 ÷ 12.90   _____

28. Divide $1,290.45 ÷ 128.50   _____

PERCENT KEY PROBLEMS

Most hand-held calculators have a special percent (%) key that eliminates the need for converting a percent (sometimes called the rate) into a decimal.

EXAMPLE 6   Find 20% of 80

Depress clear key

Enter 80   Depress   ×
Enter 20   Depress   %

Read the answer   16

NOTE: With *some* calculators, it may be necessary to depress the percent key, *then* the equals key (=) in order to obtain the final answer.

Percent Tryout Problems

29. Find 12% of 500   _____60_____

30. Find 25% of 1,000   _____

31. Find 15% of 800   _____

32. Find 32.5% of 2,000   _____

33. Find 24.6% of 820   _____

34. Find 56.7% of 110   _____

35. Find 12.3% of 220   _____

MEMORY KEY PROBLEMS

Most hand-held calculators can store individual answers with storage keys that are labeled M+ for storing positive answers and M– for storing negative answers.

EXAMPLE 7   Determine the grand total for the following problems:

2 × 3
4 × 5
6 × 7

Depress clear key

Enter 2   Depress   ×
Enter 3   Depress   M+
Enter 4   Depress   ×
Enter 5   Depress   M+
Enter 6   Depress   ×
Enter 7   Depress   M+
          Depress   MRC Key

Read the answer   68

It is also possible to subtract one answer from another answer by using memory keys.

EXAMPLE 8   Determine the grand total for the following problems:

4 × 6
3 × 7
−8 × 2

Depress clear key

Enter 4   Depress   ×
Enter 6   Depress   M+
Enter 3   Depress   ×
Enter 7   Depress   M+
Enter 8   Depress   ×
Enter 2   Depress   M−
          Depress   MRC Key

Read the answer   29

Memory Tryout Problems

36.   3 × 9
          14 × 2
          16 × 3   *103*

37. 12 × 62
        31 × 40
        28 × 56

38. 112 × 4.5
     89 × 2.6
   –12 × 3.4   _____

39. 120 × 5.6
   –31 × 4.1
   –10 × 2.5   _____

40.   23.4 × 5.2
   –11.4 × 2.5
   – 5.6 × 2.2   _____

MULTIPLE OPERATION PROBLEMS

Many business activities involve the combination of several mathematical functions to obtain the answer. For example, you may have to add to determine an intermediate answer and then subtract to determine the final answer.

EXAMPLE 9   Determine the correct answer for the following problem.

44 + 55 – 61

Depress clear key

Enter 44   Depress   +
Enter 55   Depress   –
Enter 61   Depress   =

Read the answer   38

EXAMPLE 10   Determine the correct answer for the following problem.

20 × 45 + 110

Depress clear key

Enter 20   Depress   ×
Enter 45   Depress   +
Enter 110   Depress   =
Read the answer   1,010

| EXAMPLE 11 | Determine the correct answer for the following problem. |

1,400 ÷ 20 − 25

Depress clear key

Enter 1,400   Depress   ÷
Enter   20   Depress   −
Enter   25   Depress   =

Read the answer   45

Multiple Operation Tryout Problems

41. 2,400 − 120 + 40 =   _2,320_

42. 456 + 240 − 300 =   _____

43. 2,900 − 400 + 350 =   _____

44. 130 × 34 − 285 =   _____

45. 3.45 × 2.26 + 10.40   _____

Percent: Portion, Rate, and Base

CHAPTER PREVIEW

This chapter begins with a discussion of simple conversions for percents, fractions, and decimals. Then it shifts attention to three formulas that can be used to find portion, rate, and base. The ability to work with these formulas is especially important today because it will enable you to solve many real-world business application problems.

MATH TODAY

INVESTORS "JUST DO IT" FOR NIKE

According to Moody's Investors Service, Nike, Inc., has established a sound foundation for long-term financial prosperity and increased profits. Why did a respected source go out on a limb and make these recommendations?

For starters, Nike is a leader in the design, manufacture, and distribution of athletic shoes and sports apparel for men, women, and children. Incorporated in 1968, Oregon-based Nike, Inc., started small but enjoyed phenomenal growth during the 1980s and 1990s. In 1992, the last year for which complete financial results are available, the company generated $3.9 billion in revenues—up from $1.2 billion in 1988. That's an annual increase of 45 percent. And the fact that the firm is almost debt free and has about $200 million in cash reserves only enhances its investment potential. Investors also realize the potential of a firm that holds more than 30 percent of the $6 billion US market for athletic shoes.

Nike's domestic sales of over 400 different types of athletic products account for 80 percent of the firm's sales revenues. International sales of the firm's products in over 60 different countries account for approximately 20 percent of sales revenues.

But before you run out and buy Nike stock, be warned. The US market for athletic shoes is expanding at about 5 percent a year—down from the phenomenal 15 percent annual growth experienced during the 1980s. According to the experts, this trend will affect Nike's short-term sales figures and profits.

To combat this problem, Nike has cut expenses, has reduced the workforce by approximately 7 percent, and is diversifying into retailing, entertainment, and sports management. Finally, Nike continues to spend millions on advertising. According to Nike's chairman and CEO, Philip Knight, the firm is banking on endorsements from basketball's Michael Jordan, baseball's Bo Jackson, football's Jerry Rice, and to a lesser extent several hundred other professional athletes to maintain Nike's number one position.

Source: For more information, see Gary Hoover, ed., *Hoover's Handbook of American Business 1994* (Austin, TX.: The Reference Press, Inc., 1993), pp. 814–815; *Moody's Handbook of Common Stocks,* (New York, NY: Moody's Investors Service, Fall 1993), Dori Jones Yang, "Nike's New Treads," *Business Week,* October 4, 1993, p. 38; David E. Thigpen, "Is Nike Getting Too Big for Its Shoes?" *Time,* April 26, 1993, p. 55; and Jim Impoco, "Nike Goes to the Full-Court Press," *U.S. News & World Report,* April 19, 1993, pp. 48–50.

UNIT 25 | CONVERTING PERCENTS, DECIMALS, AND FRACTIONS

LEARNING OBJECTIVES   *After completing this unit, you will be able to:*

1. convert percent to decimals.
2. convert decimals to percents.
3. convert common fractions to percents.
4. convert percents to common fractions.

Take a moment and look back at the financial data used to describe Nike, Inc. Nike sales increased at an annual rate of over 40 percent, Nike had more than 30 percent of the $6 billion US market for athletic shoes, domestic sales accounted for 80 percent of Nike's sales revenues, and so on. In fact, there are seven—count them—financial references that deal with percents. Obviously, the ability to work with percents is one of the most important skills a business person can have. We begin our discussion with material on converting percents, decimals, and fractions. Then we discuss how formulas for portion, rate, and base can be used to solve real-world business problems.

CONVERTING PERCENTS TO DECIMALS

A **percent** is a specific number that represents a part of 100. For example, 75 percent means 75 parts of 100. Percents are like decimals and fractions, which also represent parts of 100. Thus, it is possible to convert one number form to another number form. For example, to convert percents to decimals, first drop the percent sign and then move the decimal *two* places to the left.

EXAMPLE 1   Convert 24% to a Decimal

24% = 24% =

24% = .24% = .24

STEP 1   Drop the percent sign.

STEP 2   Move the decimal point two places to the left.

In Example 1, it is assumed that there is a decimal point to the right of the 4. If the percent is a single-digit whole number, it may be necessary to add a zero *before* the single digit and then move the decimal point two places to the left. For example, 4.2 percent becomes 0.042.

EXAMPLE 2   Convert 4.2% to a Decimal

4.2% = 4.2%

4.2% = .042% = .042

STEP 1   Drop the percent sign.

STEP 2   Move the decimal point two places to the left.

It is also possible to convert a percent that contains a fraction to a decimal. In Example 3, notice that $6\frac{3}{8}$ percent has been changed to the decimal 0.06375.

EXAMPLE 3   Convert $6\frac{3}{8}$% to a Decimal

$$6\frac{3}{8}\% \ = 8)\overline{3.000}^{\ .375} = 6.375\%$$
$$\underline{24}$$
$$60$$
$$\underline{56}$$
$$40$$

STEP 1   Convert the fraction to its decimal equivalent: Divide the numerator (3) by the denominator (8).

6.375% = 6.375%

6.375% = .06375% = .06375

STEP 2   Drop the percent sign.

STEP 3   Move the decimal point two places to the left.

> **CALCULATOR TIP**
>
> Depress clear key
> Enter 3   Depress ÷
> Enter 8   Depress =
> Read the answer 0.375

CONVERTING DECIMALS TO PERCENTS

To convert decimals to percents, reverse the process: First move the decimal point two places to the *right,* and then add a percent sign at the end of the number.

EXAMPLE 4   Convert 0.35 to a percent

0.35 = 0.35.   **STEP** 1   Move the decimal point two places to the right.

0.35 = 0.35% = 35%   **STEP** 2   Add a percent sign.

If the decimal has only a single digit to the right of the decimal point, place a zero after the single digit, then move the decimal point. For example, 0.2 becomes 20 percent.

CONVERTING COMMON FRACTIONS TO PERCENTS

It is possible to change a common fraction to a percent, but first you must convert the fraction to a decimal. Then, follow the preceding steps to convert the decimal to a percent. In other words:

1. Divide the top number by the bottom number to obtain a decimal.
2. Move the decimal point *two* places to the right.
3. Add a percent sign at the end of the number.

EXAMPLE 5   Convert $\frac{4}{5}$ to a Percent

$$\begin{array}{r} 0.8 \\ 5{\overline{)4.0}} \\ \underline{4\ 0} \end{array}$$

STEP 1   Divide the top number (4) by the bottom number (5).

0.8 = 0.80   **STEP** 2   Move the decimal point two places to the right.

0.8 = 0.80% = 80%   **STEP** 3   Add a percent sign.

CALCULATOR TIP

Depress clear key
Enter 4   Depress ÷
Enter 5   Depress =
Read the answer 0.8

CONVERTING PERCENTS TO COMMON FRACTIONS

To change a percent to a common fraction, first you must convert the percent to a decimal. Then you can convert the decimal to a common fraction. In other words:

1. Drop the percent sign.
2. Move the decimal point two places to the *left.*
3. Convert the decimal to a common fraction.
4. If necessary, reduce the fraction to its lowest terms.

EXAMPLE 6   Convert 75% to a Common Fraction

75% = 75   **STEP** 1   Drop the percent sign.

75% = .75 = .75   **STEP** 2   Move the decimal point two places to the left.

75% = .75 = $\frac{75}{100}$ **STEP** 3   Convert the decimal to a common fraction.

$\frac{75}{100} = \frac{3}{4}$ **STEP** 4   If necessary, reduce the fraction to lowest terms.

In step 3, .75 represents "75 hundredths"; therefore, the denominator in the fraction is 100.

HINT: You may want to review the methods used to reduce fractions in Unit 18 and the material on converting fractions to decimals in Unit 24. For all problems in Unit 25 that do not work out evenly, carry your answers to three decimal places.

Convert the Following Percents to Decimals

1. 23% = _____.23_____   **2.** 15% = _____

3. 10% = _____   **4.** 12.5% = _____

5. 11.4% = _____   **6.** 4% = _____

7. 3.1% = _____   **8.** $6\frac{7}{8}$% = _____

9. $14\frac{1}{2}$% = _____   **10.** $2\frac{1}{5}$% = _____

Convert the Following Decimals to Percents

11. .33 = _____33%_____   **12.** .22 = _____

13. .12 = _____   **14.** .55 = _____

15. .4 = _____   **16.** .442 = _____

17. .115 = _____   **18.** .021 = _____

19. .6 = _____   **20.** .014 = _____

Convert the Following Common Fractions to Percents

21. $\frac{1}{4} =$ _____ 25% _____   **22.** $\frac{2}{5} =$ _____

23. $\frac{7}{10} =$ _____   **24.** $\frac{3}{20} =$ _____

25. $\frac{3}{8} =$ _____   **26.** $\frac{5}{6} =$ _____

27. $\frac{3}{4} =$ _____   **28.** $\frac{4}{9} =$ _____

29. $\frac{3}{10} =$ _____   **30.** $\frac{1}{6} =$ _____

Convert the Following Percents to Common Fractions and Reduce them when Possible

31. 20% = _____ $\frac{1}{5}$ _____   **32.** 10% = _____

33. 12% = _____   **34.** 8% = _____

35. 45% = _____   **36.** 2% = _____

37. 55% = _____   **38.** 50% = _____

39. 3.2% = _____   **40.** 0.6% = _____

After completing this unit, it is possible for you to convert percents to decimals or fractions. It is also possible to convert decimals or fractions to percents. In order to practice these conversions, complete the following grid for the numbers given.

	Percent	Decimal	Common Fraction
41.	45%	.45	$\frac{45}{100} = \frac{9}{20}$
42.			$\frac{1}{4}$
43.		.44	
44.	9%		
45.		.05	
46.			$\frac{625}{1000} = \frac{5}{8}$
47.		.034	
48.	$2\frac{1}{2}$%		
49.		.03	
50.	$3\frac{3}{10}$%		

| UNIT 26 | FINDING THE PORTION |

LEARNING OBJECTIVE   *After completing this unit, you will be able to:*

1. Find the portion in a mathematical problem when given the rate and the base.

Many business applications like cost markup, markdown, discounts, percent of increase, and retail markup are based on the portion, rate, and base formulas. Before we examine these three formulas, we must define the three terms necessary to complete the formulas. **Base** is the whole, or total, amount. The base in a problem is *always* equal to 100 percent. The base is *always* the number that follows the word *of* in a mathematical problem. For example, take the problem, "Find 20% of 500." The 500 follows the word *of* and is thus the whole amount, or base.

The **rate** is a percent; it indicates a specific part of the base in a mathematical problem. It helps to remember that the words *rate* and *percent* are used interchangeably in most mathematical problems. Because it is a percent, the rate is always followed by a percent sign (%). In the problem in the preceding paragraph, 20% is the rate.

The **portion** (sometimes referred to as percen*tage*) is what you get when you multiply the rate and base together. The portion gets its name because it is a "portion," or part, of the base.

FINDING THE PORTION

When working a portion problem, you will always be given the rate and the base. You can find the portion by multiplying the rate times the base. Here is how the formula is stated:

$$\text{Portion} = \text{Rate} \times \text{Base}$$

Rather than having to memorize formulas to work portion, rate, or base problems, many students find it easier to work with a visual learning aid. You can use the circle illustrated in Example 1 to find the specific formula you need. Simply cover the *unknown* part with your finger. Then perform the indicated operation. For example, if portion is covered, multiply rate times base.

| EXAMPLE 1 | Find 25% of 600 |

$\text{Portion} = \text{Rate} \times \text{Base}$ STEP [1]   State the formula.

CALCULATOR TIP

Depress clear key
Enter .25   Depress ×
Enter 600   Depress =
Read the answer 150

$\text{Portion} = 25\% \times \text{Base}$ STEP [2]   Identify the rate. It is 25%, or the number with the percent sign.

$\text{Portion} = 25\% \times 600$ STEP [3]   Identify the base. It is 600, the number preceded by the word *of.*

| Portion = .25 × 600 | **STEP** 4 | Convert the percent to a decimal. |
| Portion = 0.25 × 600 = 150 | **STEP** 5 | Multiply rate (.25) × base (600). |

When analyzing the problem, note three things: (a) 25% is the rate because it is the number with a percent sign; (b) the word *of* indicates multiplication and tells you to multiply the rate times the base; and (c) 600 is the *base* because it is the whole, or total, amount and follows the word *of.* Once you have determined the rate and the base, you can find the portion by multiplying the rate and base.

For all problems in Unit 26 that do not work out evenly, carry your answers to two decimal places.

Find the Portion in the Following Problems

1. Find 20% of 420

.20 × 420 = 84

2. Find 12% of 112

3. Find 15% of 1,300

4. Find 36% of 985

5. Find 2% of 410

6. Find $4\frac{1}{2}$% of 3,460

7. Find 11.45% of 2,004

8. Find 0.35% of 10,560

9. Rocky Mountain Corporation mailed out 24,600 advertisements for a new resort development in Colorado. Thirty-three percent of those receiving advertisements responded. What portion of potential customers responded?

P = R × B
P = 33% × 24,600
P = .33 × 24,600 = 8,118

10. Old Tyme Wicker Shop received a shipment of 450 wicker baskets. If 6 percent of the baskets were defective, how many baskets were defective?

11. Twenty-four percent of Acme Manufacturing's employees earn in excess of $40,000 a year. If the firm has 700 employees, what portion of the employees earn more than $40,000?

12. Makim Barkley earns $2,210 a month. If his employer withholds 26.4 percent of his salary for deductions, what portion of his salary goes to deductions?

13. In problem 12, what portion of Makim's salary is his take-home pay (after deductions)?

14. The Aim Charter Mutual Fund charges 4 1/2 percent for all investments under $10,000. How much would an investor pay to invest $6,500?

15. Mike Matlock receives social security benefits. In January, all social security recipients received a 3.2 percent cost-of-living increase. If Mike Matlock's benefits total $660 a month *before* the increase, how much is his increase?

16. In problem 15, what is the total of Mike Matlock's social security benefits per month after the cost-of-living increases?

17. On January 15, Judy Cole purchases Federal Express common stock for $50 a share. A year later, the stock had increased in value by $16\frac{1}{4}$ percent. What is the dollar amount of the increase?

18. In problem 17, what is the value of the stock after the 16.25 percent increase?

Personal Math Problem

Bob Hartman works for American Purchasing Exchange and earns $17,680 a year. Because both sales revenues and profits have increased this year, the firm has agreed to reward all employees with a 15 percent year-end bonus, payable on December 31.

19. What is the bonus amount that Mr. Hartman will receive on December 31?

20. If you were Bob Hartman, what would you do with this one-time, year-end

bonus? _____

_____

_____

_____

_____

_____

_____

| UNIT 27 | FINDING THE RATE |

LEARNING OBJECTIVES   *After completing this unit, you will be able to:*

1. find the rate in a mathematical problem when given the portion and base.
2. work percent-of-increase or -decrease problems.

HINT: Before completing this unit, you may want to review the following definitions:

1. The *base* is the whole, or total, amount; it represents 100 percent.
2. The *rate* is always the number followed by a percent sign.
3. The *portion* is the product of multiplying the rate and the base.

FINDING THE RATE

The following formula can be used to find the rate:

$$\text{Rate} = \text{Portion} \div \text{Base}$$

When working a rate problem, you will always be given the portion and the base. A typical rate problem might read, "Find what percent 30 is of 120." Neither number given in the problem has a percent sign because the rate (often referred to as the percent) is what you are trying to find. One hundred twenty is the base because it is the whole, or total, amount and follows the word *of*. Thirty is the portion because it is a part of 120. Now, you can find the rate by dividing the portion by the base.

EXAMPLE 1   30 Is What Percent of 120

Rate = Portion ÷ Base   **STEP** ☐1   Since neither number in the problem has a percent sign, you are looking for the rate. State the rate formula.

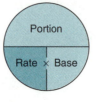

CALCULATOR TIP
Depress the clear key
Enter 30   Depress ÷
Enter 120   Depress =
Read the answer 0.25

Rate = Portion ÷ 120   **STEP** ☐2   Identify the base. It is 120.

Rate = 30 ÷ 120   **STEP** ☐3   By process of elimination, identify the portion. It is 30.

Rate = 30 ÷ 120 = .25   **STEP** ☐4   Divide the portion (30) by the base (120).

Rate = 0.25 = 25%   **STEP** ☐5   Convert the decimal answer to a percent.

PERCENT OF INCREASE AND DECREASE

Often, a businessperson must calculate the percent of increase or decrease for profits, sales, and expenses. To calculate the percent of increase or decrease, follow these steps:

1. Find the *dollar* amount of increase or decrease by subtracting the smaller number from the larger number.
2. Use the rate formula to find the *rate (percent) of increase or decrease*.
3. Convert the decimal answer to a percent.

EXAMPLE 2   If 1993 Sales Were $125,000 and 1994 Sales Were $150,000, What Is the Percent of Increase?

$150,000
−125,000
$ 25,000   **STEP** ☐1   Subtract the smaller number from the larger number.

Rate = Portion ÷ Base   **STEP** ☐2   Since none of the numbers in the problem has a percent sign, you are looking for the rate. State the rate formula.

Rate = Portion ÷ $125,000   **STEP** ☐3   Identify the base, which is the dollar amount for the older time period. The base is $125,000.

Rate = $25,000 ÷ $125,000	**STEP** 4	By process of elimination, identify the portion. It is $25,000.
Rate = $25,000 ÷ $125,000 = 0.20 **STEP** 5		Divide the portion by the base.
Rate = 0.20 = 20%	**STEP** 6	Convert the decimal answer to a percent.

In a percent of increase and decrease problem, the *base* is *always* the *older* time period. When working a percent of decrease problem, the *same* steps are used, but you need to indicate that the dollar amount of change and the percent of change are *negative* amounts. Most people use either the minus sign or write the word *decrease* behind their answers.

Complete the Following Problems

For all problems in Unit 27 that do not work out evenly, carry your answers to three decimal places.

1. 12 is what percent of 48?

   12 ÷ 48 = .25 = 25%

2. 25 is what percent of 200?

3. 112 is what percent of 240?

4. 22 is what percent of 44?

5. 345 is what percent of 360?

6. 11.5 is what percent of 45?

7. 4.5 is what percent of 20?

8. 13.6 is what percent of 198?

August Sales	September Sales	Amount of Change	Percent of Increase or Decrease
$225,600.00	$190,857.60	**9.** *−$34,742.40*	**10.** *−15.4%*

$34,742.40 ÷ 225,600 = .154 = −15.4%

March Profits	April Profits	Amount of Change	Percent of Increase or Decrease
$ 92,245.40	$104,350.00	**11.** _____	**12.** _____

13. On Monday, Mayfair's Clothing Store sold 64 appliances. On Tuesday, eight of the appliances were returned. What is the rate of appliances that were returned?

8 ÷ 64 = .125 = 12.5%

14. Phoenix-based Manufacturers' Sales, Inc., does business in two states: Arizona and New Mexico. During the month of April, the firm's telemarketing sales force made 4,500 sales calls. If 2,000 calls were to New Mexico residents, what percent of total sales calls were to New Mexico residents?

15. In problem 14 what percent of total sales calls were to Arizona residents?

16. At the beginning of 1994, the average home in the San Antonio area was valued at $86,840. If the value of the average home increased $7,555 by the end of the year, what is the rate of increase?

17. In problem 16, what is the value of the average home at the end of the twelve-month period?

18. During 1993, profits for Hancock Hardware were $76,210. During 1994, profits declined to $58,910. What is the percent of decrease for profits?

19. In 1993, total sales for a local Ford dealership were $784,500. In 1994, sales were $856,000. What is the percent of increase for total sales?

20. The salary expense for Atlanta Office Supply was $8,459 during the month of October. During the month of November, salary expense was $7,910. What is the percent of decrease for salary expense?

UNIT 28	FINDING THE BASE

LEARNING OBJECTIVE

After completing this unit, you will be able to:

1. find the base in a mathematical problem when given the portion and rate.

HINT: Before completing this unit, you may want to review the following definitions:

1. The *base* is the whole, or total, amount; it represents 100 percent.
2. The *rate* is always the number followed by a percent sign.
3. The *portion* is the product of multiplying the rate and the base.

FINDING THE BASE

The following formula can be used to find the base:

Base = Portion ÷ Rate

When working a base problem, you will always be given the portion and rate. A typical base problem might read, "156 is 65% of what number?" Just like solving portion or rate problems, you need to identify which number is the portion, which is the rate, and which is the base. Notice that 65% is the rate in this problem because it is the number followed by the percent sign. The words *of what number* tell you that the problem is asking you to find the base. Finally, because the rate and base have already been identified, 156 is the portion. Now you can find the base by dividing the portion by the rate.

EXAMPLE 1   156 is 65% of What Number?

Base = Portion ÷ Rate   **STEP** 1   The words *of what number* indicate you are looking for the base. State the base formula.

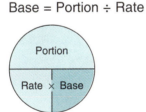

Base = Portion ÷ 65%   **STEP** 2   Identify the rate. It is 65%, or the number with the percent sign.

Base = 156 ÷ 65%   **STEP** 3   By the process of elimination, identify the portion. The portion is 156.

Base = 156 ÷ 0.65   **STEP** 4   Convert the percent to a decimal.

Base = 156 ÷ 0.65 = 240   **STEP** 5   Divide the portion (156) by the rate (0.65).

CALCULATOR TIP

Depress the clear key
Enter 156   Depress ÷
Enter .65   Depress =
Read the answer 240

Complete the Following Problems

For all problems in Unit 28 that do not work out evenly, carry your answers to two decimal places.

1. 22 is 40% of what number?

22 ÷ .40 = 55

2. 56 is 10% of what number?

3. 5 is $10\frac{1}{4}$% of what number?

4. 114 is 5% of what number?

5. 246 is 80% of what number?

6. 456 is 2% of what number?

7. 23.4 is 9.5% of what number?

8. 15.6 is 9.4% of what number?

9. If soccer equipment sales during the month of April were $28,680, or 30 percent of total sales, for Omaha Sports Supply, what is the firm's total sales revenues?

$28,680 ÷ .30 = $95,600

10. Total deductions from Kelly Mathew's weekly paycheck amounted to $130, which is 24 percent of her total earnings for the week. How much did she earn before payroll deductions?

11. In problem 10, what was Kelly Mathew's take-home pay after deductions?

12. During the month of September, Richard Slezak paid $88.50 on his credit card account, which is $31\frac{1}{2}$ percent of his total credit card debt. Before the payment, how much did he owe the credit card company?

13. To finish an apartment complex, McPherson Construction Company placed an order for kitchen sinks from a Cleveland manufacturer. Because the manufacturer's employees were on strike, the firm could ship only 112 kitchen sinks. This shipment represents 70 percent of the total order. How many sinks did McPherson Construction originally order from the manufacturer?

Critical Thinking Problem

While working as a real estate agent for Cypress Springs Realty, Amy Gonzales sold property valued at $1,200,000 during 1994. On January 1, 1995, Ms. Gonzales set a goal of increasing her annual sales by 20 percent.

14. To achieve a 20 percent increase in sales, how much property should Ms. Gonzales sell during 1995?

15. During the first three months of 1995, Ms. Gonzales sold property valued at $315,000. Assuming that Ms. Gonzales's sales remain constant for the remainder of 1995, what will her sales total at the end of 1995?

16. If Ms. Gonzales's sales remain constant for the remainder of 1995, will Ms. Gonzales obtain her goal of increasing sales by 20 percent?

Yes _____ No _____

 17. Based on the above information, what would you do if you were Ms. Gonzales? _____

_____

_____

UNIT 29	PORTION, RATE, AND BASE: PULLING IT ALL TOGETHER

LEARNING OBJECTIVE   *After completing this unit, you will be able to:*

1. work portion, rate, and base problems.

Because portion, rate, and base problems are used so often in business applications, this unit provides additional problems to reinforce each type of problem. As illustrated in Units 26, 27, and 28, there are three separate formulas to use when you work portion, rate, and base problems. Each individual formula is presented below.

$$P = R \times B$$
$$R = P \div B$$
$$B = P \div R$$

To complete a portion problem, you must multiply the rate times the base. To complete a rate problem, you must divide the portion by the base. To complete a base problem, you must divide the portion by the rate. You can also use the model shown below:

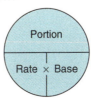

In this model, the line between portion and base and rate indicates division. To find the specific formula you need, simply cover the *unknown* part with your finger. Then perform the indicated operation. For example, if portion is covered, multiply rate times base. If rate is covered, divide portion by base. If base is covered, divide portion by rate.

EXAMPLE 1   Find 42% of 600

Portion = Rate × Base   **STEP** 1   Use the model to determine the correct formula for portion.

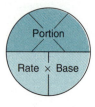

Portion = 0.42 × 600   **STEP** 2   Multiply the rate times base. Note that 42% has been converted to a decimal 0.42.
Portion = 252

EXAMPLE 2   Find What Percent 39 Is of 50

Rate = Portion ÷ Base   **STEP** 1   Use the model to determine the correct formula for rate.

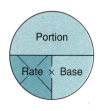

Rate = 39 ÷ 50   **STEP** 2   Divide the portion by the base.
Rate = 0.78

Rate = 0.78 = 78%   **STEP** 3   Convert the decimal answer to a percent.

EXAMPLE 3   40 is 20% of What Number?

Base = Portion ÷ Rate   **STEP** 1   Use the model to determine the correct formula for base.

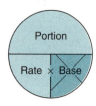

Base = 40 ÷ 0.20   **STEP** 2   Divide the portion by the rate. Note that 20% has been converted to the decimal 0.20.
Base = 200

Find the Portion, Rate, or Base for the Following Problems

For all problems in Unit 29 that do not work out evenly, carry your answers to two decimal places

HINT: Before you answer questions 1 through 5, you may want to review the material in Unit 25 on converting percents, decimals, and fractions.

1. Convert 29.7% to a decimal   *.297*

2. Convert 7% to a decimal

3. Convert $5\frac{7}{10}$% to a decimal

4. Convert 0.231 to a percent

5. Convert .04 to a fraction

6. Find 8.9% of 220.

$P = 0.089 \times 220 = 19.58$

7. 115 is what percent of 300?

8. Find 7.8% of 410.

9. 112 is 21% of what number?

10. 56 is 56% of what number?

11. Find 92% of 5,540.

12. 12 is 39.4% of what number?

13. 240 is what percent of 900?

14. Find $32\frac{9}{20}$% of 11,456.

15. 45 is what percent of 910?

16. Watson Manufacturing Company produces electrical components. On one particular part—a $\frac{1}{4}$ horse fan motor—the firm earns $24.75 profit, which represents a $16\frac{1}{2}$ percent return on the selling price of the fan motor. What is the selling price?

$B = P \div R$
$B = \$24.75 \div 16.5\%$
$B = \$24.75 \div .165$
$B = \$150$

17. Benson Manufacturing produces small electrical parts for the computer industry. Based on past experience, the firm expects 1.5 percent of all parts manufactured to be defective. If the firm manufactures 248,000 parts during the month of May, what portion of the parts will be defective?

18. Al Small works for a Century 21 real estate office. He receives a 2.75 percent commission for all sales. If Al sells real estate valued at $780,500, how much will he receive in commissions?

19. Barbara Fox contributes $216 of her monthly salary to her company's retirement plan. If she earns $3,200 a month, what percent of her monthly salary is contributed to the retirement fund?

20. John Gray spends $8\frac{1}{8}$ percent of his take-home pay on food. If his monthly food allotment is $140, what is his monthly take-home pay?

21. According to a partnership agreement, Jack Kaplan is to receive 44.2 percent of all profits earned by Kaplan and Weinstock Chemicals. Nancy Weinstock is to receive 55.8 percent of all profits. If the partnership earned $1,456,700 in 1994, what is Kaplan's share?

22. In problem 21, what is Weinstock's share of profits?

23. Deep Blue Pool Supply earned $39,120 in 1993. In 1994, the firm earned $56,700. What is the percent of increase for this firm?

24. A piece of real estate property increased $8,500—an 8.5 percent increase in value—in 1994. What is the value of the real estate after the increase?

25. A piece of manufacturing equipment cost $98,450. If this equipment depreciates 12.5 percent a year, what is the amount of annual depreciation?

INSTANT REPLAY

UNIT	IMPORTANT POINTS TO REMEMBER	EXAMPLES
UNIT 25 Converting Percents, Decimals, and Fractions	To convert a percent to a decimal, (a) drop the percent sign and (b) move the decimal point 2 places to the left. To convert a decimal to a percent, (a) move the decimal point two places to the right and (b) add a percent sign. To convert a common fraction to a percent, (a) divide the numerator by the denominator to obtain a decimal, (b) move the decimal point two places to the right, and (c) add a percent sign. To change a percent to a common fraction, (a) drop the percent sign, (b) move the decimal point two places to the left, (c) convert the decimal to a common fraction, and (d) reduce the common fraction if necessary.	$35\% = 0.35$ $0.46 = 46\%$ $\frac{5}{8} = 0.625 = 62.5\%$ $65\% = 0.65 = \frac{65}{100} = \frac{13}{20}$
UNIT 26 Finding the Portion	*Base* is the whole, or total, amount and is equal to 100 percent. *Rate* is a percent, so it is always the number with a percent sign. The *portion* is the product of multiplying the rate and the base. The following formula can be used to find the portion: Portion = Rate × Base	Find 12% of 60. Portion = Rate × Base Portion = 12% × 60 Portion = 0.12 × 60 Portion = 7.2
UNIT 27 Finding the Rate	The rate is always the number followed by a percent sign. The following formula can be used to find the rate: Rate = Portion ÷ Base	14 is what percent of 85? Rate = Portion ÷ Base Rate = 14 ÷ 85 Rate = 0.165 = 16.5%
UNIT 28 Finding the Base	The base is the whole, or total, amount; it represents 100 percent. The following formula can be used to find the base: Base = Portion ÷ Rate	90 is 60% of what number? Base = Portion ÷ Rate Base = 90 ÷ 60% Base = 90 ÷ 0.60 Base = 150

continued from page 133

UNIT	IMPORTANT POINTS TO REMEMBER	EXAMPLES
UNIT 29 Portion, Rate, and Base: Pulling It All Together	The portion, rate, and base formulas can be illustrated by the following model:	Find 10% of 650. Portion = Rate × Base Portion = 10% × 650 Portion = .10 × 650 Portion = 65

$$\frac{\text{Portion}}{\text{Rate} \times \text{Base}}$$

Portion = Base × Rate
Rate = Portion ÷ Base
Base = Portion ÷ Rate

To find the specific formula you need, simply cover the unknown part with your finger. Then perform the mathematical operation that remains.

45 is what % of 900?
Rate = Portion ÷ Base
Rate = 45 ÷ 900
Rate = 0.05 = 5%

56 is 10% of what number?
Base = Portion ÷ Rate
Base = 56 ÷ 10%
Base = 56 ÷ 0.10
Base = 560

CHAPTER 4	MASTERY QUIZ

DIRECTIONS   *Round each answer to two decimal places or hundredths. (Each answer counts 5 points.)*

1. Convert 25.5% to a decimal.   _____
2. Convert 0.854 to a percent.   _____
3. Convert $\frac{3}{4}$ to a percent.   _____
4. Convert 40% to a fraction.   _____
5. What is the formula for portion?   _____
6. What is the formula for rate?   _____
7. What is the formula for base?   _____
8. Find 15% of 780.   _____
9. 86 is what percent of 132?   _____
10. 2,510 is 32% of what number?   _____

11. Landry Real Estate sold a total of 82 properties during the month of June. If 15 of the 82 properties sold were townhomes, what percent of the total do townhomes represent?

12. During the month of April, Chuck Burns paid $120 on his Visa Bank Card, which is 15 percent of his total credit card debt. Before the payment, how much did he owe the credit card company?

13. During the month of August, Jill Clayborne sold merchandise valued at $164,000. If Ms. Clayborne is paid $2\frac{1}{2}$ percent commission, how much did she earn during August?

14. K mart Corporation distributed 450,000 advertising circulars to people in a large metropolitan area during the first week of October. Based on previous marketing research, $1\frac{1}{2}$ percent of those who receive the circulars will shop in a local K mart store within five days. Based on the research findings, how many customers will K mart have within five days after the circulars are distributed?

15. In 1994, Kapoor Accounting Service paid $9,600 for rental expense. In 1993, the firm paid $10,800. What is the percent of increase or decrease for rent expense?

16. Juan Delgado earns $2,970 a month. If his employer withholds 22.6 percent of his salary for deductions, what is the amount of his deductions?

17. In problem 16, what is the amount of Juan Delgado's take-home pay after deductions?

18. Grand Isle Men's Store sold merchandise valued at $124,210 during September. During the same month, sales returns totaled $4,968.40. What percent of total sales were returned?

19. During the last 12-month period, a piece of real estate increased $4,400—a 3.2 percent increase in value. What is the value of the real estate after the increase?

20. Last year, Business Supply, Inc., sold merchandise and services worth $2,345,610. If the firm earns 12.4 percent profit on each sales dollar, what is the dollar amount of profit for this firm?

CHAPTERS 1–4   CUMULATIVE TEST

SPECIAL NOTE TO STUDENTS: To help you build a foundation for success in math, we have included a cumulative test at the end of every three or four chapters in the text. If you do not score at least 80 on this cumulative test, you may want to review the type of problems that you missed.

DIRECTIONS: *If necessary, round each answer to two decimal places or hundredths. (Each answer counts 5 points.)*

1. Convert $1,346.54 to written form _____

_____

2. Add 34.5 + 92.46 + 112.47   _____

3. Add $1,245.68 + $89.32 + $6.75   _____

4. Subtract 3,112.52 – 993.87   _____

5. Subtract $672.11 – $563.33   _____

6. Multiply 1,288 × 300   _____

7. Multiply $45.60 × 7.84   _____

8. Multiply $24,568 × 1,000   _____

9. Divide 5,610 ÷ 15   _____

10. Divide 166.45 ÷ 23.4   _____

11. Reduce $\frac{36}{144}$ _____

12. Add $\frac{1}{5} + \frac{3}{4} + \frac{5}{8}$ _____

13. Subtract $12\frac{1}{4} - 8\frac{5}{18}$ _____

14. Multiply $\frac{2}{3} \times \frac{9}{16} \times \frac{15}{20}$ _____

15. Divide $8\frac{1}{4} \div 2\frac{3}{8}$ _____

16. Convert $21\frac{3}{5}$% to a decimal   _____

17. Convert $\frac{2}{5}$ to a percent   _____

18. 966 is what percent of 2,800   _____

19. Find 22% of 4,500   _____

20. 162 is 40% of what number   _____

Banking Records, Credit Cards, and Installment Purchases

We begin this chapter with a discussion of checks and deposit slips. Then we discuss how to maintain accurate account balances with the help of check stubs or a check register. Next, we describe the process used to reconcile a bank statement. We also examine credit card charges paid by the merchant and the consumer. The chapter closes with a discussion of finance charges that result from installment purchases.

COMMUNITY CREDIT UNION—A BANK THAT MEETS CUSTOMER NEEDS

To meet customer needs, Dallas-based Community Credit Union—like other financial institutions—offers checking accounts, savings accounts, automated teller machines, and other services. And yet, as one member says, "There's something different at Community Credit Union." Take, for example, the number of different checking accounts available to members.

Types of Accounts	Minimum Opening Deposit	Balance Required to Avoid Service Charge	Service Charge	Interest
Express Checking If you deposit at least $600 per month using direct deposit or payroll deduction, choose this FREE account.	$25	N/A	FREE	N/A
Personal Checking If you want to maintain a minimum balance and earn interest, this is the best choice for you.	$200	$500	$9	Based on low balance; interest is paid monthly when balance is $1000 or more.
Money Market Checking If you write only a few checks per month but want to earn market-based interest rates, choose this investment. Unlimited over the counter withdrawals. $5 charge for each additional check over three check minimum.	$2500	$2500	$10	Calculated on daily balance; interest is paid monthly when balance is $2500 or more.
Ultimate Checking Enjoy this free, interest-bearing checking account when you have an Ultimate Savings Account.	$100	N/A	N/A	Based on low balance; interest is paid monthly when balance is $500 or more.
Economy Checking Save money with this basic account for budget-minded members. Write up to 10 checks per month free, pay 65¢ per check for each additional check.	$100	N/A	$1.95	N/A
Prelude Checking For students age 13–21 who are ready for their own account.	$50	$200	$2	N/A
Business Checking See New Accounts Representative for details.				

While many competitors offer just one type of checking account—a take it or leave it approach to banking—Community Credit Union offers seven different checking accounts. And each type of account is customized to meet the business's or individual's needs. This dedication to meeting customer needs may help explain why Community Credit Union has been able to expand from one location to six locations during a period of time when there have been record bank failures in the United States.

Source: Adapted from *Checking* (Richardson, TX: Community Credit Union, February 1994).

UNIT 30	DEPOSIT SLIPS AND CHECKS

LEARNING OBJECTIVES *After completing this unit, you will be able to:*

1. complete deposit slips.
2. identify three different types of endorsements.
3. Write checks correctly.

To meet the needs of their customers, Community Credit Union offers seven different checking accounts. Some have service charges, some pay interest, and some offer extra services that make banking easier. Obviously, there are at least three factors to consider before opening a checking account.

First, many banks have monthly service charges. Although it may be possible to avoid service charges if the customer maintains a minimum balance, monthly service charges usually range from $5 to $15 at most banks. There are also fees for check printing, overdraft fees, and fees for stop-payment orders.

Second, many financial institutions now offer interest-bearing checking accounts. Interest rates on checking accounts are approximately the same as the interest rates for savings accounts. Before paying interest, however, many financial institutions may impose certain restrictions, including the following:

- A minimum balance.
- Fees for accounts whose balances fall below a set minimum balance.
- A maximum number of checks that may be written each month free of any service charges.

Third, many financial institutions offer special services that include 24-hour teller machines, banking over the telephone, storage of canceled checks on microfilm, and overdraft protection. **Overdraft protection** is an automatic loan made to customers to cover the amount of checks written in excess of the balance in their checking account.

When opening a checking account, the account holder must sign a signature card. A **signature card** is a record of the official signature(s) of the person(s) who are authorized to write checks on an account. Once you have opened a checking account, it is now possible to make deposits and write checks.

DEPOSIT SLIPS

A **deposit slip** is a form you complete when you deposit money in a bank account. Note that in Example 1, checks, paper currency, and coins are listed separately. Once listed in proper order, all entries must be added to obtain the total for the deposit slip.

EXAMPLE 1   Sample Deposit Slip

CHECKS AND OTHER ITEMS ARE RECEIVED FOR DEPOSIT SUBJECT TO THE TERMS AND CONDITIONS OF THIS BANK'S COLLECTION AGREEMENT NOW IN EFFECT.

Woodside Hardware
909 Birch Drive
Dallas, TX 75248

USE BACK SIDE FOR LISTING ADDITIONAL CHECKS

IF YOU ARE BANKING BY MAIL PLEASE REFER TO INSTRUCTIONS ON REVERSE SIDE

RECORD CHECKS FOR DEPOSIT	DOLLARS	CENTS
	419	50
1	233	10
	56	24
CURRENCY	177	00
COINS	1	65
TOTAL DEPOSIT	1 887	49

CALCULATOR TIP

Enter   419.50 Depress +
Enter   1,233.10 Depress +
Enter   56.24 Depress +
Enter   177.00 Depress +
Enter   1.65 Depress =
Read the answer $1,887.49

STEP [1]   List individual checks.

STEP [2]   List currency.

STEP [3]   List coins.

STEP [4]   Add each entry to obtain the total deposit amount.

THREE DIFFERENT TYPES OF ENDORSEMENTS

Before you can deposit checks in a bank account, you must endorse them— that is, sign them on the back. A signature on the back of a check is called an **endorsement.** Under current federal regulations, an endorsement must be placed on the back within 1½ inches of the left edge of the check. The rest of the back of the check is reserved for bank endorsements. You can endorse a check three different ways: by using a blank endorsement, a restrictive endorsement, or a special endorsement. Each type of endorsement, along with the parts of a check, is illustrated in Example 2.

EXAMPLE 2 Blank, Restrictive, and Special Endorsements

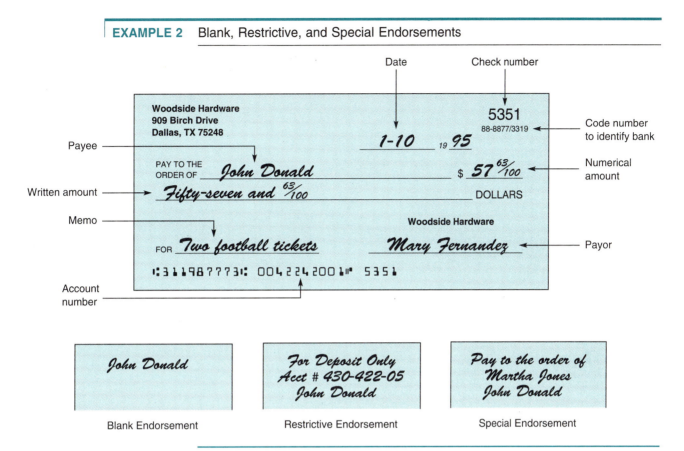

A *blank endorsement* contains only the signature of the payee (the person to whom the check is written). It is quick and easy but dangerous because it makes the check payable to anyone who gets possession of it—legally or otherwise. A *restrictive endorsement* states the purpose for which the check is to be used. For example, the words *For Deposit Only* in Example 2 means that this check must be deposited in the specified account. A *special endorsement* identifies the person or firm to whom the check is payable. The words *Pay to the Order of Martha Jones* in Example 2 means that the only person who can cash, deposit, or negotiate this check is Martha Jones, because John Donald has endorsed or signed it over to her. Like the restrictive endorsement, the special endorsement protects the check in case it is lost or stolen.

WRITING A CHECK

A **check** is a written order for a bank or other financial institution to pay a stated dollar amount to the business or person indicated on the face of the check. The steps necessary for proper check writing are illustrated in Example 3.

EXAMPLE 3   How to Write a Check

STEP [1]   Record the current date.

STEP [2]   Write the name of the payee (the person or business that is receiving the payment).

STEP [3]   Record the amount of the check in figures.

STEP [4]   Write the amount of the check in words.

STEP [5]   Sign the check.

HINT: You may want to review the material in Unit 2 on writing numerical amounts with decimals before completing this unit.

Complete the Deposit Slips and Checks That Are Provided; Identify Different Endorsements

1. On January 22, R. I. Singh deposited the following checks: $345.13, $26.78, $567.10, and $3,210.45. What is the total for this deposit?

$4,149.46

2. On March 10, Peter A. Gonzales deposited the following checks: $1,245.60; $356.78; $240.50; and $10.42. What is the total for this deposit?

3. On April 3, Mary Barnes, the bookkeeper for Home Decorators, Inc., deposited the following amounts: currency, $124.00, and coins, $3.41. In addition, there were five checks included in the deposit. Individual amounts for the checks were $4,235.60, $23.49, $345.60, $31.76, and $1,455.63. Using this information, complete the following deposit slip:

RECORD CHECKS FOR DEPOSIT	DOLLARS	CENTS
CURRENCY		
COINS		
USE BACK SIDE FOR LISTING ADDITIONAL CHECKS — **TOTAL DEPOSIT**		

CHECKS AND OTHER ITEMS ARE RECEIVED FOR DEPOSIT SUBJECT TO THE TERMS AND CONDITIONS OF THIS BANK'S COLLECTION AGREEMENT NOW IN EFFECT.

Home Decorators, Inc.
133 Newton Drive
Boston, MA 02108

IF YOU ARE BANKING BY MAIL PLEASE REFER TO INSTRUCTIONS ON REVERSE SIDE

4. "For Deposit Only, Bob Harris" written on the back of a check is a _restrictive_ endorsement.

5. "Dianne Martin" written on the back of a check is a _____ endorsement.

6. "Pay to the Order of Debbie Brown; Bill Jackson" is a _____ endorsement.

7. "Hawthorne, Inc." written on the back of a check is a _____ endorsement.

8–10. As the bookkeeper for Oklahoma Service Company, you are responsible for paying the monthly bills. During the first week of May, the following three bills were received:

Interstate Delivery Service for $945.90
Franklin Wholesale for $3,456.70
A-1 Manufacturing for $10,456.30

Using the three blank checks provided, complete the checks needed to pay these bills. Use May 3, 1994, for the date and sign your name as payor.

Oklahoma Service Company
411 N. W. 23rd
Oklahoma City, OK 76214

455

_____ 19 _____

PAY TO THE
ORDER OF _____ $ _____

_____ DOLLARS

Oklahoma Service Company

FOR _____   _____

⑆091901 480⑆ 8530002118314 1⑈ 0455

Oklahoma Service Company
411 N. W. 23rd
Oklahoma City, OK 76214

456

_____ 19 _____

PAY TO THE
ORDER OF _____ $ _____

_____ DOLLARS

Oklahoma Service Company

FOR _____   _____

⑆091901 480⑆ 8530002118314 1⑈ 0456

Oklahoma Service Company
411 N. W. 23rd
Oklahoma City, OK 76214

457

_____ 19 _____

PAY TO THE
ORDER OF _____ $ _____

_____ DOLLARS

Oklahoma Service Company

FOR _____   _____

⑆091901 480⑆ 8530002118314 1⑈ 0457

UNIT 31	CHECK STUBS AND CHECK REGISTERS

LEARNING OBJECTIVES   *After completing this unit, you will be able to:*

1.   record a check or deposit on a check stub or in a check register.
2.   determine the correct balance for a checking account.

To avoid making banking errors, you must be careful when determining the balance in your checkbook. Both individuals and businesses use either a check stub or a check register to keep track of their current balance. The information needed to complete the check stub or check register should be completed *before* the check is written. For example, Betty Habib completes check number 2345 for $319.20 to pay her apartment rent. To determine her current balance, she completes the check stub illustrated in Example 1.

EXAMPLE 1	Recording a Check on a Check Stub

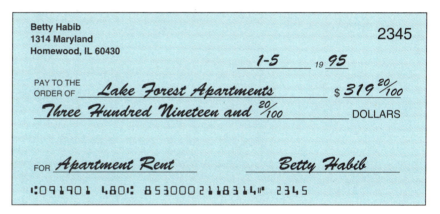

CALCULATOR TIP

Depress clear key
Enter   1,775.40 Depress −
Enter   319.20 Depress =
Read the answer $1,456.20

STEP 1   Enter the current date on the stub.

STEP 2   Enter the number of the check on the stub.

STEP 3   Enter the name of the payee (the person or business receiving the check).

STEP 4   Record the dollar amount of the check.

STEP 5   Subtract the check amount from the beginning balance (balance forward) to determine the new balance.

The difference between recording a check and recording a deposit is that a deposit must be *added* to the previous balance. On January 13, Jim Begg deposited $875.50 in his checking account. What is his new balance?

EXAMPLE 2   Recording a Deposit in a Check Register

CODE OR NUMBER	DATE	DESCRIPTION OF TRANSACTION		PAYMENT/DEBIT OR FEE (–)		T (√)	DEPOSIT OR CREDIT (+)		BALANCE FORWARD		
										2,575	18
2346	¹/₁₁/₉₅	To	Metro Lighting Inc	229	50				Pay't or Dep		
		For	Fixtures						Bal	2,345	68
①	¹/₁₃/₉₅	To	Deposit ②				③		Pay't or Dep		
		For					875	50	Bal ④	3,221	18
		To							Pay't or Dep		
		For							Bal		
		To							Pay't or Dep		
		For							Bal		
		To							Pay't or Dep		
		For							Bal		

STEP ☐1   Record the current date.

STEP ☐2   Write the word *Deposit.*

STEP ☐3   Record the dollar amount of the deposit.

STEP ☐4   Add the deposit to the previous balance to obtain the new balance.

If you use a check stub (like the one illustrated in Example 1), you can use the same steps to record a deposit.

HINT: Before completing this unit, you may want to review the material in units 6 and 8 on addition and subtraction of numbers that contain decimals.

Complete the Following Problems

1. On November 30, Joe Bates wrote check number 2105 for $234.50. If his previous balance was $3,220.43, what is his new balance after he records this check?

   $2,985.93

2. On May 14, Jan Perez, the bookkeeper for Red Baron Wrecking, Inc., wrote check number 1142 for $1,986.32. If the firm's previous balance was $17,774.81, what is the new balance after Ms. Perez records this check?

3. On April 10, Joyce Clark deposited $1,456.00 in her checking account. If her previous balance was $1,154.30, what is her new balance after she records this deposit?

4. At the beginning of October, Debbie Radford's checking account had a $1,495.40 balance. During the month, she made a $2,345.50 deposit and wrote checks for $21.19, $56.70, $34.50, $125.67, and $546.30. What is her balance at the end of the month?

5. Given the information written on the following check, complete the check stub.

DATE		Bal. For'd
		2,145. 60
To the Order of		Total
		Amt. this Check
		Balance

```
Sandra Gomez                                              440
1456 Burbank
Los Angeles, CA 90562
                                        4-15    19 95

PAY TO THE
ORDER OF   Internal Revenue Service        $ 1,235 43/100

One Thousand, Two Hundred Thirty-Five and 43/100 DOLLARS


FOR  1994 Taxes                        Sandra Gomez

⑆091901 480⑆ 85300021183140 0440
```

6. On May 10, Joan Norwood wants to purchase a new coat from the Dillard's department store for $149.95. Assume that you are Joan, and complete the check necessary to pay for the coat. Also, complete the check stub to record this transaction.

DATE		Bal. For'd
		1,233 24
To the Order of		Total
		Amt. this Check
		Balance

```
Joan Norwood                                             1441
294 Reading
Dallas, TX 75233
                                         _____ 19 ____

PAY TO THE
ORDER OF _____  $ _____

_____ DOLLARS


FOR _____      _____

⑆091901 480⑆ 85300021183140 1441
```

7. Given the information below, complete the following check register.

CODE OR NUMBER	DATE		DESCRIPTION OF TRANSACTION	PAYMENT/DEBIT OR FEE (–)		T (√)	DEPOSIT OR CREDIT (+)		BALANCE FORWARD 1,345	60
1025	3/1	To	Jones Lighting	115	13				Pay't or Dep	
		For							Bal	
1026	3/2	To	Farm & Home Savings	754	00				Pay't or Dep	
		For							Bal	
	3/3	To	Deposit				2,100	00	Pay't or Dep	
		For							Bal	
1027	3/9	To	Cash	125	00				Pay't or Dep	
		For							Bal	
	3/11	To	Deposit				410	00	Pay't or Dep	
		For							Bal	
1028	3/15	To	South-Western Bell	63	20				Pay't or Dep	
		For							Bal	
1029	3/20	To	All-Cities Electric	71	32				Pay't or Dep	
		For							Bal	

8. Given the information below, complete the following check register.

CODE OR NUMBER	DATE	DESCRIPTION OF TRANSACTION	PAYMENT/DEBIT OR FEE (–)		T (√)	DEPOSIT OR CREDIT (+)		BALANCE FORWARD 2,463	21
	5/2	To Deposit For				1,000	00	Pay't or Dep Bal	
3341	5/3	To Woodside Apts For	415	00				Pay't or Dep Bal	
3342	5/3	To Lone Star Gas For	31	90				Pay't or Dep Bal	
3343	5/3	To Metro Water Service For	25	16				Pay't or Dep Bal	
3344	5/3	To GMAC For	310	35				Pay't or Dep Bal	
3345	5/10	To Cash For	140	00				Pay't or Dep Bal	
	5/15	To Deposit For				1,131	45	Pay't or Dep Bal	

Personal Math Problem

Bob Patton's problems started when he received a phone call from the office manager of Tom Thumb Supermarket. The office manager told Bob that a $32 check he had written for groceries had been returned for nonsufficient funds (NSF). She asked Bob to come in and pick up the "bounced" check. In addition, she told him he would have to pay a $15 charge to the store. According to Bob's bank records, he still had more than enough money in his account to cover the check.

9. A portion of Bob's check register is reproduced below. Is the last balance correct? If not, what should his balance be?

CODE OR NUMBER	DATE	DESCRIPTION OF TRANSACTION	PAYMENT/DEBIT OR FEE (–)		T (√)	DEPOSIT OR CREDIT (+)		BALANCE FORWARD 245	60
764	3/5	To City of New Castle For (Water)	35	10				Pay't or Dep Bal 210	50
	3/7	To Deposit For				140	00	Pay't or Dep Bal 450	50
765	3/11	To Circuit City For	224	30				Pay't or Dep Bal 226	20
766	3/16	To Trinity Savings - Cash For	100	00				Pay't or Dep Bal 126	20
767	3/17	To Tom Thumb Supermarket For	32	00				Pay't or Dep Bal 94	20
		To For						Pay't or Dep Bal	
		To For						Pay't or Dep Bal	

10. In your own words, describe why individuals or businesses should avoid "bouncing" checks. _____

_____

_____

_____

_____

FIGURE 32–1. Monthly Statement from the Community Credit Union.

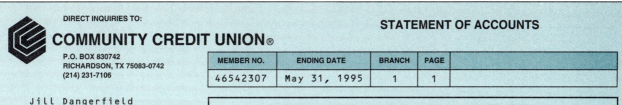

DATE	TRANSACTION DESCRIPTION		AMOUNT		BALANCE
	PERSONAL CHECKING		PREVIOUS BALANCE		4,968.93
MAY01	DIVIDEND		2.91		4,971.84
MAY03	SHARE DRAFT	4862	5.01–		4,966.83
MAY03	SHARE DRAFT	4864	216.18–		4,750.65
MAY03	SHARE DRAFT	4859	346.38–		4,404.27
MAY04	SHARE DRAFT	4863	20.59–		4,383.68
MAY04	SHARE DRAFT	4866	31.81–		4,351.87
MAY04	SHARE DRAFT	4867	37.41–		4,314.46
MAY04	SHARE DRAFT	4865	88.41–		4,226.05
MAY05	SHARE DRAFT	4861	20.00–		4,206.05
MAY05	SHARE DRAFT	4874	25.00–		4,181.05
MAY05	SHARE DRAFT	4860	650.29–		3,530.76
MAY06	DEPOSIT		63.81		3,594.57
MAY06	SHARE DRAFT	4878	21.38–		3,573.19
MAY06	SHARE DRAFT	4877	24.47–		3,548.72

| UNIT 32 | BANK STATEMENT RECONCILIATION |

LEARNING OBJECTIVE   *After completing this unit, you will be able to:*

1. reconcile a checkbook balance with the balance contained in a monthly bank statement.

Each month you receive a detailed bank statement of transactions that affect your checking account (See Figure 32–1).

This statement shows the dollar amounts subtracted from your account (checks and bank charges). The statement also shows the dollar amounts *added* to your account (deposits and special collections). The balance shown in your checkbook is *usually* different from the balance on the bank statement. Four things can explain the difference between these two balances:

1. *Outstanding checks:* checks you have written that have not been presented to or paid by the bank.
2. *Deposits in transit:* deposits you have made that the bank has not yet processed.
3. *Bank charges:* charges made by the bank that you are not aware of, such as monthly activity or service charges, check printing charges, or nonsufficient funds charges.
4. *Other transactions:* amounts that must be added to or subtracted from your checkbook balance. These include the following:

Transaction	Effect on Your Account
(a) Checks you received from another person or business and deposited by you in your account that are returned for non-sufficient funds (NSF)	Reduces your account balance
(b) Fees for NSF checks	Reduces your account balance
(c) Fees for using an automated teller machine (ATM)	Reduces your account balance
(d) Bills paid automatically with money taken from your checking account	Reduces your account balance
(e) A payroll deposit made automatically to your account	Increases your account balance
(f) Interest on your checking account	Increases your account balance

Because of these differences, it is important to compare your checkbook balance with the bank statement promptly. To determine whether your checkbook balance is correct, you can perform a bank reconciliation—that is, you reconcile your balance with the bank's balance. As part of the bank statement, most banks include a form like the one in Example 1 to make the bank reconciliation process a little easier. Notice that the directions needed to complete a bank reconciliation are included as part of the form.

EXAMPLE 1   Bank Reconciliation for Clayton's Dry Cleaners

On March 31, Clayton's Dry Cleaners had a $2,699.10 checkbook balance. The following information was included in the firm's most recent bank statement:

Bank statement balance—$2,334.56   Activity charge—$13.45
Interest earned—$5.20   Check printing—$15.00

A comparison of the bank statement with the firm's check stubs showed the following: deposits in transit—$564.35, $521.35, and $456.70; and outstanding checks—$113.47, $356.78, $76.51, and $654.35.

Bank balance	_____	**1**	Checkbook balance	_____	**6**
Add deposits not shown	_____	**2**	Add interest earned, if any	_____	**7**
Subtotal	_____	**3**	Subtotal	_____	**8**
Subtract outstanding checks	_____	**4**	Subtract activity fees and other bank charges	_____	**9**
Balance	_____	**5**	Balance	_____	**10**

Do the two balances agree? ____ yes ____ no **11**

Bank Balance Side

STEP	1	Record the balance shown on the bank statement.
STEP	2	Total deposits in transit.
STEP	3	Add the bank balance to total deposits in transit.
STEP	4	Total all outstanding checks not shown on the statement.
STEP	5	Subtract the total of all outstanding checks from the subtotal.

Checkbook Balance Side

STEP	6	Record the balance from the checkbook.
STEP	7	Record interest earned, if any.
STEP	8	Add interest earned to the checkbook balance.
STEP	9	Record other bank charges.
STEP	10	Subtract activity fees and other bank charges from the subtotal.
STEP	11	Compare the two balances.

Once the reconciliation process is completed, the total on the bank balance side should agree with the total on the checkbook balance side. If you cannot reconcile your checkbook balance with the bank statement—that is, if you cannot find a way to make the totals agree—a mistake has been made. You might need to check all your addition and subtraction, first in the reconciliation and then in your checkbook register (or your check stubs).

Complete the Following Problems

In the space below, indicate if the transaction will increase or decrease your checking account balance.

Transaction	Increase	Decrease
1. A bank activity charge	_____	*X*
2. Interest on your account	_____	_____
3. A bill paid automatically	_____	_____
4. A check printing charge	_____	_____
5. A check returned for NSF	_____	_____
6. A charge for using an automated teller	_____	_____
7. A fee for an NSF check	_____	_____
8. An automatic payroll deposit	_____	_____

9. On the last day of each month, Mary Turner's $2,543.60 paycheck is automatically deposited into her checking account. On January 30, Mary Turner had a $2,346.51 balance in her checkbook. What is her checkbook balance on January 31?

$4,890.11

10. On Monday, Alvin Peters's checkbook balance was $1,874.53. On Tuesday, Alvin got a notice from his bank that a $119 check he had deposited had been returned for nonsufficient funds. His bank also notified him that there is a $3 charge for all returned checks. After he adjusts his account for the returned check and NSF fee, what is his new checkbook balance?

11. During the month of April, Detroit Printing Company wrote the following checks that have not been presented to the bank for payment: $45.10, $130.20, $17.84, $12.40, $1,765.30, and $567.89. What is the total of the firm's outstanding checks?

12. When Alex Castle examined his bank statement, he found that the bank had not processed three deposits. The amounts for these deposits were $1,245, $672.50, and $252.67. What is the total of his deposits in transit?

13. On May 31, Julie Mathis had a $2,603.68 checkbook balance. The following information was included in her bank statement:

Bank statement balance—$1,789.04
Activity charge—$7.50
Check printing—$9.00.

After checking all the deposits and checks noted in her statement, she found she had a $2,345.60 deposit in transit and outstanding checks in the amounts of $14.56, $167.80, $1,245.10, and $120. Using the following form, prepare a bank reconciliation:

Bank balance	a. _____	Checkbook balance	f. _____
Add deposits not shown	b. _____	Add interest earned, if any	g. _____
Subtotal	c. _____	Subtotal	h. _____
Subtract outstanding checks	d. _____	Subtract activity fees and other bank charges	i. _____
Balance	e. _____	Balance	j. _____
Do the two balances agree? k. ____ yes ____ no			

14. On June 30, Dreyfuss Hardware had a $2,247.75 checkbook balance. The following information was included in the firm's bank statement: bank statement balance, $4,388.50; interest earned, $11.15; activity charge, $24.50; and a nonsufficient funds charge, $34.00. A review of the statement against the company checkbook showed the following: deposits in transit, $1,333.00, $1,132.40, and $996.52; outstanding checks, $123.45, $67.89, $345.60, $1,234.50, $1,567.00, and $2,311.58. Using the following form, prepare a bank reconciliation:

Bank balance	a. _____	Checkbook balance	f. _____
Add deposits not shown	b. _____	Add interest earned, if any	g. _____
Subtotal	c. _____	Subtotal	h. _____
Subtract outstanding checks	d. _____	Subtract activity fees and other bank charges	i. _____
Balance	e. _____	Balance	j. _____
Do the two balances agree?	k. ____ yes ____ no		

15. On October 31, the law firm of Jacobs & Lin had a $12,627.52 checkbook balance. The following information was included in the firm's bank statement:

Bank statement balance—$10,954.67
Activity charge—$22.30
Check printing—$49.50
A check returned for NSF—$200.00

A comparison of the bank statement with the firm's check stubs showed the following: deposits in transit, $1,345.81 and $4,567.82; and outstanding checks, $256.74, $344.56, $1,256.89, and $2,654.39. Using the following form, prepare a bank reconciliation:

Bank balance	a. _____	Checkbook balance	f. _____
Add deposits not shown	b. _____	Add interest earned, if any	g. _____
Subtotal	c. _____	Subtotal	h. _____
Subtract outstanding checks	d. _____	Subtract activity fees and other bank charges	i. _____
Balance	e. _____	Balance	j. _____
Do the two balances agree?	k. ____ yes ____ no		

UNIT 33 | CREDIT CARD TRANSACTIONS

LEARNING OBJECTIVES   *After completing this unit, you will be able to:*

1. understand why merchants accept credit cards.
2. calculate the credit card fee paid by merchants.
3. calculate the finance charge paid by consumers on credit card accounts.

"Charge it!" If those two words sound familiar, it is no wonder. Over 75 million Americans use credit cards to pay for everything from tickets on American Airlines to Zenith computers. And the number of credit cardholders increases every month. In fact, most Americans receive at least two or three credit card applications in the mail every month. Why have credit cards become so popular?

WHY MERCHANTS ACCEPT CREDIT CARDS

Credit is immediate purchasing power that is exchanged for a promise to repay the amount at a later date. Today, merchants accept credit cards because *some* customers simply cannot afford to pay cash for their purchases.

Once the merchant accepts a credit card for payment of goods or services, the charge slip can be deposited in a bank or other financial institution. As a result, the merchant has immediate access to the amount deposited. In return for processing the merchant's credit card transactions, the bank charges a fee that ranges between $1\frac{1}{2}$ percent and 5 percent.

Calculating bank fees for processing credit card transactions is a practical application of the portion formula (Portion = Rate × Base) presented in Unit 26. The bank fee (portion) is found by multiplying the total of credit card transactions (base) by the bank fee stated as a percent (rate).

EXAMPLE 1   Calculate the Bank Fee for Credit Card Transactions

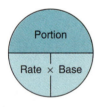

Advanced Travel Service deposits credit card transactions that total $4,650. The Interstate Bank charges 4 percent to process this merchant's credit transactions. Find the bank fee.

$4,650	Total credit card transactions
× .04	Bank fee stated as a percent
$186.00	Dollar amount of the bank fee

CALCULATOR TIP

Depress clear key
Enter 0.04   Depress   ×
Enter 4,650   Depress   =
Read the answer $186

In this example, note the bank fee stated as a percent (4%) has been changed to the decimal 0.04. To calculate the amount that the merchant receives, one additional step is necessary: subtract the bank fee from the total credit card charges.

EXAMPLE 2 | Determine the Amount the Merchant in Example 1 Receives

$4,650	Total credit card charges
− 186	Dollar amount of the bank fee
$4,464	Dollar amount the merchant receives

CREDIT CARD FEES FOR CONSUMERS

There are two basic questions to ask when comparing the cost of credit cards. First, is there an annual fee? While some banks and other financial institutions use "no annual fees" to entice consumers, most charge annual fees that range from $15 to $35. Obviously, the lower the annual fee, the better.

The second question also deals with cost. How much are the finance charges and how are they calculated? This question is more difficult to answer than the first question. Banks and other financial institutions that issue credit cards must tell consumers how they calculate finance charges. Typically, the dollar amount of the finance charge is calculated by multiplying either the previous month's balance or the average daily balance by the finance charge stated as a percent. Both methods are a practical application of the portion formula (Portion = Rate × Base) presented in Unit 26.

To use the previous balance method, multiply the outstanding balance at the end of the previous month (base) by the finance charge stated as a percent (rate).

EXAMPLE 3 | Calculate the Finance Charge Based on the Previous Month's Balance

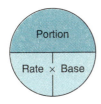

At the end of last month, Larry Angelo had an $840 balance on his bank card. Each month, the bank charges a monthly rate of $1\frac{1}{2}$ percent based on the previous month's balance. What is the finance charge?

$ 840	Previous month's balance
×.015	Finance charge stated as a percent
$12.60	Dollar amount of finance charge

To use the second method, you must first determine the average daily balance. All credit card companies use a computer to maintain a customer's average daily balance, but you should know that it is the sum of the daily account balances throughout the monthly billing period divided by the number of days in the current billing period. Notice that the average daily balance amount is also included at the bottom of the monthly Mastercard statement illustrated in Figure 33–1. Once the average daily balance (base) and the finance charge stated as a percent (rate) are known, multiplication can be used to find the dollar amount of finance charge (portion).

FIGURE 33–1 Cardmember Statement for First USA Mastercard

Cardmember Statement **First USA**

032 7MA 7 15 7
17122000 1836

MasterCard

ACCOUNT NUMBER	PAST DUE AMOUNT	NEW BALANCE	MINIMUM PAYMENT DUE	PAYMENT DUE DATE	WRITE AMOUNT OF PAYMENT
	0.00	1155.00	23.00		

Please make check payable to **First USA MasterCard.**
Send top portion of statement with payment in enclosed envelope.
Please send payment to address shown below.

FIRST USA MASTERCARD
P.O. BOX 740085
ATLANTA, GA 30374-0085

Detach here ▼

For customer service 24 hours a day, call (Outside U.S.A. call 800-524-8200) 800-955-9900

PAGE OF

ACCOUNT NUMBER	TOTAL CREDIT LINE	CASH ADVANCE CREDIT LINE	AVAILABLE CREDIT	AVAILABLE PORTION FOR CASH ADVANCES	PAYMENT DUE DATE	STATEMENT DATE
	5000	2500	3845	2500		

TRANS DATE	POST DATE	REFERENCE NUMBER	MERCHANT NAME OR TRANSACTION DESCRIPTION	AMOUNT
09 13	09 13	7043177LJ1LT4QSAG	THE HOME DEPOT 544 RICHARDSON TX	15.40
09 15	09 15	8041019LMSD3RHZ5H	K MART 00007438 DALLAS TX	11.80
09 15	09 15	7042574LKABVEQD35	HAN-CHU RESTAURANT DALLAS TX	15.91
09 18	09 18	7041019LR49VXAMYX	RED LOBSTER 00000901 RICHARDSON TX	19.12
09 20	09 20	7042574LTABVEQD1M	HAN-CHU RESTAURANT DALLAS TX	20.74
			PERIODIC RATE *FINANCE CHARGE*	23.09

PREVIOUS BALANCE	+ PURCHASES, FEES AND ADJUSTMENTS	– CASH ADVANCES	**+ FINANCE CHARGES**	– PAYMENTS AND CREDITS	NEW BALANCE
753.52	578.39	0.00	14.33	200.00	1155.00

CARDMEMBER NEWS

Send payments to: FIRST USA MASTERCARD P.O. BOX 740085 ATLANTA, GA 30374-0085

FINANCE CHARGE SUMMARY

Average Daily Balance subject to Finance Charge

Purchases and Cash Advances → 1023.45

Average Daily Balance

Monthly Periodic Rate 1.40%
Nominal ANNUAL PERCENTAGE RATE 16.90%
Effective ANNUAL PERCENTAGE RATE 16.90%

Send billing and other inquiries to:
First USA Bank, P.O. Box 8650
Wilmington, DE 19899-8650

*Cash Advance Credit Line is a Portion of Your Total Credit Line

See reverse side for important information.

First USA Bank
Member FDIC

| EXAMPLE 4 | Calculate the Finance Charge Based on the Average Daily Balance |

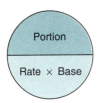

Portion

Rate × Base

The average daily balance for Marty Green's credit card account is $552. Each month, the bank charges a monthly rate of 1¼ percent based on the average daily balance. What is the finance charge?

$ 552	Average daily balance
×.0125	Finance charge stated as a percent
$ 6.90	Dollar amount of finance charge

CALCULATOR TIP

Depress clear key
Enter 0.0125   Depress ×
Enter 552   Depress =
Read the answer $6.90

HINT: You may want to review the material on converting fractions to decimals in Unit 24 and the material on finding portion in Unit 26 before completing this unit.

Complete the Following Problems

1. In your own words, describe why merchants accept credit cards. _____
_____
_____
_____

2. In your own words, describe why consumers use credit cards. _____
_____
_____

3. T   F   A bank usually charges the merchant between 1½ percent and 5 percent to process credit card transactions.

4. T   F   To get most credit cards, all consumers must pay an annual fee between $35 and $50.

Total Credit Card Charges	Bank Fee Stated as a Percent	Dollar Amount of Merchant's Bank Fee
$12,469.42	2%	5. _____$249.39_____
$ 342.60	3¼%	6. _____
$ 1,890.00	5%	7. _____
$ 9,456.20	4½%	8. _____

Previous Month's Balance	Monthly Finance Charge Stated as a Percent	Dollar Amount of Finance Charge
$ 460.20	1%	9. _____
$1,208.40	1½%	10. _____
$2,490.75	1¾%	11. _____

Average Daily Balance	Monthly Finance Charge Stated as a Percent	Dollar Amount of Finance Charge
$2,400.00	1¼%	12. _____
$ 762.10	1½%	13. _____
$ 942.34	1⅓%	14. _____

Critical Thinking Problem

Pat and Andrea Fuentes own Fuentes' Antiques. Since the store opened six months ago, they have always insisted that customers pay by check or cash. While their "check or cash" policy seems to be working well, they are concerned that they may be losing sales. Andrea did talk to their banker and was told that the bank would be glad to process the firm's credit card transactions, but the bank does charge small firms $4\frac{3}{4}$ percent for the service.

15. If customers bought merchandise valued at $10,500 and used credit cards to pay for the merchandise, what is the dollar amount of the bank fee?

16. If you were the Fuentes, would you drop the "check or cash" policy and allow customers to pay with credit cards? Why? _____

_____

_____

17. Charles Garrett owns Garrett Jewelry Store. Because of a special sales promotion, the store sold merchandise totaling $12,566 last weekend. Cash sales totaled $5,110 while credit sales totaled $7,456. If his bank charges 3 percent to process credit card transactions, what is the amount of the bank fee?

18. In problem 17, what is the dollar amount the merchant receives after the bank fee?

19. At the end of last month, Kate Hoover's credit card balance was $2,345. If her bank charges a monthly rate of 1⅕ percent, what is the finance charge?

20. The average daily balance for Hilda Southmore's credit card account is $946.32. Each month, the bank charges a monthly rate of 1 1/8 percent based on the average daily balance. What is the finance charge?

UNIT 34	INSTALLMENT PURCHASES

LEARNING OBJECTIVES   *After completing this unit, you will be able to:*

1. find the total installment price for an installment purchase.
2. determine the finance charge for an installment purchase.
3. calculate monthly payment amounts.
4. find the annual percentage rate for an installment purchase.

When consumers purchase automobiles, major appliances, and many other expensive items, they often choose to buy on an installment plan (buy on time). An **installment purchase** allows the consumer to make regular payments over an extended period of time to pay for merchandise.

DETERMINING THE TOTAL INSTALLMENT PRICE

The cost of an installment purchase is usually higher than the cost of a cash purchase because retailers charge interest or assess a finance charge. The **total installment price** is the total of all payments *plus* the down payment, if any. You can determine the total installment price by multiplying the payment amount by the number of payments then adding the down payment.

EXAMPLE 1   Find the Total Installment Price

A microwave oven has a cash price of $299. If consumers use an installment purchase plan, they must make a down payment of $60 and make 10 equal monthly payments of $32. Determine the total installment price.

CALCULATOR TIP

Enter 32   Depress ×
Enter 10   Depress =
Read the answer $320
Enter 320 Depress +
Enter 60   Depress =
Read the answer $380

$32 × 10 = $320   **STEP** 1   Multiply the payment amount ($32) by the number of payments (10).

$320
+ 60
$380   **STEP** 2   Add the down payment.

The **finance charge** is the difference between the total installment price and the cash price. To calculate the finance charge, subtract the cash price from the total installment price.

EXAMPLE 2   Find the Finance Charge When the Cash Price is $299 and the Total Installment Price is $380

CALCULATOR TIP

Depress clear key
Enter 380   Depress −
Enter 299   Depress =
Read the answer $81

$380   total installment price
−299   cash price
$ 81   finance charge

DETERMINING THE MONTHLY PAYMENT AMOUNT

To calculate the monthly payment amount, first subtract the down payment from the cash price. Second, add the finance charge to the answer from step 1. Third, divide the step 2 answer by the number of payments.

EXAMPLE 3   Determine the Monthly Payment Amount

A stereo system has a cash price of $995. If consumers choose an installment purchase plan, they must make a down payment of $175 and pay a finance charge of $155. Assuming a consumer pays for the stereo in 12 equal monthly installments, what is the payment amount?

CALCULATOR TIP

Depress clear key
Enter 995   Depress   −
Enter 175   Depress   =
Read the answer $820
Enter 820   Depress   +
Enter 155   Depress   =
Read the answer $975
Enter 975   Depress   ÷
Enter 12   Depress   =
Read the answer $81.25

$995
−175
$820

STEP 1   Subtract the down payment ($175) from the cash price ($995).

$820
+155
$975

STEP 2   Add the finance charge ($155) to the answer from step 1.

$975 ÷ 12 = $81.25

STEP 3   Divide the answer from step 2 by the number of payments.

DETERMINING THE ANNUAL PERCENTAGE RATE

With an installment purchase, the borrower (the buyer) makes regular payments (usually on a monthly basis) to pay off a loan. And even though a portion of the loan is repaid with each payment, the borrower still pays a finance charge based on the entire borrowed amount.

To inform consumers about the *true* financial cost of borrowing money, the US Congress enacted the Truth-in-Lending Act in 1969. This law requires the lender to tell the borrower what the true annual percentage rate (APR) is for an installment loan. The **annual percentage rate** is the cost—stated as a percent—for a loan on a yearly basis. Once consumers know the APR for a loan, they can compare the cost of one loan with the cost of other loan options.

Calculating the APR requires the use of complicated formulas that often only approximate the APR. A better approach to calculating APR is to use tables published by the government. A portion of an annual percentage rate table is illustrated in Table 34–1 on pages 166–167. To use the APR table, find the number of monthly payments in the column on the far left. Then, follow across that row to find the amount closest to the finance charge per $100 of the amount financed. The easiest way to explain this is by an example.

EXAMPLE 4   Calculate the Annual Percentage Rate (APR)

The cash price for a washing machine is $450, or it can be financed for 24 months with a down payment of $100 and a finance charge of $60. What is the APR?

$450
−100
$350

STEP 1   To determine the amount financed, subtract the down payment ($100) from the cash price ($450).

$60 ÷ $350 = 0.1714   **STEP** 2   Divide the finance charge ($60) by the amount financed ($350). Carry answer to four decimal places.

0.1714 × 100 = 17.14   **STEP** 3   Because the APR table is based on $100 units, multiply the answer from step 2 by 100.

The APR (based on Table 34–1) is 15.75%   **STEP** 4   Using the APR table, go down the left column to locate the correct number of payments (24). Then move across that row to find the amount *closest* to the answer in step 3 (17.22). Follow that column up to the very top of the table to find the APR.

Complete the Following Problems

If necessary, round your answer to two decimal places.

1. In your own words, describe how you would calculate the total installment price. _____

_____

_____

2. In your own words, describe how you would calculate the finance charge for an installment purchase. _____

_____

_____

3. What does APR stand for? _____

_____

Find the Total Installment Price

Monthly Payment	Number of Monthly Payments	Down Payment	Total Installment Price
$50	10	$ 80	**4.** $50 × 10 = $500 $500 + $80 = $580
$75	18	$250	**5.** _____
$92	24	$400	**6.** _____
$24	8	$ 50	**7.** _____

Find the Monthly Payment Amount

Cash Price	Down Payment Amount	Finance Charge	Number of Payments		Monthly Payment Amount
$ 375	$ 50	$ 80	12	8.	
$1,200	$100	$190	10	9.	
$2,450	–0–	$300	24	10.	
$3,600	$720	$520	18	11.	

12. Harry Swanson saw a newspaper advertisement for a new air conditioner. The cash price was $339. Since Harry was short of cash, he decided to take advantage of the retailer's installment plan. To use the installment plan, he had to make a down payment of $50 and 10 monthly payments of $38. What was the total installment price?

$38 x 10 = $380
$380 + $50 = $430

13. A new fax machine has a cash price of $419. If consumers use an installment purchase plan, they must make a down payment of $69 and make 8 equal monthly payments of $48.75. Determine the total installment price.

14. In problem 13, what is the finance charge?

15. Best Buy Electronics offers a 27-inch color television set for $799. If consumers choose an installment purchase plan, they must make a down payment of $99 and pay a finance charge of $110. Assuming the customer pays for the television set in 24 equal monthly installments, what is the payment amount?

16. The cash price for a new ice maker is $1,050, or it can be financed for 36 months with a down payment of $200 and a finance charge of $220. What is the annual percentage rate? (Use Table 34–1.)

17. The cash price for a new television is $675, or it can be financed for 18 months with no down payment and a finance charge of $80. What is the annual percentage rate? (Use Table 34–1.)

18. The cash price for a compact disc player is $199, or it can be financed for 12 months with no down payment and a finance charge of $19. What is the annual percentage rate? (Use Table 34–1.)

19. The cash price for a fax machine is $419, or it can be financed for 18 months with a down payment of $41 and a finance charge of $45. What is the annual percentage rate? (Use Table 34–1.)

Personal Math Problems

Sandra Watson is trying to decide whether to pay cash or use an installment purchase plan to buy a computer. One local retailer offers a personal computer, monitor, and printer for $2,499, or it can be financed with a $299 down payment, a finance charge of $195, and 12 monthly payments of $199.50.

20. If Ms. Watson chooses to pay for this purchase over 12 months, what is the amount that will be financed?

21. Assuming that Ms. Watson uses the installment purchase plan, what is the APR for this transaction?

22. If you were Ms. Watson would you pay cash or finance this purchase? Why?

_____

_____

_____

_____

TABLE 34–1   Annual Percentage Rate (Finance Charge per $100 of Amount Financed)

Number of Payments	Annual Percentage Rate (Finance Charge per $100 of Amount Financed)															
	14.00%	14.25%	14.50%	14.75%	15.00%	15.25%	15.50%	15.75%	16.00%	16.25%	16.50%	16.75%	17.00%	17.25%	17.50%	17.75%
1	1.17	1.19	1.21	1.23	1.25	1.27	1.29	1.31	1.33	1.35	1.37	1.40	1.42	1.44	1.46	1.48
2	1.75	1.78	1.82	1.85	1.88	1.91	1.94	1.97	2.00	2.04	2.07	2.10	2.13	2.16	2.19	2.22
3	2.34	2.38	2.43	2.47	2.51	2.55	2.59	2.64	2.68	2.72	2.76	2.80	2.85	2.89	2.93	2.97
4	2.93	2.99	3.04	3.09	3.14	3.20	3.25	3.30	3.36	3.41	3.46	3.51	3.57	3.62	3.67	3.73
5	3.53	3.59	3.65	3.72	3.78	3.84	3.91	3.97	4.04	4.10	4.16	4.23	4.29	4.35	4.42	4.48
6	4.12	4.20	4.27	4.35	4.42	4.49	4.57	4.64	4.72	4.79	4.87	4.94	5.02	5.09	5.17	5.24
7	4.72	4.81	4.89	4.98	5.06	5.15	5.23	5.32	5.40	5.49	5.58	5.66	5.75	5.83	5.92	6.00
8	5.32	5.42	5.51	5.61	5.71	5.80	5.90	6.00	6.09	6.19	6.29	6.38	6.48	6.58	6.67	6.77
9	5.92	6.03	6.14	6.25	6.35	6.46	6.57	6.68	6.78	6.89	7.00	7.11	7.22	7.32	7.43	7.54
10	6.53	6.65	6.77	6.88	7.00	7.12	7.24	7.36	7.48	7.60	7.72	7.84	7.96	8.08	8.19	8.31
11	7.14	7.27	7.40	7.53	7.66	7.79	7.92	8.05	8.18	8.31	8.44	8.57	8.70	8.83	8.96	9.09
12	7.74	7.89	8.03	8.17	8.31	8.45	8.59	8.74	8.88	9.02	9.16	9.30	9.45	9.59	9.73	9.87
13	8.36	8.51	8.66	8.81	8.97	9.12	9.27	9.43	9.58	9.73	9.89	10.04	10.20	10.35	10.50	10.66
14	8.97	9.13	9.30	9.46	9.63	9.79	9.96	10.12	10.29	10.45	10.62	10.78	10.95	11.11	11.28	11.45
15	9.59	9.76	9.94	10.11	10.29	10.47	10.64	10.82	11.00	11.17	11.35	11.53	11.71	11.88	12.06	12.24
16	10.20	10.39	10.58	10.77	10.95	11.14	11.33	11.52	11.71	11.90	12.09	12.28	12.46	12.65	12.84	13.03
17	10.82	11.02	11.22	11.42	11.62	11.82	12.02	12.22	12.42	12.62	12.83	13.03	13.23	13.43	13.63	13.83
18	11.45	11.66	11.87	12.08	12.29	12.50	12.72	12.93	13.14	13.35	13.57	13.78	13.99	14.21	14.42	14.64
19	12.07	12.30	12.52	12.74	12.97	13.19	13.41	13.64	13.86	14.09	14.31	14.54	14.76	14.99	15.22	15.44
20	12.70	12.93	13.17	13.41	13.64	13.88	14.11	14.35	14.59	14.82	15.06	15.30	15.54	15.77	16.01	16.25
21	13.33	13.58	13.82	14.07	14.32	14.57	14.82	15.06	15.31	15.56	15.81	16.06	16.31	16.56	16.81	17.07
22	13.96	14.22	14.48	14.74	15.00	15.26	15.52	15.78	16.04	16.30	16.57	16.83	17.09	17.36	17.62	17.88
23	14.59	14.87	15.14	15.41	15.68	15.96	16.23	16.50	16.78	17.05	17.32	17.60	17.88	18.15	18.43	18.70
24	15.23	15.51	15.80	16.08	16.37	16.65	16.94	17.22	17.51	17.80	18.09	18.37	18.66	18.95	19.24	19.53
25	15.87	16.17	16.46	16.76	17.06	17.35	17.65	17.95	18.25	18.55	18.85	19.15	19.45	19.75	20.05	20.36

26	21.19	20.87	20.56	20.24	19.93	19.62	19.30	18.99	18.68	18.37	18.06	17.75	17.44	17.13	16.82	16.51
27	22.02	21.69	21.37	21.04	20.71	20.39	20.06	19.74	19.41	19.09	18.76	18.44	18.12	17.80	17.47	17.15
28	22.86	22.52	22.18	21.84	21.50	21.16	20.82	20.48	20.15	19.81	19.47	19.14	18.80	18.47	18.13	17.80
29	23.70	23.35	22.99	22.64	22.29	21.94	21.58	21.23	20.88	20.53	20.18	19.83	19.49	19.14	18.79	18.45
30	24.55	24.18	23.81	23.45	23.08	22.72	22.35	21.99	21.62	21.26	20.90	20.54	20.17	19.81	19.45	19.10
31	25.40	25.02	24.64	24.26	23.88	23.50	23.12	22.74	22.37	21.99	21.61	21.24	20.87	20.49	20.12	19.75
32	26.25	25.86	25.46	25.07	24.68	24.28	23.89	23.50	23.11	22.72	22.33	21.95	21.56	21.17	20.79	20.40
33	27.11	26.70	26.29	25.88	25.48	25.07	24.67	24.26	23.86	23.46	23.06	22.65	22.25	21.85	21.46	21.06
34	27.97	27.54	27.12	26.70	26.28	25.86	25.44	25.03	24.61	24.19	23.78	23.37	22.95	22.54	22.13	21.72
35	28.83	28.39	27.96	27.52	27.09	26.66	26.23	25.79	25.36	24.94	24.51	24.08	23.65	23.23	22.80	22.38
36	29.70	29.25	28.80	28.35	27.90	27.46	27.01	26.57	26.12	25.68	25.24	24.80	24.35	23.92	23.48	23.04
37	30.57	30.10	29.64	29.18	28.72	28.26	27.80	27.34	26.88	26.42	25.97	25.51	25.06	24.61	24.16	23.70
38	31.44	30.96	30.49	30.01	29.53	29.06	28.59	28.11	27.64	27.17	26.70	26.24	25.77	25.30	24.84	24.37
39	32.32	31.83	31.34	30.85	30.36	29.87	29.38	28.89	28.41	27.92	27.44	26.96	26.48	26.00	25.52	25.04
40	33.20	32.69	32.19	31.68	31.18	30.68	30.18	29.68	29.18	28.68	28.18	27.69	27.19	26.70	26.20	25.71
41	34.08	33.56	33.04	32.52	32.01	31.49	30.97	30.46	29.95	29.44	28.92	28.41	27.91	27.40	26.89	26.39
42	34.97	34.44	33.90	33.37	32.84	32.31	31.78	31.25	30.72	30.19	29.67	29.15	28.62	28.10	27.58	27.06
43	35.86	35.31	34.76	34.22	33.67	33.13	32.58	32.04	31.50	30.96	30.42	29.88	29.34	28.81	28.27	27.74
44	36.76	36.19	35.63	35.07	34.51	33.95	33.39	32.83	32.28	31.72	31.17	30.62	30.07	29.52	28.97	28.42
45	37.66	37.08	36.50	35.92	35.35	34.77	34.20	33.63	33.06	32.49	31.92	31.36	30.79	30.23	29.67	29.11
46	38.56	37.96	37.37	36.78	36.19	35.60	35.01	34.43	33.84	33.26	32.68	32.10	31.52	30.94	30.36	29.79
47	39.46	38.86	38.25	37.64	37.04	36.43	35.83	35.23	34.63	34.03	33.44	32.84	32.25	31.66	31.07	30.48
48	40.37	39.75	39.13	38.50	37.88	37.27	36.65	36.03	35.42	34.81	34.20	33.59	32.98	32.37	31.77	31.17
49	41.29	40.65	40.01	39.37	38.74	38.10	37.47	36.84	36.21	35.59	34.96	34.34	33.71	33.09	32.48	31.86
50	42.20	41.55	40.89	40.24	39.59	38.94	38.30	37.65	37.01	36.37	35.73	35.09	34.45	33.82	33.18	32.55
51	43.12	42.45	41.78	41.11	40.45	39.79	39.17	38.46	37.81	37.15	36.49	35.84	35.19	34.54	33.89	33.25
52	44.04	43.36	42.67	41.99	41.31	40.63	39.96	39.28	38.61	37.94	37.27	36.60	35.93	35.27	34.61	33.95
53	44.97	44.27	43.57	42.87	42.17	41.48	40.79	40.10	39.41	38.72	38.04	37.36	36.68	36.00	35.32	34.65
54	45.90	45.18	44.47	43.75	43.04	42.33	41.63	40.92	40.22	39.52	38.82	38.12	37.42	36.73	36.04	35.35
55	46.83	46.10	45.37	44.64	43.91	43.19	42.47	41.74	41.03	40.31	39.60	38.88	38.17	37.46	36.76	36.05
56	47.77	47.02	46.27	45.53	44.79	44.05	43.31	42.57	41.84	41.11	40.38	39.65	38.92	38.20	37.48	36.76
57	48.71	47.94	47.18	46.42	45.66	44.91	44.15	43.40	42.65	41.91	41.16	40.42	39.68	38.94	38.20	37.47
58	49.65	48.87	48.09	47.32	46.54	45.77	45.00	44.23	43.47	42.71	41.95	41.19	40.43	39.68	38.93	38.18
59	50.60	49.80	49.01	48.21	47.42	46.64	45.85	45.07	44.29	43.51	42.74	41.96	41.19	40.42	39.66	38.89
60	51.55	50.73	49.92	49.12	48.31	47.51	46.71	45.91	45.11	44.32	43.53	42.74	41.95	41.17	40.39	39.61

INSTANT REPLAY

UNIT	IMPORTANT POINTS TO REMEMBER	EXAMPLES
UNIT 30 Deposit Slips and Checks	A *deposit slip* is a form used for depositing money in a bank account. Before checks can be deposited, they must be endorsed. Today, three types of endorsements can be used: blank, restrictive, and special. A *check* is a written order for a bank or other financial institution to pay a stated dollar amount to the business or person indicated on the face of the check.	"John Dexter"—blank endorsement "For Deposit Only, Acct. # 234-521-07, John Dexter"—restrictive endorsement "Pay to the order of Nancy Gregg; John Dexter"—special endorsement
UNIT 31 Check Stubs and Check Registers	The information needed to complete the check stub or check register should be completed before the check is written. To record a deposit, add the deposit amount to the previous balance. To record a check, subtract the amount of the check from the previous balance.	Previous balance $1,560.34 Plus deposits + 678.00 Minus this check − 123.40 New balance $2,114.94
UNIT 32 Bank Statement Reconciliation	A bank statement shows the dollar amounts subtracted from your account. The statement also shows the dollar amounts added to your account. To reconcile your bank statement, deposits in transit are added to the balance on the bank statement. Then, outstanding checks are subtracted from the balance on the bank statement. On the checkbook side, monthly activity charges are subtracted from the checkbook balance. Then, any other bank charges (nonsufficient funds charges, check printing charges, etc.) are subtracted from the checkbook balance. Adjustments for deposits made automatically and interest paid on your account are added to the checkbook balance.	*Bank Statement* Balance $1,421.69 Deposits in transit + 905.42 Subtotal $2,327.11 Outstanding checks − 436.70 Balance $1,890.41 *Checkbook* Balance $1,903.11 Interest + 8.80 Subtotal $1,911.91 Activity fees and other bank charges − 21.50 Balance $1,890.41

continued from page 168

UNIT	IMPORTANT POINTS TO REMEMBER	EXAMPLES

UNIT 33
Credit Card Transactions

Banks charge a fee that ranges between $1\frac{1}{2}$ percent and 5 percent to process a merchant's credit card transactions. Consumers also pay finance charges. This finance charge is calculated by multiplying the finance charge stated as a percent by either the previous month's balance or the average daily balance.

$3,000	total credit card sales
× .03	bank fee (percent)
$ 90	dollar amount of bank fee
$ 900	previous month's balance
× .015	finance charge (percent)
$13.50	bank finance charge
$ 600	average daily balance
×.012	finance charge (percent)
$7.20	bank finance charge

UNIT 34
Installment Purchases

An *installment purchase* allows the consumer to make regular payments over an extended period. The *total installment price* is the total of all payments plus the down payment. The *finance charge* is the difference between the total installment price and the cash price. To calculate the monthly payment amount, first subtract the down payment from the cash price. Next, add the finance charge to the answer. Then divide the answer by the number of payments. The Truth-in-Lending Act requires that lenders disclose the annual percentage rate (APR) to consumers. When the government's tables are used, the APR is determined by the number of monthly payments and the finance charge per $100 of the amount financed.

$ 50	monthly payment
× 10	payments
$500	
+100	down payment
$600	total installment price
$600	total installment price
−499	cash price
$101	finance charge
$1,095	cash price
− 200	down payment
$ 895	
$ 895	
+ 150	finance charge
$1,045	

$1,045 ÷ 20 payments = $52.25 monthly payment

Find the APR for a 20-month loan when the finance charge is $110 and the financed amount is $850. (Use Table 34–1.)

$110 ÷ $850 × 100 = 12.94

Interest Rate from Table 34–1 = 14.25%

NOTES

CHAPTER 5 | MASTERY QUIZ

DIRECTIONS *Round each answer to two decimal places or hundredths.*

1. On July 3, Nancy Green deposited the following amounts: coins, $1.98; paper currency, $110.00; checks, $456.72. What was the total deposit amount?

2. "Jason Bart" is what type of endorsement?

3. "Pay to the order of Nell Hampton; Mike Lamp" is what type of endorsement?

4. "For Deposit Only, Baxter Company" is what type of endorsement?

5. On March 15, Peggy Bennett wants to purchase a new fuel pump for her car from Chief Auto Parts. The pump will cost $39.95. Assume that you are Peggy and complete the following check and check stub.

DATE		Bal. For'd 685.40
To the Order of		Total
		Amt. this Check
		Balance

Peggy Bennett
135 Back Bay Drive
Boston, MA 02108

439

_____ 19 _____

PAY TO THE
ORDER OF _____ $ _____

_____ DOLLARS

FOR _____   _____

⑆091901 480⑆ 8530002118314⑊ 0439

6. Given the following information, complete the check register.

CODE OR NUMBER	DATE	DESCRIPTION OF TRANSACTION	PAYMENT/DEBIT OR FEE (−)	T (√)	DEPOSIT OR CREDIT (+)	BALANCE FORWARD 8,721 35	
1123	5/1	To Nation's Repair Serv / For	145 30			Pay't or Dep	
						Bal	
		To Deposit / For			310 00	Pay't or Dep	
						Bal	

7. During the month of August, George Self wrote the following checks that have not been presented to the bank for payment: $16.70, $176.40, $2,341.10, and $56.78. What is the total of George's outstanding checks?

8–13. On November 30, Charlotte Goodson had a $1,402.03 checkbook balance. Included in Ms. Goodson's bank statement:

Bank statement balance—$1,234.58
Interest earned—$10.84
Activity charge—$7.50

A comparison of that statement with her checkbook revealed the following:

Deposits in transit—$176.00, $563.40
Outstanding checks—$12.39, $187.62, $345.00, $23.60

Using the following form, prepare a bank reconciliation:

Bank balance	_____		Checkbook balance	_____
Add deposits not shown	8. _____		Add interest earned, if any	_____
Subtotal	9. _____		Subtotal	12. _____
Subtract outstanding checks	10. _____		Subtract activity fees and other bank charges	_____
Balance	11. _____		Balance	13. _____
Do the two balances agree? _____ yes _____ no				

14. Last week, Majestic Dry Cleaners' credit card sales totaled $2,348. If the firm's bank charges $3\frac{3}{4}$ percent to process credit card transactions, what is the amount of the bank fee?

15. Last month, Mike Irwin's credit card balance was $342. If the bank charges a monthly rate of $1\frac{1}{4}$ percent calculated on the previous month's balance, what is the finance charge?

16. The average daily balance for Julie Tran's credit card account is $711. If the bank charges a monthly rate of $1\frac{2}{3}$ percent calculated on the average daily balance, what is the finance charge?

17. A stereo has a cash price of $459. If consumers use an installment purchase plan, they must make a down payment of $50 and make 12 equal monthly payments of $40. What is the total installment price?

18. In problem 17, what is the finance charge?

19. A used automobile has a cash price of $3,450. If consumers choose an installment purchase plan, they must make a down payment of $1,450 and pay a finance charge of $220. Assuming a consumer pays for the automobile in 12 equal monthly installments, what is the payment amount?

20. The cash price for a portable television is $449, or it can be financed for 18 months with a down payment of $49 and a finance charge of $49. What is the APR? (Use Table 34–1.)

Payroll

In this chapter, we look at a topic that concerns both employers and employees—payroll calculations. First, we look at how employees are paid. In this unit, we also examine how to determine overtime pay amounts. Next, we discuss commission—another method used to pay employees. Then, we describe the procedures used to determine the deduction amounts for both Social Security and income taxes. Finally, we examine taxes that must be paid by the employer.

ADP—THE NATION'S PAYROLL CLERK

New Jersey–based Automatic Data Processing, Inc., (ADP) processes payroll checks for more than 15 million employees that work for 275,000 US companies. In fact, they process payroll checks for more than 12 percent of the nation's workforce. Founded in 1949 with just eight accounts, ADP is now the largest payroll processing company in the United States. What's more, ADP is the darling of Wall Street. ADP has increased earnings per share by 10 percent or more for every quarter for the past 128 quarters—or for 32 years. Total annual sales revenues are in excess of $2.2 billion, and net profits exceed $294 million. What factors account for ADP's success?

For starters, ADP provides payroll services that take the burden off their customers. Today, the procedures for determining the correct payroll amount for each employee, withholding federal income tax, and making tax deposits for social security, medicare, federal unemployment, and state unemployment are complicated. And there are penalties if payroll calculations are done incorrectly. By contrast, ADP's "Easy Pay" computer program is designed to simplify the entire payroll process. Both small and large businesses can input information to ADP by phone, by facsimile, or by personal computer. Pay periods can be set up on a weekly, biweekly, or monthly basis. And ADP expertly prepares and submits all government reports.

Another reason for ADP's success is management's commitment to employee motivation. All employees are called "associates," and more than half own ADP stock. And because managers work hard, they motivate by example. In fact, Josh Weston, ADP's chairman and chief executive officer, visits each of the company's 50 data-processing locations at least once a year. Also, employees are trained to provide immediate solutions to problems. Even Mr. Weston answers his own telephone, and *all* 275,000 ADP clients have his direct-line telephone number!

Source: A special thanks to Automatic Data Processing, Inc., One ADP Boulevard, Roseland, NJ 07068 for providing much of the background information for this Math Today case. In addition, the following sources were used: Gary Hoover, *Hoover's Handbook of Common Stocks, 1994* (The Reference Press, 1993), pp. 226–27; *Moody's Handbook of Common Stocks, Winter 1993–1994* (New York: Moody's Investor's Service); Peter Nulty, "Making Money Like Clockwork," *Fortune,* September 30, 1993, pp. 80–81; and William Stern, "They Make Money Paying Us," *Forbes,* January 4, 1993, p. 99.

UNIT 35   DETERMINING GROSS PAY

LEARNING OBJECTIVE   *After completing this unit, you will be able to:*

1.   determine gross pay for hourly employees.

Each of Automatic Data Processing's 20,000 employees is committed to providing the best payroll service available. Ironically, it is difficult to appreciate the service provided by ADP unless the reader has ever received an *incorrect* payroll check. To calculate payroll amounts *correctly,* at least two factors must be considered. First, the employee's gross pay amount must be calculated. **Gross pay** is the amount of money an employee earns before deductions. Second, the correct amounts for Social Security, income tax, and various other deductions must be subtracted from gross pay to determine an employee's net pay. **Net pay** is an employee's actual take-home pay—what is left after all deductions have been subtracted from gross pay. This unit shows how to calculate gross pay; the following units discuss how to figure deductions and how to calculate net pay.

CALCULATING GROSS PAY FOR SALARIED EMPLOYEES

A **salary** is a specific amount of money paid for an employee's work during a calendar period, regardless of the actual number of hours worked. Today, most professional and managerial positions are salaried. How often employees are paid determines how much money they receive each pay period. Common pay periods include weekly, biweekly, semimonthly, or monthly.

Number of Paychecks Each Year

Weekly	52 paychecks
Biweekly	26 paychecks
Semimonthly	24 paychecks
Monthly	12 paychecks

To determine gross pay for a salaried employee, divide the annual salary amount by the number of paychecks the employee receives. June Ericson's annual salary is $28,000, and she is paid semimonthly. What is her gross pay each pay period?

EXAMPLE 1 Gross Pay Calculation for a Salaried Employee

$28,000 ÷ 24 = $1,166.67 — Divide the annual salary amount by 24 (the number of pay periods for employees paid semimonthly).

CALCULATOR TIP

Depress clear key
Enter 28,000 Depress ÷
Enter 24 Depress =
Read the answer $1,166,67

Because the salary amount in Example 1 represents dollars and cents, the answer is rounded off to two decimal places.

CALCULATING GROSS PAY FOR HOURLY WORKERS

Today, a large number of employees are paid an hourly wage. To calculate gross pay, multiply an employee's hourly wage rate by the number of hours worked. During the first week of March, John Peterson worked 40 hours. If John earns $8.20 an hour, what is his gross pay?

EXAMPLE 2 Gross Pay Calculation

$8.20	hourly wage
× 40	number of hours worked
$328.00	gross pay

CALCULATOR TIP

Depress clear key
Enter 8.20 Depress ×
Enter 40 Depress =
Read the answer $328

CALCULATING OVERTIME PAY AMOUNTS

The Fair Labor Standards Act covers the majority of US employees and requires that if certain employees work more than 40 hours in one week, they must be paid time-and-a-half (1.5 times the hourly wage). For overtime purposes, employees are classified as exempt or nonexempt. Exempt employees receive a salary regardless of the number of hours worked. Nonexempt employees—usually those paid on an hourly basis—are paid overtime when they work more than 40 hours a week. Some firms pay overtime rates when employees work more than eight hours in one day, and some firms pay "double time"—twice the hourly rate—for Sunday or holiday work.

To calculate time-and-a-half overtime, multiply the regular hourly wage by 1.5. Then multiply the overtime rate by the number of extra hours worked. During the second week of March, John Peterson worked 45 hours for $8.20 an hour. His gross pay for that particular week is $389.50—See Example 3.

EXAMPLE 3	Overtime Pay Calculation

CALCULATOR TIP

Depress clear key
Enter 8.20   Depress ×
Enter 40   Depress =
Read the answer $328
Enter 8.20   Depress ×
Enter 1.5   Depress =
Read the answer $12.30
Enter 12.30   Depress ×
Enter 5   Depress =
Read the answer $61.50
Enter 328   Depress +
Enter 61.50   Depress =
Read the answer $389.50

$8.20
× 40 hours
$328.00

STEP 1   Calculate the regular pay amount. Multiply the hourly wage by the number of hours worked.

$8.20
× 1.5
$12.30

STEP 2   Determine the overtime rate. Multiply the hourly wage by $1\frac{1}{2}$—the mixed fraction for time-and-a-half—or 1.5, its decimal equivalent.

$12.30
×   5 hours
$61.50

STEP 3   Calculate the overtime pay. Multiply the overtime rate ($12.30) by the number of overtime hours (5).

$328.00
+ 61.50
$389.50

STEP 4   Determine the gross pay. Add the regular pay amount ($328) and the overtime pay amount ($61.50).

HINT: You may want to review Unit 21, Adding Fractions, before completing this unit.

Complete the Following Problems

 1. In your own words, define gross pay. _____
_____

 2. In your own words, define net pay. _____
_____

Use the Following Information to Answer Problems 3–6

Mary Martinez earns $32,000 a year. What is the dollar amount of her gross pay for each of the following pay periods?

3. If paid weekly, Ms. Martinez's gross pay is _____ *$615.38* _____.

4. If paid biweekly, Ms. Martinez's gross pay is _____.

5. If paid semimonthly, Ms. Martinez's gross pay is _____.

6. If paid monthly, Ms. Martinez's gross pay is _____.

7. Linda Brown worked the following hours: Monday, $8\frac{1}{4}$ hours; Tuesday, $7\frac{1}{2}$ hours; Wednesday, 8 hours; Thursday, 9 hours; Friday, $7\frac{3}{4}$ hours; and Saturday, 3 hours. How many hours did she work during this week?

8. During the first week of October, Jack Leslie worked 38 hours. If he earns $5.85 an hour, what is his gross pay?

Use the Following Information to Answer Problems 9–11

During the fourth week of November, Kelly Wallace worked $44\frac{1}{2}$ hours. Kelly's hourly wage is $6.10.

9. What is Kelly's regular pay? _____

10. What is Kelly's overtime pay? _____

11. What is Kelly's gross pay? _____

Use the Following Information to Answer Problems 12–15

During the first week in January, Kirk Chance worked the following hours: Monday, 9 hours; Tuesday, off; Wednesday, $8\frac{1}{2}$ hours; Thursday, 10 hours; Friday, $9\frac{1}{4}$ hours; Saturday, $8\frac{3}{4}$ hours.

12. How many hours did Kirk work during this one week period? _____

13. If Kirk earns $7.00 an hour, what is his regular pay amount? _____

14. What is Kirk's overtime pay amount? _____

15. What is Kirk's gross pay amount?   _____

16. The following table is a partial payroll journal for Ferrell Wholesale Supply. Assume that this company complies with the Fair Labor Standards Act and pays overtime to any employee who works in excess of 40 hours in one week. Complete the payroll journals, determining the gross pay amount for all employees.

Ferrell Wholesale Supply Week Ending					
Employee Name	Hourly Rate	Hours Worked	Regular Pay	Overtime Pay	Gross Pay
June Sanderson	$5.90	40	*$236.00*	*0*	*$236.00*
Jim Barnes	$5.70	43			
Bob North	$5.70	36			
Nancy Strong	$7.00	44			
Vera Hess	$6.50	$41\frac{1}{2}$			
John Planter	$6.10	$46\frac{1}{4}$			

UNIT 36   CALCULATING COMMISSIONS

LEARNING OBJECTIVE   *After completing this unit, you will be able to:*

1. calculate commissions based on sales.

Many salespeople—for example, those who sell merchandise, services, real estate, or life insurance—are paid on a commission basis. **Commission** is a method of paying employees based on a percentage of the value of merchandise or services sold. Payroll calculations that involve commission are a practical application of the portion formula (Portion = Base × Rate) presented in Unit 26. The dollar amount of the commission (the portion) is found by multiplying the total amount of sales (the base) by the rate of commission (the rate). *Commission is paid only on actual sales. Amounts for returned merchandise, freight charges, and storage fees are deducted from total sales* ***before*** *commission is calculated.*

EXAMPLE 1 Calculating Commission

Dan Scott sold computers valued at $56,000. Based on a 15 percent commission, Dan's gross pay is $8,400.

$56,000	total sales (base)
× 0.15	rate of commission (rate)
$8,400.00	commission amount (portion)

CALCULATOR TIP

Depress clear key
Enter 56,000 Depress ×
Enter 0.15 Depress =
Read the answer $8,400

Remember that when you multiply percents, you need to change the percent to a decimal or use your calculator's percent key. In Example 1, the 15 percent has been converted to 0.15 before multiplying.

COMMISSION PLUS SALARY

Some employers pay their salespeople a commission *plus* a salary. For example, an employee might earn a monthly salary of $1,500 and then a commission on top of that. To figure gross pay, first determine the dollar amount of the commission, then add it to the salary amount.

EXAMPLE 2 Calculating Commission Plus a Salary Amount

Patricia Polacco earns a monthly salary of $1,000. In addition, she earns an 8 percent commission on sales. This month she has total sales of $24,500. What is her gross pay?

CALCULATOR TIP

Depress clear key
Enter 24,500 Depress ×
Enter 0.08 Depress =
Read the answer $1,960
Enter 1,960 Depress +
Enter 1,000 Depress =
Read the answer $2,960

$24,500	**STEP** 1	Multiply total sales by the rate of commission.
× 0.08		
$1,960.00		
$1,960.00	**STEP** 2	Add the dollar amount of the commission to the salary amount to obtain gross pay.
+1,000.00		
$2,960.00		

Note that the rate of commission (8 percent) has been changed to 0.08 before multiplying in step 1.

FINDING GROSS PAY WHEN AN EMPLOYEE RECEIVES A DRAW

A **draw** is an advance against commissions paid to a salesperson. To figure gross pay when there is a draw, first determine the dollar amount of the commission, *then* subtract the draw amount.

EXAMPLE 3　Calculating Commission Minus a Draw Amount

Billy Janesco received a $750 draw on April 10. He also sold packaging material valued at $64,200 during the month of April. If he is paid a 5 percent commission, what is his gross pay at the end of April?

$64,200
×　.05
$3,210.00

STEP 1　Multiply total sales by the rate of commission.

$3,210.00
−　750.00
$2,460.00

STEP 2　Subtract the dollar amount of the draw from the commission amount to obtain gross pay.

GRADUATED COMMISSION SCALES

To encourage a salesperson to work harder, some firms use a graduated commission scale. When a graduated commission scale (sometimes called a sliding scale) is used, the rate of commission increases as the salesperson's sales increase. For example, a salesperson might earn a 6 percent commission on sales up to $10,000 and an 8 percent commission on additional sales. To figure gross pay, first determine the commission earned on the sales base ($10,000). Next figure the commission earned on the sales in excess of the sales base. Then, add the two commission amounts.

EXAMPLE 4　Calculating Commission Based on a Graduated Scale

Bill Monroe is paid 5 percent commission on all sales up to $12,000 and 7 percent on all sales in excess of $12,000. During the month of April, Bill sold merchandise valued at $41,000. What is his gross pay?

$12,000
×　0.05
$600.00

STEP 1　Multiply the sales base ($12,000) by the rate of commission (5 percent).

$41,000
−12,000
$29,000

STEP 2　Subtract the sales base from total monthly sales to determine sales in excess of the base amount.

$29,000
×　0.07
$2,030.00

STEP 3　Multiply sales in excess of the base amount ($29,000) by the appropriate rate of commission (7 percent).

$600.00
+2,030.00
$2,630.00

STEP 4　Add the commission amounts.

HINT: You may want to review the material on portion in Unit 26 before completing this unit.

Complete the Following Problems

Total Sales	Commission Rate	Commission Amount
$24,000	10%	1. $24,000 × .10 = $2,400
$ 8,600	8%	2. _____
$42,100	12%	3. _____
$50,360	5%	4. _____
$26,800	7%	5. _____

6. During the month of May, Jan Michaels sold office equipment valued at $23,500. If Jan is paid a 12 percent commission, how much commission did she earn?

$23,500 × .12 = $2,820

7. Mark Sammons works as a salesperson for Madison's Home Furnishings. During the first week of June, Mark sold furniture valued at $10,430. If he is paid a $4\frac{1}{2}$ percent commission, how much commission did he earn?

8. Peter Cervantes is paid a monthly salary of $650 plus 4 percent commission on sales. If he sold merchandise valued at $86,400 during the month of December, what was his gross pay for the month?

9. Maxwell Manufacturing pays all sales personnel $1,000 a month plus $3\frac{1}{4}$ percent commission on all of their individual sales. During the month of November, Mary Jensen sold plastic products worth $104,200. Her sales returns totaled $1,290. What was her gross pay for the month of November?

10. Mike Nottoway works for Cleveland Printing, Inc. For the month of December, Mr. Nottoway received an $800 draw. He also sold printing services totaling $78,640. If he is paid 4.1 percent commission, what is his gross pay amount at the end of December?

11. Jennifer's Sports Wear pays all salespeople 3 percent on sales up to $18,000 and $5\frac{1}{2}$ percent commission on sales in excess of $18,000. If Miriam Goldsmith sold merchandise valued at $43,264 during April, what is her gross pay for April?

12. Baltimore-based Pilgrim's Manufacturing Company pays all salespersons 2.5 percent commission on sales up to $25,000 and 4.3 percent commission on sales in excess of $25,000. Also, some employees are paid monthly salaries in addition to commissions. Using the following table, determine the gross pay for each employee, then determine the totals for salaries, sales and gross pay.

Pilgrim's Manufacturing Company Month ended May 19XX			
Salesperson	**Salary**	**Sales**	**Gross Pay**
Sid Bonner	-0-	$64,310	$ 2,315.33
Betty Mason	$1,000	$73,210	
Bill Wooley	-0-	$49,750	
Wallace Oakley	-0-	$51,635	
Connie Ling	$1,500	$61,111	
Dan Patrillo	-0-	$69,200	
Peggy Horton	-0-	$54,390	
Totals			

Personal Math Problem

Betty Matlock is currently unemployed, but she has been offered two different jobs. One company—Mountain Vernon Carpets—has offered to pay her $6\frac{3}{4}$ percent commission on all sales. The second company—Down East Publishing—has offered her a $1,500 monthly salary plus a $3\frac{1}{4}$ percent commission on all sales.

13. If she sold merchandise or services valued at $40,000 and worked for Mountain Vernon Carpets, what is her gross pay?

14. If she sold merchandise or services valued at $40,000 and worked for Down East Publishing, what is her gross pay?

15. If you were Ms. Matlock, which job would you choose? Why? _____

_____

_____

| UNIT 37 | COMPUTING DEDUCTIONS FOR SOCIAL SECURITY (FICA) |

LEARNING OBJECTIVE   *After completing this unit, you will be able to:*

1. calculate the deduction amount for Social Security.

Today, there are many deductions that reduce an employee's take-home pay. One in particular—the Federal Insurance Contributions Act (FICA)—provides funding for retirement, disability, Medicare, and death benefits for eligible employees. Both the employer and the employee pay this tax. The employer withholds the employee's share from his or her salary and sends it to the federal government along with the employer's share. **(NOTE: The employer's contribution is equal to the employee's contribution.)**

At the time of publication, the FICA tax was broken into two components: Social Security and Medicare. The employees' tax rate for Social Security was 6.2 percent on the first $60,600 earned. The employees' tax rate for Medicare was 1.45 percent. Currently, there is no wage ceiling for Medicare. Self-employed individuals pay 12.4 percent on the first $60,600 earned. Self-employed individuals also pay 2.9 percent for Medicare. Again, there is no wage ceiling for Medicare for self-employed individuals.

Deductions for Social Security and Medicare are a practical application of the portion formula (Portion = Base × Rate) presented in Unit 26. For example, the Social Security tax amount (portion) is found by multiplying the gross pay (base) by the Social Security tax rate (6.2 percent or 0.062). The Medicare tax amount (portion) is found by multiplying the gross pay (base) by the Medicare tax rate (1.45 percent or 0.0145).

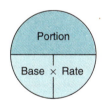

Alexander Wallace earned $560 during the first week of July. How much of his salary will be deducted for the Social Security tax and the Medicare tax?

EXAMPLE 1   Calculating the Social Security and Medicare Tax Amount

CALCULATOR TIP

Depress clear key
Enter 560   Depress ×
Enter 0.062   Depress =
Read the answer $34.72
Enter 560   Depress ×
Enter 0.0145   Depress =
Read the answer $8.12

$560 × .062 = $34.72   **STEP** [1]   Multiply the wages subject to Social Security by 6.2 percent.

$560 × .0145 = $8.12   **STEP** [2]   Multiply the wages subject to Medicare by 1.45 percent.

Social Security taxes are deducted only from the first $60,600 an employee earns. That is, no Social Security taxes are withheld from the portion of the wages that exceeds the wage ceiling. Medicare taxes are deducted on all wages an individual earns.

| EXAMPLE 2 | Calculating the Social Security and Medicare Tax Amounts When Gross Pay Exceeds the Ceiling Amount |

Amy Polaski has earned $59,300 during the first ten months of the calendar year. During the month of November, Amy earns $3,500. What are her Social Security and Medicare tax deductions for November?

CALCULATOR TIP

Depress clear key
Enter 60,600   Depress   −
Enter 59,300   Depress   =
Read the answer $1,300
Enter 1,300   Depress   ×
Enter 0.062   Depress   =
Read the answer $80.60
Enter 3,500   Depress   ×
Enter 0.0145   Depress   =
Read the answer $50.75

$60,600
−59,300
$ 1,300

STEP 1 — Subtract year-to-date wages ($59,300) from the wage ceiling ($60,600) to determine the amount of wages still subject to Social Security.

$1,300 × 0.062 = $80.60

STEP 2 — Multiply the wages subject to Social Security ($1,300) by the Social Security tax rate (6.2 percent).

$3,500 × .0145 = $50.75

STEP 3 — Multiply the November salary ($3,500) by the Medicare tax rate (1.45 percent).

Once an employee's wages have passed the wage ceiling for Social Security, she or he no longer has to pay Social Security taxes until the next calendar year begins.

HINT: Since FICA tax rates and the wage amounts subject to FICA do change, you may want to check with the IRS or the Social Security Administration for the most current tax rates. Both government agencies have toll-free telephone numbers.

Complete the Following Problems

1. ⓣ F   The Federal Insurance Contributions Act provides funding for the Social Security program.

2. T   F   The FICA tax is paid only by employees.

3. T   F   The Social Security tax rate is 4.5 percent on the first $57,600.

4. T   F   The Medicare tax rate is 1.45 percent.

NOTE: For all problems in this unit, assume a 6.2 percent tax rate and a wage ceiling of $60,600 for Social Security. Also assume a 1.45 percent tax rate for Medicare.

5. Mathew Smith earns $360 a week. What is his weekly Social Security tax deduction?

   $360 × .062 = $22.32

6. Noreen Chambers earns $570 a week. What is Noreen's weekly Medicare tax deduction?

7. Jane Winston works for Morningstar Publications. She earns $6.70 an hour. During the first week of January, she worked 38 hours. What is her gross pay amount?

8. For problem 7, what is Jane's Social Security deduction?

9. For problem 7, what is Jane's Medicare deduction?

10. Mark Sanders works for Financial Securities, Inc. He earns $8.40 an hour. During the first week of February, he worked 42 hours. Assuming that Financial Securities pays time-and-a-half for all hours in excess of 40 hours a week, what is Mark's gross pay?

11. For problem 10, what is Mark's Social Security tax deduction?

12. For problem 10, what is Mark's Medicare tax deduction?

13. For problem 10, what is the *total* amount that Mark pays for FICA tax deductions?

14. Walter Knox's cumulative earnings to date for the year are $58,750. During the month of October, Walter earned $4,250. What is Walter's Social Security tax deduction for the month of October?

15. For problem 14, what is Walter's Medicare tax deduction for the month of October?

16. Sarah Swanson's cumulative earnings to date for the year are $54,970. During the month of December, Sarah earned $4,975. What is Sarah's Social Security tax deduction for the month of December?

17. For problem 16, what is Sarah's Medicare deduction for the month of December?

| UNIT 38 | COMPUTING DEDUCTIONS FOR INCOME TAX |

LEARNING OBJECTIVES   *After completing this unit you will be able to:*

1. calculate the withholding amount for income tax.
2. determine net pay for employees.

An employer is required by law to deduct federal income tax withholding from each employee's paycheck. Both marital status and withholding allowances (sometimes called exemptions) affect the amount of federal income tax withheld.

Today, business firms can choose from two methods when calculating the deduction for federal income withholding: the percentage method or the wage bracket table method. Instructions and tables for both methods are provided in the Internal Revenue Service's *Circular E—Employer's Tax Guide.*

THE WAGE BRACKET TABLE METHOD

Since most employers—especially small business owners—prefer the wage bracket table method, we will discuss that method first. Wage bracket tables are available for single and married employees who are paid weekly, biweekly, semimonthly, and monthly. Table 38–1 (p. 188) shows a portion of the tax table for single taxpayers paid on a monthly basis; Table 38–2 (p. 190) shows a portion of the tax table for married taxpayers paid on a weekly basis.

EXAMPLE 1   Calculating Federal Income Tax Withheld Using the Wage Bracket Table Method

Find the federal income tax withholding amount for Theo Lea, who is single, claims two exemptions, and earns $1,985 a month.

STEP ☐1   Determine if the worker is single or married and paid weekly or monthly, and refer to the appropriate tax table. Theo is single and paid on a monthly basis. Use Table 38–1.

STEP 2   Using the two columns on the left side of the tax table, find the employee's wage amount by moving down the column to the appropriate line. Theo earns $1,985 each month, which is "at least" $1,960 "but less than" $2,000.

STEP 3   Read across to the correct column for the number of withholding allowances (exemptions) claimed by the worker. Theo claims two withholding allowances, so read across to the appropriate column. His withholding amount is $204.

THE PERCENTAGE METHOD

If you do not use the wage-bracket method to determine how much to withhold, you can use the percentage method. Withholding allowance tables and tax tables for the percentage method for both single and married taxpayers are available from the federal government. Table 38–3 (p. 192) shows a table used to determine withholding allowances; Table 38–4 (p. 193) shows tables for the percentage method of withholding for employees paid weekly, biweekly, semimonthly, or monthly.

EXAMPLE 2   Calculating Federal Income Tax Withholding Using the Percentage Method

Find the federal income tax withholding amount for Leah Bell, who is single, claims two withholding allowances, and earns $1,240 semimonthly.

CALCULATOR TIP

Depress clear key
| Enter 102.08 | Depress × |
| Enter 2 | Depress = |
Read the answer $204.16
| Enter 1,240 | Depress − |
| Enter 204.16 | Depress = |
Read the answer $1,035.84
| Enter 1,035.84 | Depress − |
| Enter 1,004.00 | Depress = |
Read the answer $31.84
| Enter 31.84 | Depress × |
| Enter 0.28 | Depress = |
Read the answer $8.92
| Enter 134.55 | Depress + |
| Enter 8.92 | Depress = |
Read the answer $143.47

The dollar amount is $102.08

STEP 1   From Table 38–3 determine the dollar amount for one withholding allowance.

$102.08 × 2 = $204.16

STEP 2   Multiply the dollar amount for each withholding allowance (step 1) by the number of withholding allowances for this employee (2).

$1,240.00
− 204.16
$ 1035.84

STEP 3   Subtract the answer in step 2 from the employee's gross pay.

$1,035.84
− 1004.00
$ 31.84

$31.84 × 0.28 = $8.92

$134.55
+ 8.92
$143.47 total withholding
       amount

STEP 4   Using the section of Table 38–4 for single taxpayers paid semimonthly, determine the employee's withholding tax amount. The tax on the first $1,004 is $134.55 *plus* 28 percent of the excess over $1,004. The total tax is $134.55 plus $8.92 or $143.47.

CALCULATING NET PAY

Along with the Social Security and Medicare tax deductions (Unit 37), the deduction for federal income tax withholding must be subtracted from gross pay to determine an employee's net pay.

TABLE 38–1   SINGLE Persons—**MONTHLY** Payroll Period (for Wages Paid in 1994)

If the wages are—		And the number of withholding allowances claimed is—										
		0	1	2	3	4	5	6	7	8	9	10
At least	But less than	The amount of income tax to be withheld is—										
$ 0	$ 220	$ 0	$ 0	$ 0	$ 0	$ 0	$ 0	$ 0	$ 0	$ 0	$ 0	$ 0
220	230	2	0	0	0	0	0	0	0	0	0	0
230	240	3	0	0	0	0	0	0	0	0	0	0
240	250	5	0	0	0	0	0	0	0	0	0	0
250	260	6	0	0	0	0	0	0	0	0	0	0
260	270	8	0	0	0	0	0	0	0	0	0	0
270	280	9	0	0	0	0	0	0	0	0	0	0
280	290	11	0	0	0	0	0	0	0	0	0	0
290	300	12	0	0	0	0	0	0	0	0	0	0
300	320	14	0	0	0	0	0	0	0	0	0	0
320	340	17	0	0	0	0	0	0	0	0	0	0
340	360	20	0	0	0	0	0	0	0	0	0	0
360	380	23	0	0	0	0	0	0	0	0	0	0
380	400	26	0	0	0	0	0	0	0	0	0	0
400	420	29	0	0	0	0	0	0	0	0	0	0
420	440	32	2	0	0	0	0	0	0	0	0	0
440	460	35	5	0	0	0	0	0	0	0	0	0
460	480	38	8	0	0	0	0	0	0	0	0	0
480	500	41	11	0	0	0	0	0	0	0	0	0
500	520	44	14	0	0	0	0	0	0	0	0	0
520	540	47	17	0	0	0	0	0	0	0	0	0
540	560	50	20	0	0	0	0	0	0	0	0	0
560	580	53	23	0	0	0	0	0	0	0	0	0
580	600	56	26	0	0	0	0	0	0	0	0	0
600	640	61	30	0	0	0	0	0	0	0	0	0
640	680	67	36	6	0	0	0	0	0	0	0	0
680	720	73	42	12	0	0	0	0	0	0	0	0
720	760	79	48	18	0	0	0	0	0	0	0	0
760	800	85	54	24	0	0	0	0	0	0	0	0
800	840	91	60	30	0	0	0	0	0	0	0	0
840	880	97	66	36	5	0	0	0	0	0	0	0
880	920	103	72	42	11	0	0	0	0	0	0	0
920	960	109	78	48	17	0	0	0	0	0	0	0
960	1,000	115	84	54	23	0	0	0	0	0	0	0
1,000	1,040	121	90	60	29	0	0	0	0	0	0	0
1,040	1,080	127	96	66	35	4	0	0	0	0	0	0
1,080	1,120	133	102	72	41	10	0	0	0	0	0	0
1,120	1,160	139	108	78	47	16	0	0	0	0	0	0
1,160	1,200	145	114	84	53	22	0	0	0	0	0	0
1,200	1,240	151	120	90	59	28	0	0	0	0	0	0
1,240	1,280	157	126	96	65	34	4	0	0	0	0	0
1,280	1,320	163	132	102	71	40	10	0	0	0	0	0
1,320	1,360	169	138	108	77	46	16	0	0	0	0	0
1,360	1,400	175	144	114	83	52	22	0	0	0	0	0
1,400	1,440	181	150	120	89	58	28	0	0	0	0	0
1,440	1,480	187	156	126	95	64	34	3	0	0	0	0
1,480	1,520	193	162	132	101	70	40	9	0	0	0	0
1,520	1,560	199	168	138	107	76	46	15	0	0	0	0
1,560	1,600	205	174	144	113	82	52	21	0	0	0	0
1,600	1,640	211	180	150	119	88	58	27	0	0	0	0

continued

TABLE 38–1　SINGLE Persons—MONTHLY Payroll Period *(concluded)*

If the wages are—		And the number of withholding allowances claimed is—										
		0	1	2	3	4	5	6	7	8	9	10
At least	But less than	The amount of income tax to be withheld is—										
1,640	**1,680**	217	186	156	125	94	64	33	2	0	0	0
1,680	**1,720**	223	192	162	131	100	70	39	8	0	0	0
1,720	**1,760**	229	198	168	137	106	76	45	14	0	0	0
1,760	**1,800**	235	204	174	143	112	82	51	20	0	0	0
1,800	**1,840**	241	210	180	149	118	88	57	26	0	0	0
1,840	**1,880**	247	216	186	155	124	94	63	32	2	0	0
1,880	**1,920**	253	222	192	161	130	100	69	38	8	0	0
1,920	**1,960**	259	228	198	167	136	106	75	44	14	0	0
1,960	**2,000**	265	234	204	173	142	112	81	50	20	0	0
2,000	**2,040**	272	240	210	179	148	118	87	56	26	0	0
2,040	**2,080**	284	246	216	185	154	124	93	62	32	1	0
2,080	**2,120**	295	252	222	191	160	130	99	68	38	7	0
2,120	**2,160**	306	258	228	197	166	136	105	74	44	13	0
2,160	**2,200**	317	264	234	203	172	142	111	80	50	19	0
2,200	**2,240**	328	271	240	209	178	148	117	86	56	25	0
2,240	**2,280**	340	282	246	215	184	154	123	92	62	31	1
2,280	**2,320**	351	294	252	221	190	160	129	98	68	37	7
2,320	**2,360**	362	305	258	227	196	166	135	104	74	43	13
2,360	**2,400**	373	316	264	233	202	172	141	110	80	49	19
2,400	**2,440**	384	327	270	239	208	178	147	116	86	55	25

EXAMPLE 3　Calculating Net Pay

Jackie Thornton is married and claims two withholding allowances. Her weekly salary is $700. What is her net pay?

CALCULATOR TIP

Depress clear key
Enter 700　　Depress ×
Enter 0.062　Depress =
Read the answer $43.40
Enter 700　　Depress ×
Enter 0.0145 Depress =
Read the answer $10.15
Enter 700　　Depress −
Enter 43.40　Depress −
Enter 10.15　Depress −
Enter 73　　　Depress =
Read the answer $573.45

$700 × .062 = $43.40　　**STEP** [1]　Multiply gross pay ($700) by the Social Security tax rate (6.2 percent or 0.062).

$700 × .0145 = $10.15　**STEP** [2]　Multiply gross pay ($700) by the Medicare tax rate (1.45 percent or 0.0145).

Based on the information in Table 38–2, Jackie's deduction for federal income tax withholding is $73.　**STEP** [3]　Determine the amount for federal income tax withholding. Use the wage-bracket table method.

$700.00
− 43.40
− 10.15
− 73.00
$573.45

STEP [4]　Subtract Social Security, Medicare, and federal withholding amounts from gross pay to determine net pay.

HINT: You may want to review the material on calculating Social Security and Medicare tax amounts in Unit 37 before completing this unit.

190 CHAPTER SIX

TABLE 38–2 MARRIED Persons—WEEKLY Payroll Period (for Wages Paid in 1994)

At least	But less than	0	1	2	3	4	5	6	7	8	9	10
$ 0	$125	$ 0	$ 0	$ 0	$ 0	$ 0	$ 0	$ 0	$ 0	$ 0	$ 0	$ 0
125	130	1	0	0	0	0	0	0	0	0	0	0
130	135	2	0	0	0	0	0	0	0	0	0	0
135	140	2	0	0	0	0	0	0	0	0	0	0
140	145	3	0	0	0	0	0	0	0	0	0	0
145	150	4	0	0	0	0	0	0	0	0	0	0
150	155	5	0	0	0	0	0	0	0	0	0	0
155	160	5	0	0	0	0	0	0	0	0	0	0
160	165	6	0	0	0	0	0	0	0	0	0	0
165	170	7	0	0	0	0	0	0	0	0	0	0
170	175	8	0	0	0	0	0	0	0	0	0	0
175	180	8	1	0	0	0	0	0	0	0	0	0
180	185	9	2	0	0	0	0	0	0	0	0	0
185	190	10	3	0	0	0	0	0	0	0	0	0
190	195	11	3	0	0	0	0	0	0	0	0	0
195	200	11	4	0	0	0	0	0	0	0	0	0
200	210	12	5	0	0	0	0	0	0	0	0	0
210	220	14	7	0	0	0	0	0	0	0	0	0
220	230	15	8	1	0	0	0	0	0	0	0	0
230	240	17	10	3	0	0	0	0	0	0	0	0
240	250	18	11	4	0	0	0	0	0	0	0	0
250	260	20	13	6	0	0	0	0	0	0	0	0
260	270	21	14	7	0	0	0	0	0	0	0	0
270	280	23	16	9	2	0	0	0	0	0	0	0
280	290	24	17	10	3	0	0	0	0	0	0	0
290	300	26	19	12	5	0	0	0	0	0	0	0
300	310	27	20	13	6	0	0	0	0	0	0	0
310	320	29	22	15	8	1	0	0	0	0	0	0
320	330	30	23	16	9	2	0	0	0	0	0	0
330	340	32	25	18	11	4	0	0	0	0	0	0
340	350	33	26	19	12	5	0	0	0	0	0	0
350	360	35	28	21	14	7	0	0	0	0	0	0
360	370	36	29	22	15	8	1	0	0	0	0	0
370	380	38	31	24	17	10	3	0	0	0	0	0
380	390	39	32	25	18	11	4	0	0	0	0	0
390	400	41	34	27	20	13	6	0	0	0	0	0
400	410	42	35	28	21	14	7	0	0	0	0	0
410	420	44	37	30	23	16	9	2	0	0	0	0
420	430	45	38	31	24	17	10	3	0	0	0	0
430	440	47	40	33	26	19	12	5	0	0	0	0
440	450	48	41	34	27	20	13	6	0	0	0	0
450	460	50	43	36	29	22	15	8	0	0	0	0
460	470	51	44	37	30	23	16	9	2	0	0	0
470	480	53	46	39	32	25	18	11	3	0	0	0
480	490	54	47	40	33	26	19	12	5	0	0	0
490	500	56	49	42	35	28	21	14	6	0	0	0
500	510	57	50	43	36	29	22	15	8	1	0	0
510	520	59	52	45	38	31	24	17	9	2	0	0
520	530	60	53	46	39	32	25	18	11	4	0	0
530	540	62	55	48	41	34	27	20	12	5	0	0

If the wages are— / And the number of withholding allowances claimed is— / The amount of income tax to be withheld is—

continued

TABLE 38–2 MARRIED Persons—**WEEKLY** Payroll Period *(concluded)*

If the wages are—		And the number of withholding allowances claimed is—										
		0	1	2	3	4	5	6	7	8	9	10
At least	But less than	The amount of income tax to be withheld is—										
540	550	63	56	49	42	35	28	21	14	7	0	0
550	560	65	58	51	44	37	30	23	15	8	1	0
560	570	66	59	52	45	38	31	24	17	10	3	0
570	580	68	61	54	47	40	33	26	18	11	4	0
580	590	69	62	55	48	41	34	27	20	13	6	0
590	600	71	64	57	50	43	36	29	21	14	7	0
600	610	72	65	58	51	44	37	30	23	16	9	2
610	620	74	67	60	53	46	39	32	24	17	10	3
620	630	75	68	61	54	47	40	33	26	19	12	5
630	640	77	70	63	56	49	42	35	27	20	13	6
640	650	78	71	64	57	50	43	36	29	22	15	8
650	660	80	73	66	59	52	45	38	30	23	16	9
660	670	81	74	67	60	53	46	39	32	25	18	11
670	680	83	76	69	62	55	48	41	33	26	19	12
680	690	84	77	70	63	56	49	42	35	28	21	14
690	700	86	79	72	65	58	51	44	36	29	22	15
700	710	87	80	73	66	59	52	45	38	31	24	17
710	720	89	82	75	68	61	54	47	39	32	25	18
720	730	90	83	76	69	62	55	48	41	34	27	20
730	740	92	85	78	71	64	57	50	42	35	28	21

TABLE 38–3 Percentage Method—Amount for One Withholding Allowance

Payroll Period	One withholding allowance
Weekly	$ 47.12
Biweekly	94.23
Semimonthly	102.08
Monthly	204.17
Quarterly	612.50
Semiannually	1,225.00
Annually	2,450.00
Daily or miscellaneous (each day of the payroll period)	9.42

TABLE 38–4 Tables for Percentage Method of Withholding (for Wages Paid in 1994)

TABLE 1—WEEKLY Payroll Period

(a) SINGLE person (including head of household)—

If the amount of wages (after subtracting withholding allowances) is:

Not over $50 . . . . The amount of income tax to withhold is: $0

Over—	But not over—	The amount of income tax to withhold is:	of excess over—
$50	—$463	15%	—$50
$463	—$968	$61.95 plus 28%	—$463
$968	—$2,238	$203.35 plus 31%	—$968
$2,238	—$4,834	$597.05 plus 36%	—$2,238
$4,834	. . . .	$1,531.61 plus 39.6%	—$4,834

(b) MARRIED person—

If the amount of wages (after subtracting withholding allowances) is:

Not over $122 . . . . The amount of income tax to withhold is: $0

Over—	But not over—	The amount of income tax to withhold is:	of excess over—
$122	—$806	15%	—$122
$806	—$1,606	$102.60 plus 28%	—$806
$1,606	—$2,767	$326.60 plus 31%	—$1,606
$2,767	—$4,883	$686.51 plus 36%	—$2,767
$4,883	. . . .	$1,448.27 plus 39.6%	—$4,883

TABLE 2—BIWEEKLY Payroll Period

(a) SINGLE person (including head of household)—

If the amount of wages (after subtracting withholding allowances) is:

Not over $99 . . . . The amount of income tax to withhold is: $0

Over—	But not over—	The amount of income tax to withhold is:	of excess over—
$99	—$927	15%	—$99
$927	—$1,936	$124.20 plus 28%	—$927
$1,936	—$4,475	$406.72 plus 31%	—$1,936
$4,475	—$9,667	$1,193.81 plus 36%	—$4,475
$9,667	. . . .	$3,062.93 plus 39.6%	—$9,667

(b) MARRIED person—

If the amount of wages (after subtracting withholding allowances) is:

Not over $244 . . . . The amount of income tax to withhold is: $0

Over—	But not over—	The amount of income tax to withhold is:	of excess over—
$244	—$1,612	15%	—$244
$1,612	—$3,212	$205.20 plus 28%	—$1,612
$3,212	—$5,535	$653.20 plus 31%	—$3,212
$5,535	—$9,765	$1,373.33 plus 36%	—$5,535
$9,765	. . . .	$2,896.13 plus 39.6%	—$9,765

(continued)

192

TABLE 38-4 Tables for Percentage Method of Withholding (for Wages Paid in 1994) concluded

TABLE 3—SEMIMONTHLY Payroll Period

(a) SINGLE person (including head of household)—

If the amount of wages (after subtracting withholding allowances) is:

The amount of income tax to withhold is:

Not over $107 . . . . $0

Over—	But not over—		of excess over—
$107	—$1,004	15%	—$107
$1,004	—$2,097	$134.55 plus 28%	—$1,004
$2,097	—$4,848	$440.59 plus 31%	—$2,097
$4,848	—$10,473	$1,293.40 plus 36%	—$4,848
$10,473	. . . .	$3,318.40 plus 39.6%	—$10,473

(b) MARRIED person—

If the amount of wages (after subtracting withholding allowances) is:

The amount of income tax to withhold is:

Not over $265 . . . . $0

Over—	But not over—		of excess over—
$265	—$1,746	15%	—$265
$1,746	—$3,479	$222.15 plus 28%	—$1,746
$3,479	—$5,996	$707.39 plus 31%	—$3,479
$5,996	—$10,579	$1,487.66 plus 36%	—$5,996
$10,579	. . . .	$3,137.54 plus 39.6%	—$10,579

TABLE 4—MONTHLY Payroll Period

(a) SINGLE person (including head of household)—

If the amount of wages (after subtracting withholding allowances) is:

The amount of income tax to withhold is:

Not over $215 . . . . $0

Over—	But not over—		of excess over—
$215	—$2,008	15%	—$215
$2,008	—$4,194	$268.95 plus 28%	—$2,008
$4,194	—$9,696	$881.03 plus 31%	—$4,194
$9,696	—$20,946	$2,586.65 plus 36%	—$9,696
$20,946	. . . .	$6,636.65 plus 39.6%	—$20,946

(b) MARRIED person—

If the amount of wages (after subtracting withholding allowances) is:

The amount of income tax to withhold is:

Not over $529 . . . . $0

Over—	But not over—		of excess over—
$529	—$3,492	15%	—$529
$3,492	—$6,958	$444.45 plus 28%	—$3,492
$6,958	—$11,992	$1,414.93 plus 31%	—$6,958
$11,992	—$21,158	$2,975.47 plus 36%	—$11,992
$21,158	. . . .	$6,275.23 plus 39.6%	—$21,158

Complete the Following Problems

For all problems in this unit, assume a 6.2 percent tax rate and a wage ceiling of $60,600 for Social Security. Also, assume a 1.45 percent tax rate for Medicare.

For problems 1–3, use Tables 38–1 and 38–2.

1. Benny Rosen is married and claims four withholding allowances. If Benny's weekly salary is $722, how much should his employer withhold for federal income tax?

   $62

2. Sharon Doherty is single and claims one withholding allowance. If Sharon's monthly salary is $1,840, how much should her employer withhold for federal income tax?

3. Peter Walker is single and claims five withholding allowances. His gross pay for the month is $1,950. How much should his employer withhold for federal income tax?

For problems 4–6, use Tables 38–3 and 38–4.

4. Martin Guerra is single and claims one withholding allowance. If Martin's biweekly salary is $1,340, how much should his employer withhold for federal income tax?

5. Patti Pyle is married and claims two withholding allowances. Her semimonthly salary is $2,000. How much should her employer withhold for federal income tax?

6. Peggy Hess is married and claims three withholding allowances. Her monthly salary is $2,900. How much should her employer withhold for federal income tax?

7. Charlene Wolf works as an administrative assistant for Southwest Hospital System. She is paid $16.25 an hour. She is married and claims one withholding allowance. If she works 40 hours a week, what is her gross pay?

8. For problem 7, what is Charlene's Social Security tax deduction each week?

9. For problem 7, what is Charlene's Medicare tax deduction each week?

10. For problem 7, how much federal income tax is withheld from Charlene's weekly paycheck? (Use the wage-bracket-tax-table method.)

11. Robert Hall works for Acme Plastics and is paid $14.40 an hour. Robert is married and claims two withholding allowances. During the first week of March, Robert worked $43\frac{1}{2}$ hours. What is his gross pay (Assume time-and-a-half for overtime.)

12. For problem 11, what is Robert's Social Security tax deduction for the week?

13. For problem 11, what is Robert's Medicare tax deduction for the week?

14. For problem 11, what is Robert's federal income tax withholding for the week? (Use the wage-bracket-tax-table method.)

15. For problem 11, what is Robert's net pay?

16. A partial monthly payroll register for Carter's Electronics is provided here. Your job as payroll clerk is to use the information provided on the form to determine each employee's net wages paid. Then, determine the dollar amount for (a) total wages; (b) federal withholding; (c) Social Security tax; (d) Medicare tax; (e) total deductions; and (f) net wages paid. (Use the wage-bracket-tax-table method.)

Carter's Electronics
Month Ending 1–31–19XX

Employee's Name	Marital Status	Exemptions	Total Wages	Federal Withholding	Social Security	Medicare	Total Deductions	Net Wages Paid
Bosworth, B.	S	1	$1,560	$174	$ 96.72	$22.62	$ 293.34	$ 1,266.66
Duncan, T.	S	4	2,175					
Hartley, C.	S	0	1,210					
James, S.	S	3	1,800					
Martin, B.	S	2	1,970					
Neeley, F.	S	0	2,080					
Roberts, S.	S	1	1,900					
Taylor, M.	S	2	2,080					
Totals	XXXXX	XXXXX						

UNIT 39	EMPLOYER'S TAX RESPONSIBILITIES

LEARNING OBJECTIVES

After completing this unit, you will be able to:

1. calculate the tax amount for federal unemployment.
2. calculate the tax amount for state unemployment.

Generally, an employer is any person or business that during the current year or during the last year either (1) paid wages of $1,500 or more in any calendar quarter *or* (2) had one or more employees at any time in each of 20 different calendar weeks. As an employer, you are responsible for deducting Social Security, Medicare, and federal income tax withholding from employees' earnings. You are also responsible for depositing all monies collected from employees *and* paying the employer's share of Social Security and Medicare. Finally, you must pay federal unemployment and state employment taxes.

In order to deposit employment taxes, an employer must have an Employer Identification Number (EIN). The EIN is a nine-digit number issued by the Internal Revenue Service. In order to obtain an EIN, employers must fill out IRS form SS–4.

EMPLOYER'S QUARTERLY TAX RETURN (FORM 941)

All employers that collect Social Security taxes, Medicare taxes, and federal income tax withholding must file Form 941—the Employer's Quarterly Federal Tax Return. The dollar amounts for Mayflower's Supermarket are listed in Example 1. As discussed in Unit 37, equal amounts for Social Security and Medicare taxes are paid by the employee *and* the employer. The amount reported for federal income tax withholding is the total dollar amount withheld from individual employee's paychecks during the pay period.

EXAMPLE 1 | Calculating the Amount to Report on Form 941

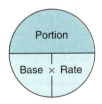

During the first calendar quarter, Mayflower's Supermarket paid wages and salaries totaling $14,600. During this same period, the federal income tax withholding totaled $3,212. The dollar amounts reported on Form 941 are as follows.

$14,600 × .124 = $1,810.40   **STEP** 1   To determine Social Security tax amount, multiply total wages ($14,600) by 12.4 percent (6.2 percent employee's tax rate *plus* 6.2 employer's rate).

$14,600 × .029 = $423.40   **STEP** 2   To determine Medicare tax amount, multiply total wages ($14,600) by 2.9 percent (1.45 percent employee's tax rate *plus* 1.45 employer's rate).

CALCULATOR TIP

Depress clear key
Enter 14,600   Depress ×
Enter 0.124   Depress =
Read the answer $1,810.40
Enter 14,600   Depress ×
Enter 0.029   Depress =
Read the answer $423.40
Enter 1,810.40 Depress +
Enter 423.40   Depress +
Enter 3,212   Depress =
Read the answer $5,445.80

Federal income tax
withholding = $3,212

STEP ☐3   List the amount for federal
income tax withholding.

$1,810.40
+   423.40
$3,212.00
$5,445.80

STEP ☐4   The total amount reported on
Form 941 is determined by
adding the amounts from
step 1, step 2, and step 3.

UNEMPLOYMENT TAXES

The Federal Unemployment Tax Act (FUTA), together with state unemployment systems, provides for payments of unemployment compensation to workers who have lost their jobs. Most employers pay both a federal and state unemployment tax. In all cases, only the employer pays this tax; the tax is not deducted from the employee's wages. At the time of publication, the FUTA tax rate was 6.2 percent of the first $7,000 paid to each employee. No federal unemployment taxes are paid on the portion of employee wages that exceed the wage ceiling.

Today, most states have enacted their own State Unemployment Tax Acts (SUTAs). In most cases, the SUTA tax rate is 5.4 percent of the first $7,000 paid to each employee. (Note: individual state unemployment tax rates and wage ceilings may vary.)

Generally, an employer can take a credit against FUTA tax amounts for amounts paid into state unemployment funds. This credit cannot be more than 5.4 percent of taxable wages. The FUTA tax rate after the credit is 0.8 percent.

The dollar amount of tax for either FUTA or SUTA is a practical application of the portion formula (Portion = Base × Rate) presented in Unit 26. For example, the FUTA tax amount (portion) is found by multiplying the total wages (base) by the FUTA tax rate.

EXAMPLE 2   Calculating FUTA and SUTA Tax Amounts

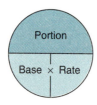

Barker Creek Construction, Inc., paid wages totaling $32,100 during the month of January. During this period, all employees had cumulative earnings of less than $7,000. The SUTA tax rate is 5.4 percent, and the FUTA tax rate is 0.8 percent.

$32,100 × .054 = $1,733.40

STEP ☐1   To determine the SUTA tax amount, multiply total wages ($32,100) by the SUTA tax rate (5.4 percent or 0.054).

$32.100 × 0.008 = $256.80

STEP ☐2   To determine the FUTA tax amount, multiply total wages ($32,100) by the FUTA tax rate (0.8 percent or 0.008).

CALCULATOR TIP

Depress clear key
Enter 32,100 Depress ×
Enter 0.054   Depress =
Read the answer $1,733.40
Enter 32,100 Depress ×
Enter 0.008   Depress =
Read the answer $256.80

Notice what happens when an employee's wages exceed the FUTA and SUTA wage ceilings.

| EXAMPLE 3 | SUTA and FUTA Tax Calculations for Earnings in Excess of the Wage Ceiling |

Nancy Weinstein has year-to-date wages of $8,900. Assuming a SUTA tax rate of 5.4 percent, a FUTA tax rate of 0.8 percent, and a wage ceiling of $7,000, her employer must pay the following unemployment taxes.

CALCULATOR TIP

Depress clear key
Enter 7,000 Depress ×
Enter 0.054 Depress =
Read the answer $378
Enter 7,000 Depress ×
Enter 0.008 Depress =
Read the answer $56

Wages subject to unemployment taxes = $7,000

STEP 1 — Only the first $7,000 is subject to unemployment taxes.

$7,000 × .054 = $378

STEP 2 — To determine the SUTA tax amount, multiply total taxable wages ($7,000) by the SUTA tax rate (5.4 percent or 0.054).

$7,000 × 0.008 = $56

STEP 3 — To determine the FUTA tax amount, multiply total taxable wages ($7,000) by the FUTA tax rate (0.8 percent or 0.008).

Employers use Internal Revenue Service Form 940, Employer's Annual Federal Unemployment (FUTA) Tax Return, to report federal unemployment tax. To report state unemployment taxes, employers must use the proper state forms.

HINT: You may want to review the material on portion in Unit 26 and the material on calculating Social Security taxes and Medicare taxes in Unit 37 before completing this unit.

Complete the Following Problems

NOTE: For all problems in this unit, assume (1) a 6.2 percent tax rate and a wage ceiling of $60,600 for Social Security; (2) a 1.45 percent tax rate for Medicare; (3) a 5.4 percent tax rate and a wage ceiling of $7,000 for SUTA; and (4) a 0.8 percent tax rate and a wage ceiling of $7,000 for FUTA.

1. T F An EIN number is issued by the Department of Commerce to qualifying employers.

2. T F Employers use IRS Form 941 to report Social Security, Medicare, and federal income tax withholding.

3. In your own words, define FUTA.

4. In your own words, define SUTA.

Total Wages or Salaries for All Employees	Employees' Social Security Amount	Employer's Social Security Amount
$ 24,500	5.	6.
$ 5,600	7.	8.
$138,900	9.	10.

Total Wages or Salaries for All Employees	Employee's Medicare Tax Amount	Employer's Medicare Tax Amount
$ 32,100	11.	12.
$ 8,700	13.	14.
$220,500	15.	16.

17. During the first calendar quarter, McPherson Tool and Manufacturing paid wages and salaries totaling $22,460. During this same period, the federal income tax withholding totaled $5,615. What is the total that all employees paid for the Social Security tax?

$22,460 × .062 = $1,392.52

18. For problem 17, how much did the employer pay for the Social Security tax?

19. For problem 17, what is the total that all employees paid for the Medicare tax?

20. For problem 17, how much did the employer pay for the Medicare tax?

21. For problem 17, what is the total amount that should be reported on Form 941?

For Problems 22–31, Assume That All Employees Had Cumulative Earnings of Less than $7,000

Total Wages or Salaries Paid by Employer	SUTA TAX Amount	FUTA TAX Amount
$ 20,500	22. *$20,500 × .054 = $1,107*	23. *$20,500 × .008 = $164*
$ 5,600	24.	25.
$134,200	26.	27.
$ 11,420	28.	29.

30. Billie Watson has year-to-date wages of $17,500. Calculate the amount of SUTA tax that her employer must pay.

31. For problem 30, what is the amount of FUTA tax that Ms. Watson's employer must pay?

Critical Thinking Problem

Jim Bateson, the owner of Bateson's Tile and Marble Company has been notified by the IRS that the company's Employer's Quarterly Tax Return (Form 941) was prepared incorrectly. After checking his accounting records, Mr. Bateson found that the company paid wages and salaries totaling $19,600. During this same calendar quarter, $4,704 was withheld from employees' paychecks for federal income tax withholding.

32. Given the above information, what is the correct amount of Social Security, Medicare, and federal income tax withholding that should have been reported to the IRS on Form 941.

33. In this situation, what should Mr. Bateson do to ensure that this situation is not repeated? _____

_____

_____

_____

INSTANT REPLAY

UNIT	IMPORTANT POINTS TO REMEMBER	EXAMPLES
UNIT 35 Determining Gross Pay	To calculate gross pay for a salaried employee, divide the annual salary by the number of paychecks the employee receives. To calculate gross pay for an hourly worker, the hourly wage rate is multiplied by the number of hours worked. To calculate time-and-a-half overtime, multiply the regular hourly wage by 1.5. Then multiply the overtime rate by the number of extra hours worked.	$20,000 ÷ 52 weeks = $384.62 weekly salary. $ 6.20 hourly wage × 40 number of hrs. $248.00 gross pay $8.00 an hour × 1.5 × number of overtime hrs. = overtime amount.

continued from page 201 UNIT	IMPORTANT POINTS TO REMEMBER	EXAMPLES	
UNIT 36 Calculating Commissions	The dollar amount of commission is found by multiplying the sales amount by the rate of commission. In addition, some firms pay their employees a salary plus commission, and some firms let salespeople draw against their future commissions. Other firms use a graduated commission scale to motivate employees.	**$25,000** **× 0.04** **$1,000**	sales rate commission
UNIT 37 Computing Deductions for Social Security (FICA)	At the time of publication, the Social Security tax was 6.2 percent on the first $60,600. The Medicare tax was 1.45 percent. The employer's contribution is equal to the employee's contribution.	**$680** **×0.062** **$42.16** **$ 680** **×0.0145** **$9.86**	gross pay Social Security tax rate Social Security tax gross pay Medicare tax rate Medicare tax
UNIT 38 Computing Deductions for Income Tax	Both marital status and withholding allowances affect the amount of federal income tax withheld. The actual dollar amount to be withheld may be determined by using the percentage method or the wage-bracket method. Instructions and tables for both methods are provided by the Internal Revenue Service in its publication *Circular E—Employer's Tax Guide*.	John Martin is single and claims two withholding allowances. He earns $2,175 a month. Based on Table 38–1, John's income tax withholding amount is $234.	
UNIT 39 Employer's Tax Responsibilities	All employers that collect Social Security, Medicare and federal income tax withholding must file IRS Form 941. In addition, employers must pay FUTA and SUTA taxes. The FUTA tax rate is 6.2 of the first $7,000 paid to each employee. Because of a credit for participating in a state system, the FUTA tax rate in that case is usually 0.8 percent. The SUTA tax rate for most states is 5.4 percent of the first $7,000 paid to each employee.	**$1,488** **+ 348** **+2,900** **$4,736** **$900** **×0.054** **$48.60** **$900** **×0.008** **$7.20**	Social Security taxes Medicare taxes federal income taxes withheld total reported on form 941 gross pay SUTA tax rate SUTA tax gross pay FUTA tax rate FUTA tax

CHAPTER 6	MASTERY QUIZ

DIRECTIONS *Round each answer to two decimal places or hundredths. (Each answer counts 5 points.)*

NOTE: For all problems in this Mastery Quiz, assume (1) a 6.2 percent tax rate and a wage ceiling of $60,600 for Social Security; (2) a 1.45 percent tax rate for Medicare; (3) a 5.4 percent tax rate and a wage ceiling of $7,000 for SUTA; and (4) a 0.8 percent tax rate and a wage ceiling of $7,000 for FUTA.

1. In your own words, define gross pay. _____

_____

2. In your own words, define net pay. _____

_____

3. Mark Jefferson worked the following hours: Monday, $9\frac{1}{4}$ hours; Tuesday, $6\frac{1}{2}$ hours; Wednesday, off; Thursday, 8 hours; Friday, $8\frac{3}{4}$ hours; and Saturday, $9\frac{1}{2}$ hours. How many hours did he work during this week?

4. Janice Scott works for Cypress Marina and is paid $6.30 an hour. Assuming that Cypress pays time-and-a-half for overtime, what is Janice's overtime pay rate?

5. Roger Evans works for Pollock Paper Company and is paid $7.10 an hour. During the first week in May, Roger worked 39 hours. What is his gross pay?

6. Rose Burley earns $10.50 an hour as a health aide. During the third week in September, Rose worked $44\frac{1}{4}$ hours. Assuming that her employer pays time-and-a-half for all hours in excess of 40, what is her gross pay?

7. Bob Upshaw is paid a monthly salary of $1,000 plus $3\frac{1}{2}$ percent commission on sales. If Bob sold merchandise valued at $49,710 during the month of March, what was his gross pay for the month?

8. Wesley's Chemical Supply pays all salespeople 5 percent on sales up to $15,000 and $7\frac{1}{2}$ percent commission on sales in excess of $15,000. If Cathy Mann sold merchandise valued at $39,710 during May, what is her gross pay for May?

9. Andy Begg's cumulative earnings for the year are $55,910. During the month of November, Andy earned $5,210. What is Andy's Social Security tax deduction for the month of November?

10. Raul Gonzales earns $2,380 a month. What is Raul's monthly Medicare tax deduction?

11. Anne Corey is single and claims two withholding allowances. If Anne's monthly salary is $2,120, how much should her employer withhold for federal income tax? (Use Table 38–1.)

12. Allison Stinson is married and claims three withholding allowances. If Allison's weekly salary is $694, how much should her employer withhold for federal income tax? (Use Tables 38–3 and 38–4.)

13. Arnie Nash works for Harley Manufacturing, Inc., and is paid $14.60 an hour. Arnie is married and claims three withholding allowances. During the second week of November, Arnie worked 45 hours. What is his gross pay amount?

14. For problem 13, what is Arnie's Social Security deduction? (Assume a 6.2 percent Social Security tax rate.)

15. For problem 13, what is Arnie's Medicare deduction? (Assume a 1.45 percent Medicare tax rate.)

16. For problem 13, what is Arnie's federal income tax withholding? (Use Table 38–2.)

17. For problem 13, what is Arnie's net pay?

18. During the first calendar quarter, Boston Dairy Products paid wages and salaries totaling $31,456. During this same period, the federal income tax withholding totaled $7,550. Assuming that no employee had cumulative wages in excess of $60,600, what is the total amount that should be reported on Form 941?

19. Jack Peters has annual wages of $16,400. Calculate the amount of SUTA tax that his employer must pay.

20. For problem 19, what is the amount of FUTA tax that Mr. Peters' employer must pay?

Discounts and Invoices

In this chapter, we explain trade discounts that manufacturers, wholesalers, and suppliers offer to retailers and other customers who purchase merchandise in large quantities. We begin with step-by-step procedures that can be used to find both discount amounts and wholesale prices. Then, we discuss cash discounts that manufacturers, wholesalers, and suppliers offer retailers who pay invoices promptly. Finally, we show how both trade discounts and cash discounts are calculated on a typical invoice.

MATH TODAY

THE RACE OF THE AIRWAVES IS ON

Since the breakup of AT&T, more companies than ever have begun offering long-distance telephone service. Today, the Big 3 (AT&T, MCI, and Sprint) compete with one another and battle for their shares of the telecommunications market. Rarely can you turn on your television without seeing their advertising. "Friends and Family" plans lure the public from one company to the other. "Come back to us for free" lures them back. Quality, efficiency, and pricing all get their share of promotion. Even Murphy Brown has gotten into the act.

To date, however, the Big 3 have concentrated on both consumers and large businesses, while overlooking the small business market. All three companies compete for consumer business by offering discounts that range between 10 and 30 percent. And large businesses—corporate customers whose telephone bills are in excess of $200,000 per month—receive discounts of 20 to 40 percent. However, most small businesses cannot qualify for these discount programs and are forced to pay higher rates for long-distance service. But there may be a change on the business horizon for at least two reasons.

First, industry experts predict that changes in technology will push prices down for all customers. For example, AT&T and 45 Asian telecommunications companies are investing over $1 billion in fiberoptic lines between the United States and Asia. This investment will not only improve the quality of service but is also expected to reduce costs by 20 to 30 percent. And a reduction in international rates will force a reduction in domestic rates. Lower international rates will also help smaller US companies compete in the world marketplace.

Second, and hard to believe, competition between large firms will increase even more in the future. In addition, more smaller telecommunications firms will throw their hats in the ring and offer long-distance services to small businesses. Competition in this competitive industry can take many forms. For example, AT&T and the National Association of Women Business Owners (NAWBO) recently announced a new agreement under which AT&T will offer discounts on long-distance services to NAWBO members. Another company, Advantech, founded by two former AT&T employees, entered into this competitive race. They decided that it was time for someone to offer lower rates to small businesses. Their company contracts with major carriers like AT&T, MCI, and Sprint for long-distance service. Because Advantech buys large amounts of long-distance service, they receive the discounts offered to large business firms. They, in turn, sell long-distance service to smaller businesses at prices that are discounted by as much as 30 percent. And the race goes on.

Source: For more information, see Carolyn M. Brown, "Career Profiles, Phillip Spencer," *Black Enterprise,* August 1993, p. 80; "Reach Out and Be a Partner," *Nation's Business,* August 1993, p. 38; and "Deals in Global Telephoning," *Kiplinger's Personal Finance Magazine,* January 1993, p. 17.

UNIT 40	CALCULATING SINGLE TRADE DISCOUNTS

LEARNING OBJECTIVES   *After completing this unit, you will be able to:*

1. calculate the trade discount amount.
2. calculate the wholesale, or net, price.

Firms like AT&T or Advantech, two firms discussed in the opening case, offer discounts to their customers to make their products or services more attractive than similar products or services offered by competitors. Discounts can also be used to encourage customers to buy in large quantities or to entice them to pay their bills promptly. Finally, manufacturers, wholesalers, and service firms allow their customers to purchase products or services at discounted prices and then the products or services are resold (retailed) at higher prices.

In fact, many manufacturers and wholesalers use catalogs to market their products. Most catalogs include a picture and description of the product along with the list price (sometimes called the suggested retail price) of the product. The **list price** is the price that the retail customer pays for the merchandise. It is the price that you pay when you purchase a product or service in a retail store. Most manufacturers and wholesalers also include a trade discount sheet in the back of a catalog that retailers can use to figure the wholesale price they pay for merchandise. By inserting different trade discount sheets, sellers can offer different discounts to different customers. And discount sheets can also be changed to increase or decrease prices without the added cost of printing new catalogs.

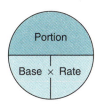

By using the information on the trade discount sheet, it is possible to calculate the trade discount amount. A **trade discount** is a deduction from the list price. The **net price** (sometimes called the wholesale price) is the list price minus the trade discount. Trade discount problems are a practical application of the portion formula presented in Unit 26. The dollar amount of the trade discount (the portion) is found by multiplying the list price (the base) by the trade discount rate (the rate).

EXAMPLE 1   Calculating the Trade Discount Amount

The list price for a swimming pool pump motor is $240. The trade discount rate is 30 percent. What is the dollar amount of the trade discount?

$240.00	list price (base)
× 0.30	trade discount rate (rate)
$ 72.00	trade discount amount (portion)

CALCULATOR TIP

Depress clear key
Enter 240   Depress ×
Enter 0.30   Depress =
Read the answer $72.00

In this example, note the trade discount rate (30 percent) has been changed to the decimal 0.30. To calculate the net price, one additional step is necessary: subtracting the trade discount amount from the list price. In Example 2, the net price is $168.

EXAMPLE 2   Find the Net Price of the Motor in Example 1

$240.00	list price
− 72.00	trade discount amount
$168.00	net price

CALCULATOR TIP

Depress clear key
Enter 240   Depress −
Enter 72   Depress =
Read the answer $168

ALTERNATE METHOD: USING COMPLEMENTS

The net price can be found directly by multiplying the list price by the **complement** of the trade discount rate. A complement is the difference between the discount and 100 percent. To find the complement, subtract the discount rate from 100 percent. For example, the complement of 20 percent is 80 percent (100 percent − 20 percent = 80 percent).

| EXAMPLE 3 | Using the Complement Method |

Find the net price for a swimming pool pump motor when the list price is $240 and the trade discount rate is 30 percent.

CALCULATOR TIP

Depress clear key
Enter 100   Depress –
Enter 30   Depress =
Read the answer 70
Enter $240 Depress ×
Enter 0.70 Depress =
Read the answer $168

100	percent	**STEP** 1	Determine the complement. Subtract the discount rate (30 percent) from 100 percent.
– 30	trade discount rate		
70	percent (complement)		

$240
×0.70
$168

STEP 2 Multiply the list price ($240) by the complement (70 percent).

In this example, note that the complement of the discount (70 percent) has been converted to the decimal 0.70.

HINT: You may want to review the material on portion in Unit 26 before completing this unit.

Complete the Following Problems

For all problems that do not work out evenly, carry your answers to two decimal places.

List Price	Trade Discount Rate	Trade Discount Amount
$299.00	25%	1. $299.00 × 0.25 = $74.75
$149.50	15%	2. _____
$795.00	22%	3. _____
$119.95	30%	4. _____
$229.95	17.5%	5. _____

Trade Discount Rate	Complement
10%	6. *100% − 10% = 90%*
15%	7.
24%	8.
32%	9.
41%	10.

List Price	Trade Discount Rate	Net (Wholesale) Amount
$75.00	35%	11. *$48.75*
		$75 $75.00 x.35 −26.25 $26.25 $48.75
$125.00	40%	12.
$49.95	12.7%	13.
$310.00	15%	14.

List Price	Trade Discount Rate	Net (Wholesale) Amount
$149.95	5.3%	**15.** _____

16. An electronic alternator produced by Precision Manufacturing has a suggested retail price of $129.50. If Precision Manufacturing offers retailers a trade discount of 28 percent, what is the trade discount amount?

$129.50 × .28 = $36.26

17. In problem 16, what is the net price?

18. A Sony compact disc player has a list price of $199.99. If Sony offers retailers a trade discount of 35 percent, what is the complement of the discount?

19. In problem 18, what is the net price?

Critical Thinking Problem

Judy Ramos is the owner and manager of Ramos Home Appliances, Inc. A local supplier—Salt Lake City Wholesale Supply—has advertised a microwave oven with a list price of $399. This wholesaler offers Ms. Ramos a 30 percent trade discount. A Detroit supplier—Northern Wholesale Appliances offers a similar microwave oven with a list price of $369 and a trade discount of 25 percent.

20. What is the net price for the microwave oven sold by Salt Lake City Wholesale Supply?

21. What is the net price for the microwave oven sold by Northern Wholesale Appliances?

22. If you were Ms. Ramos, which supplier would you choose? Why? _____

_____

_____

_____

| UNIT 41 | CALCULATING SERIES (CHAIN) DISCOUNTS |

LEARNING OBJECTIVES   *After completing this unit, you will be able to:*

1. determine the net price when a firm offers more than one trade discount.
2. calculate the total discount amount when there are a series of trade discounts.

discount (sometimes called a chain discount), which is more than one trade discount. Seasonal fluctuations, quantity of merchandise purchased, competitors' pricing of similar merchandise, geographic location, or perhaps increased promotional activity may account for changes in the number *and* size of discounts offered by a manufacturer or wholesaler. One manufacturer of air conditioning and heating products—Lennox Industries—changes its discount policy during a 12-month period to encourage more sales and reduce inventory.

MULTIPLE DISCOUNT METHOD

To compute series discounts, start with the first discount rate listed and figure the discount amount on the list price. The second discount is based on the remainder (the price after the first discount is taken). The third discount is based on the price after the second discount is taken.

EXAMPLE 1   Determining the Net Price for a Problem with Series Discounts

What is the net price for a Lennox four-ton air conditioning unit with a $1,600 list price when retailers are given a series discount of 25, 15, and 10 percent?

CALCULATOR TIP

Depress clear key
Enter 1,600   Depress ×
Enter 0.25   Depress =
Read the answer $400
Enter 1,600   Depress –
Enter 400   Depress =
Read the answer $1,200
Enter 1,200   Depress ×
Enter 0.15   Depress =
Read the answer $180
Enter 1,200   Depress –
Enter 180   Depress =
Read the answer $1,020
Enter 1,020   Depress ×
Enter 0.10   Depress =
Read the answer $102
Enter 1,020   Depress –
Enter 102   Depress =
Read the answer $918

$$\begin{array}{r} \$1,600 \\ \times\ 0.25 \\ \hline \$\ \ 400 \end{array}$$ **STEP** 1   Figure the dollar amount of the first discount by multiplying the list price ($1,600) by the first discount rate (25 percent).

$$\begin{array}{r} \$1,600 \\ -\ \ 400 \\ \hline \$1,200 \end{array}$$ **STEP** 2   Subtract the first trade discount amount ($400) from the retail price ($1,600).

$$\begin{array}{r} \$1,200 \\ \times\ 0.15 \\ \hline \$\ \ 180 \end{array}$$ **STEP** 3   Figure the dollar amount of the second discount by multiplying the first intermediate answer ($1,200) by the second discount rate (15 percent).

$$\begin{array}{r} \$1,200 \\ -\ \ 180 \\ \hline \$1,020 \end{array}$$ **STEP** 4   Subtract the second trade discount amount ($180) from the first intermediate answer ($1,200).

$$\begin{array}{r} \$1,020 \\ \times\ 0.10 \\ \hline \$\ \ 102 \end{array}$$ **STEP** 5   Figure the dollar amount of the third discount by multiplying the second intermediate answer ($1,020) by the third discount rate (10 percent).

$$\begin{array}{r} \$1,020 \\ -\ \ 102 \\ \hline \$\ \ 918 \end{array}$$ **STEP** 6   To obtain the net price, subtract the third trade discount amount ($102) from the second intermediate answer ($1,020).

In this example, all trade discount rates have been converted to decimals. When working a series discount problem, the order in which the individual trade discounts are applied does not affect the final net price. In Example 1, you could figure the 15 percent discount first, then 10 percent, then 25 percent. However, *you* **cannot** *simply add the three individual discounts and take off one larger discount.* Your net price will be wrong because each individual trade discount is calculated on a smaller intermediate answer.

ALTERNATE METHOD: USING COMPLEMENTS

To avoid using six steps as in Example 1, many retailers, wholesalers, and students prefer using the complement method, which reduces the number of steps to three. Proceed as you do when figuring single discounts by the complement method (Unit 40). To use the complement method, multiply the price at each stage of calculation by the complement of the trade discount.

EXAMPLE 2   Using the Complement Method

Find the net price for the air conditioning unit in Example 1.

CALCULATOR TIP

Depress clear key
Enter 1,600   Depress ×
Enter 0.75   Depress =
Read the answer $1,200
Enter 1,200   Depress ×
Enter 0.85   Depress =
Read the answer $1,020
Enter 1,020   Depress ×
Enter 0.90   Depress =
Read the answer $918

$1,600 × 0.75 $1,200	**STEP** 1	Multiply the list price ($1,600) by the complement (75 percent) of the first discount.
$1,200 × .085 $1,020	**STEP** 2	Multiply the first intermediate answer ($1,200) by the complement (85 percent) of the second discount.
$1,020 × 0.90 $ 918	**STEP** 3	Multiply the second intermediate answer ($1,020) by the complement (90 percent) of the third discount.

In this example, all complements have been converted to decimals. Regardless of the method used, you can determine the total amount of trade discounts by subtracting the net price from the list price.

EXAMPLE 3   Calculating the Total of All Trade Discounts

CALCULATOR TIP

Depress clear key
Enter 1,600   Depress –
Enter 918   Depress =
Read the answer $682

$1,600	list price
– 918	net price
$ 682	total of all trade discounts

ALTERNATE METHOD: SINGLE-EQUIVALENT DISCOUNT METHOD

With each of the first two methods presented in this unit, you find the net price. Then, to determine the total of all trade discounts, you must subtract the net price from the list price, as illustrated in Example 3. Some employees prefer to use the single-equivalent discount method to determine the total of all the trade discounts.

Notice in the example below, three steps are used to complete the problem. First, the complement of each individual trade discount is multiplied together. Second, the answer from step 1 is subtracted from the number 1. Third, the list price is multiplied by the answer from step 2.

EXAMPLE 4 | Using the Single-Equivalent Discount Method

CALCULATOR TIP

Depress clear key
Enter 0.80   Depress ×
Enter 0.85   Depress ×
Enter 0.90   Depress =
Read the answer 0.612
Enter 1.000   Depress −
Enter 0.612   Depress =
Read the answer 0.388
Enter 700   Depress ×
Enter 0.388   Depress =
Read the answer $271.60

What is the total of all trade discounts for a two-horse air compressor with a $700 list price when retailers are given a series discount of 20, 15, and 10 percent?

$0.80 \times 0.85 \times 0.90 = 0.612$ **STEP** [1]   Multiply the complements. (Use three decimal places.)

$$\begin{array}{r} 1.000 \\ -0.612 \\ \hline 0.388 \end{array}$$ **STEP** [2]   Subtract the answer in step 1 (0.612) from the number 1.

$700 \times 0.388 = 271.60 **STEP** [3]   To determine the total of all trade discounts, multiply the list price ($700) by the answer from step 2 (0.388).

As illustrated in Example 5, you can determine the net price by subtracting the total of all trade discounts from the list price.

EXAMPLE 5 | Calculating the Wholesale Price

CALCULATOR TIP

Depress clear key
Enter 700   Depress −
Enter 271.60   Depress =
Read the answer $428.40

$$\begin{array}{ll} \$700.00 & \text{list price} \\ -271.60 & \text{total of all trade discounts} \\ \hline \$428.40 & \text{net price} \end{array}$$

HINT: Before working the following series discount problems, you may want to review the material on single trade discounts in Unit 40.

Complete the Following Problems

Unless otherwise indicated, round your answers to two decimal places.

1. What is the single-equivalent discount for an invoice that has series discounts of 25, 10, and 5 percent? (Use 5 decimal places.)

$$\begin{array}{ccc} 1.00 & 1.00 & 1.00 \\ -.25 & -.10 & -.05 \\ \hline .75 & .90 & .95 \end{array}$$

$.75 \times 90 \times .95 = .64125$
$1.00000 - .64125 = .35875 = 35.875\%$

2. What is the single-equivalent discount for an invoice that has series discounts of 35 and 15 percent? (Use 4 decimal places.)

3. What is the single-equivalent discount for an invoice that has series discounts of 10, 10, and 10 percent? (Use 3 decimal places.)

NOTE: For the remaining problems in this unit, you may use the multiple discount method, the complement method, or the single-equivalent discount method.

List Price	Trade Discounts		Net Price
$500.00	20%, 15%, 10%	**4.**	$306

$$
\begin{array}{ccc}
1.00 & 1.00 & 1.00 \\
-.20 & -.15 & -.10 \\
\hline
.80 & .85 & .90
\end{array}
$$

$500 × 0.80 = $400
$400 × 0.85 = $340
$340 × 0.90 = $306

List Price	Trade Discounts		Net Price
$2,300.00	25%, 15%, 5%	**5.**	
$150.00	30%, 15%	**6.**	
$1,995.95	28%, 14%, 8%	**7.**	
$74.50	25%, 12%	**8.**	
$325.00	10%, 5%, 5%	**9.**	

List Price	Trade Discounts	Net Price	Total of Trade Discounts
$29.50	35%, 15%, 10%	**10.** _____	**11.** _____
$115.00	20%, 7%	**12.** _____	**13.** _____
$49.25	23%, 12%, 4%	**14.** _____	**15.** _____

16. An electric pencil sharpener has a list price of $19.99. The manufacturer offers retailers series discounts of 15 and 12 percent. What is the net price for one of the pencil sharpeners?

$$1.00 - .15 = .85 \qquad 1.00 - .12 = .88$$

$19.99 × 0.85 = $16.9915
$16.9915 × 0.88 = $14.95

17. Appliances Unlimited, Inc., received an invoice for a shipment of televisions. The total amount of 30 televisions was $7,490. The manufacturer offers series discounts of 22, 13, and 10 percent. What is the net price for this invoice?

18. In problem 17 what is the total of all trade discounts?

CHAPTER SEVEN

Critical Thinking Problem

Hilda Thomas, the purchasing specialist for Ford Industrial Products, usually orders office supplies from Midwest Office Supply, but recently Midwest has shipped inferior merchandise. Now she is trying to decide if she should change suppliers. She has obtained price quotations for a four-drawer file cabinet from Midwest and one other supplier. Midwest Office Supply offers a four-drawer file cabinet with a list price of $324.40 with series discounts of 30 and 15 percent. Rocky Mountain Office Supply offers a four-drawer file cabinet with a list price of $305 and series discounts of 10, 10, and 10 percent.

19. What is the net price for the file cabinet offered by Midwest Office Supply?

20. What is the net price for the file cabinet offered by Rocky Mountain Office Supply?

21. If you were Ms. Thomas, which supplier would you choose? Why? _____

_____

_____

_____

UNIT 42	CALCULATING CASH DISCOUNTS

LEARNING OBJECTIVES   *After completing this unit, you will be able to:*

1. calculate the dollar amount of cash discount.
2. calculate the net cash price.

When retailers purchase merchandise on a credit basis, many manufacturers and wholesalers use cash discounts to encourage prompt payment. A **cash discount** is a small discount offered to a customer for paying an invoice within a specified number of days.

Notice on the invoice illustrated in Figure 42–1 that the terms of payment indicate what the cash discount is and how soon the bill must be paid to take advantage of the discount. If the terms are stated as $\frac{2}{10}$, $\frac{N}{30}$, that means that the buyer can take a 2 percent discount if the bill is paid within 10 days of the invoice date; otherwise, the entire amount must be paid in full within 30 days. If the terms are $\frac{2}{10}$, $\frac{1}{30}$, $\frac{N}{60}$, the buyer can take a 2 percent discount if the bill is paid within 10 days of the invoice date, *or* can take a 1 percent discount if the bill is paid between day 11 and day 30, *or* can choose to pay the entire amount if the bill is paid between day 31 and day 60.

INVOICE

FROM *All-Star Sports Equipment*
411 W. 39th St., Bethany, OK 73008

Nov. 20 19 *XX*

TO _____

ADDRESS _____

CITY _____

TERMS *2/10, N/30*   ORDER NO. _____

Quantity	Description	Unit Cost		Total	
144	*Warm-up Suits*	30	00	4,320	00
144	*Knit Sweaters*	20	00	2,880	00
		List Price		7,200	00
	Less	25%, 10%		2,340	00
		Net Price		4,860	00
	Less	2/10, N/30		97	20
		Amount Due		4,762	80

FIGURE 42–1 An Invoice for Merchandise Sold by All-Star Sports Equipment

Cash discounts are always calculated on the wholesale price after trade discounts have been calculated. Cash discounts are an application of the portion formula presented in Unit 26. The dollar amount of the cash discount (the portion) is found by multiplying the net price (the base) by the cash discount rate (the rate).

EXAMPLE 1 Calculating a Cash Discount Amount

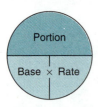

Portion

Base × Rate

On November 10, West Coast Clothes ordered sportswear from All-Star Sports Equipment. The net price for the clothing was $4,860—see Figure 42–1. All-Star Sports Equipment offers cash discount terms of $\frac{2}{10}$, $\frac{N}{30}$. What is the dollar amount of the cash discount if the bill is paid within 10 days?

$4,860	net price (base)
× 0.02	cash discount rate
$97.20	cash discount amount (portion)

CALCULATOR TIP

Depress clear key
Enter 4,860   Depress ×
Enter 0.02   Depress =
Read the answer $97.20

In this example, the cash discount rate (2 percent) has been converted to the decimal 0.02. The total amount due is the net price on the invoice minus the cash discount amount.

EXAMPLE 2 Find the Total Amount Due in Example 1

$4,860.00	net price
− 97.20	cash discount amount
$4,762.80	total amount due

CALCULATOR TIP

Depress clear key
Enter 4,860   Depress −
Enter 97.20   Depress =
Read the answer $4,762.80

ALTERNATE METHOD: USING COMPLEMENTS

You can find the total amount due directly by multiplying the invoice amount by the *complement* of the cash discount rate. In Example 3, the net price on an invoice is $10,500. The manufacturer offers a cash discount of $\frac{3}{20}$, $\frac{N}{45}$. Find the total amount due, assuming the invoice is paid within 20 days of the invoice date.

EXAMPLE 3 Find the Net Cash Price Using the Complement Method

CALCULATOR TIP

Depress clear key
Enter 100   Depress −
Enter 3   Depress =
Read the answer 97
Enter $10,500   Depress ×
Enter 0.97   Depress =
Read the answer $10,185

100	percent
− 3	cash discount rate
97	percent (complement)

STEP 1 Determine the complement. Subtract the cash discount rate (3 percent) from 100 percent.

$10,500	
× 0.97	
$10,185	

STEP 2 Multiply the net price ($10,500) by the complement (97 percent).

In this example, the complement (97 percent) has been converted to the decimal 0.97.

THE RECEIPT OF GOODS (ROG) DATING METHOD

The receipt of goods (ROG) method allows customers to wait until the merchandise is received before paying for the merchandise. For example, the terms $\frac{2}{10}$, $\frac{N}{30}$ ROG mean that the buyer can take a 2 percent discount if the bill is paid within 10 days *after* the receipt of goods; otherwise, the entire amount must be paid in full within 30 days *after* the goods are received.

EXAMPLE 4   Calculating a Cash Discount with ROG Dating

On March 3, Boise Lighting, Inc., ordered a shipment of lamps from Castleberry Electrical Supply. An invoice for $3,900 is dated March 15 and has cash discount terms of $\frac{2}{10}$, $\frac{N}{30}$ ROG. The merchandise was received June 6. If the invoice is paid on June 10, what is the total amount due?

June 6 + 10 days June 16	**STEP** 1	Determine the last date for the cash discount. Begin counting on June 6—the date the merchandise was received.
$3,900 × .02 = $78.	**STEP** 2	Since the payment is made within the discount period, calculate the cash discount amount.
$3,900 – 78 $3,822	**STEP** 3	Calculate the total amount due.

THE END OF MONTH (EOM) DATING METHOD

The end of month (EOM) method allows customers to wait until the beginning of the next month before paying for merchandise. For example, the terms $\frac{3}{15}$, $\frac{N}{60}$ EOM mean that the buyer can take a 3 percent discount if the bill is paid during the first 15 days of the next month. If payment is made after the 15th day of the next month, the entire amount must be paid.

EXAMPLE 5   Calculating a Cash Discount with EOM Dating

On June 5, Carlton Furniture ordered a shipment of foam rubber padding from Jacksonville Foam Works. An invoice for $7,100 is dated June 18 and has cash discount terms of $\frac{3}{15}$, $\frac{N}{60}$ EOM. If the invoice is paid on July 12, what is the total amount due?

The last day to take a cash discount is July 15.	**STEP** 1	Determine the last date for the cash discount.
$7,100 × .03 = $213	**STEP** 2	Since the payment is made within the discount period, calculate the cash discount amount.
$7,100 – 213 $6,887	**STEP** 3	Calculate the total amount due.

For invoices dated *after* the 25th day of any month, it is a common business practice to extend the payment period another month. Thus, the discount period extends into the *second* month. For example, if an invoice is dated September 26 with cash discount terms of $\frac{2}{10}$, $\frac{N}{30}$ EOM, the cash discount may be taken if payment is made by November 10.

REMINDER: To take a cash discount, the invoice must be paid within the discount period.

Complete the Following Problems

For all problems that do not work out evenly, carry your answers to two decimal places.

For problems 1 through 10, assume that the invoice amount is paid within the discount period.

Wholesale Price	Cash Discount Terms		Cash Discount Amount
$225.00	$\frac{2}{10}$, $\frac{N}{30}$	1.	*$4.50*
			$225.00 x 0.02 = $4.50
$1,250.00	$\frac{3}{5}$, $\frac{N}{45}$	2.	
$730.50	$\frac{2}{20}$, $\frac{N}{60}$	3.	
$3,750.00	$\frac{1}{30}$, $\frac{N}{60}$	4.	

Wholesale Price	Cash Discount Terms		Cash Discount Amount		Total Amount Due
$10,950	$\frac{2}{10}$, $\frac{N}{30}$	5.		6.	*$10,731*
					$10,950 x .98 = $10,731
$2,500	$\frac{1}{20}$, $\frac{N}{40}$	7.		8.	
$430	$\frac{3}{10}$, $\frac{N}{20}$	9.		10.	

11. Mathews Construction Company received an invoice dated January 3 for $12,350 from one of their suppliers. The supplier offers the following cash discount terms: $\frac{2}{10}$, $\frac{1}{20}$, $\frac{N}{30}$. If Mathews Construction pays the invoice on January 18, what is the cash discount amount?

    $12,350 × .01 = $123.50

12. In problem 11, what is the total amount due?

13. Phoenix Cardboard, Inc., received an invoice dated June 15 for $3,780 from one of its suppliers. The supplier offers cash discount terms of $\frac{2}{10}$, $\frac{N}{30}$. If Phoenix Cardboard pays the invoice on June 28, is the firm entitled to the cash discount?

14. In problem 13, how much should Phoenix Cardboard pay the supplier?

15. On May 10, Chicago Waste Removal ordered a shipment of new plastic waste containers. The company received an invoice for $1,750 dated May 20. The seller offers cash discount terms of $\frac{1}{15}$, $\frac{N}{30}$ ROG. The merchandise was received on July 6. If the invoice is paid on July 18, what is the total amount due?

16. Norvel Manufacturing received a $2,400 invoice for plastic boxes dated May 3. The seller offers cash discount terms of $\frac{2}{10}$, $\frac{N}{30}$ ROG. The merchandise was received on June 14. If the invoice is paid on June 28, what is the total amount due?

17. On March 18, Kansas City Postal Equipment, Inc., received an invoice dated March 15 for $4,100 with cash discount terms of $\frac{2}{10}$, $\frac{N}{60}$ EOM. What is the last date that the cash discount can be taken?

18. In problem 17, what is the amount that should be paid if the invoice is paid on April 2?

19. On November 5, New Wave Software received an invoice for computer disks. The invoice total was $12,400 and dated November 2 with cash discount terms of $\frac{2}{20}$, $\frac{N}{60}$ EOM. If the invoice is paid on December 18, what is the total amount due?

20. On April 30, Whitmore Video Productions received an invoice dated April 28 for video tape. The invoice total was $932 with cash discount terms of $\frac{2}{10}$, $\frac{N}{30}$ EOM. If the invoice is paid on June 8, how much should Whitmore Video pay the supplier?

Critical Thinking Problem

The office manager for Phoenix-based Thomas Manufacturing is trying to decide which supplier offers the best price for a personal computer. Computer World offers the computer for $1,279 with cash discount terms of $\frac{1}{10}$, $\frac{N}{60}$. Best Office Equipment—a mail order supplier located in Detroit—offers the same computer for $1,295 with cash discount terms of $\frac{3}{10}$, $\frac{N}{60}$.

21. Assuming that Thomas takes advantage of the cash discount, what is the price for the computer offered by Computer World?

22. Assuming that Thomas takes advantage of the cash discount, what is the price for the computer offered by Best Office Equipment?

23. If you were the office manager for Thomas, which supplier would you choose? Why? _____

_____

UNIT 43   COMPLETING INVOICES

LEARNING OBJECTIVE   *After completing this unit, you will be able to:*

1. complete invoices that contain trade discounts and cash discounts.

Writing up an invoice requires three sets of calculations: (a) figuring the total for each type of merchandise, (b) figuring the trade (single or series) discount, and (c) figuring the cash discount. Let's look at each calculation in the sequence.

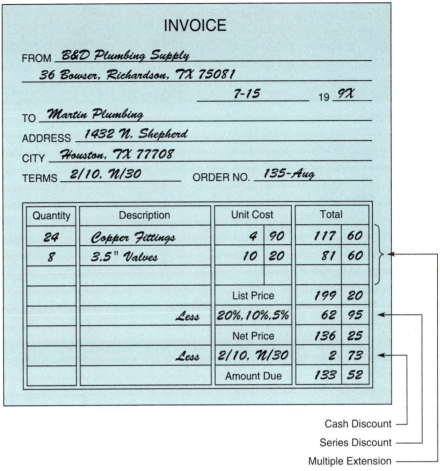

FIGURE 43–1   An Invoice for Merchandise Sold by B & D Plumbing Supply

1. Calculate the total for each type of merchandise by multiplying the quantity of each item by the cost of each item. After multiplying, write each answer in the total column on the right side of the invoice. Next, add each answer to determine the list price of all merchandise purchased (see Figure 43–1).

EXAMPLE 1   Find the Total of All Merchandise Purchased in Figure 43–1

$24 \times \$4.90 = \$117.60$	**STEP** 1	Multiply the quantity of the first item (24) by the unit price ($4.90).
$8 \times \$10.20 = \$81.60$	**STEP** 2	Multiply the quantity of the second item (8) by the unit price ($10.20).
$\begin{array}{r}\$117.60 \\ +\ 81.60 \\ \hline \$199.20\end{array}$	**STEP** 3	Add the answers from step 1 and step 2.

CALCULATOR TIP

Depress clear key
Enter 24   Depress ×
Enter 4.90   Depress =
Read the answer $117.60
Enter 8   Depress ×
Enter 10.20   Depress =
Read the answer $81.60
Enter 117.60   Depress +
Enter 81.60   Depress =
Read the answer $199.20

2. Enter the list price for all merchandise purchased on the appropriate line of the invoice. Then, apply the trade discount (single or series) to the list price. As explained in Unit 41, most people use the complement method to calculate this discount.

EXAMPLE 2   Calculate the Series Discounts and Net Price Shown in Figure 43–1

CALCULATOR TIP

Depress clear key
Enter 199.20   Depress ×
Enter 0.80   Depress =
Read the answer $159.36
Enter 159.36   Depress ×
Enter 0.90   Depress =
Read the answer $143.424
Enter 143.424   Depress ×
Enter 0.95   Depress =
Read the answer $136.25

$199.20
× 0.80
$159.36

$159.36
× 0.90
$143.424

$143.424
×   0.95
$136.253 = $136.25

STEP [1]   Multiply the list price ($199.20) by the complement of the first discount (80 percent).

STEP [2]   Multiply the first intermediate answer ($159.36) by the complement of the second discount (90 percent).

STEP [3]   Multiply the second intermediate answer ($143.424) by the complement of the third discount (95 percent).

Enter the net price on the appropriate line of the invoice. Then, subtract the net price ($136.25) from the list price ($199.20) to determine the total dollar amount of the series discounts. Enter this dollar amount on the line between the list price and the net price.

3. Apply the cash discount to the net price if the invoice is to be paid within the discount period.

EXAMPLE 3   Find the Cash Discount Amount for the Invoice Shown in Figure 43–1

CALCULATOR TIP

Depress clear key
Enter 136.25   Depress ×
Enter 0.02   Depress =
Read the answer $2.73

$136.25   net price
×   0.02   cash discount rate
$ 2.725   cash discount amount
$ 2.725 = $2.73

In this example, the cash discount amount has been rounded off to two decimal places. Finally, calculate the total amount due by subtracting the cash discount amount ($2.73) from the net price ($136.25). The total amount due is $133.52.

HINT: Before working the following problems, you may want to review (a) Unit 40, Trade Discounts, (b) Unit 41, Series Discounts, and (c) Unit 42, Cash Discounts.

Complete the Following Problems

For all problems that do not work out evenly, carry your answers to *two* decimal places.

1. Evans Electronics bought the following merchandise from Tri-Cities Office Supply: (a) 144 ballpoint pens at $1.29 each, (b) 96 mechanical pencils at $2.99 each, and (c) 5 Cross pen sets at $29.95 each. What was the total value of the merchandise purchased from Tri-Cities Office Supply?

    $622.55   *144 @ $ 1.29 ea = 185.76*
                    96 @ $ 2.99 ea = 287.04
                    5 @ $29.95 ea = 149.75
                                $622.55

2. Vantage Point Apartments received an invoice for $378.50 from Nations Electrical Supply. Nations Electrical offers trade discounts of 30, 10, and 5 percent. What is the net price for this invoice?

3. In problem 2, what is the total of all trade discounts?

4. Health Foods, Inc., received an invoice dated April 12 for $1,230 from one of its suppliers. The supplier offers cash discount terms of $\frac{2}{10}$, $\frac{N}{30}$. If the invoice is paid on April 18, how much should Health Foods pay the supplier?

5. Boston Metal Works received an invoice dated May 2 for $5,400 from one of its suppliers. The supplier offers cash discount terms of $\frac{3}{15}$, $\frac{N}{30}$. If the invoice is paid on May 20, how much should Boston Metal Works pay the supplier?

6. Jackson Furniture received an invoice dated February 5 for $12,450 from one of its suppliers. The supplier offers cash discount terms of $\frac{2}{10}$, $\frac{N}{60}$ EOM. If the invoice is paid on March 8, how much should Jackson Furniture pay the supplier?

Bob Harris, an accountant for Kentucky Chemical Supply, must complete the following invoice:

INVOICE

FROM *Kentucky Chemical Supply*

109 Louisville Ave., Louisville, KY

10-10   19 XX

TO *Forest Lane Manufacturing*

ADDRESS *13000 Forest Lane*

CITY *Columbus, Ohio 43216*

TERMS 2/20, N/30   ORDER NO. *10319*

Quantity	Description	Unit Cost		Total	
35	*Cleaning Solvent*	39	50		
	(55 gallon drums)				
12	*Cleaning Kits*	49	90		
		List Price			
	Less	30%, 15%			
		Net Price			
	Less	2/20, N/30			
		Amount Due			

Use this invoice to answer problems 7, 8, and 9.

7. What is the list price for the merchandise included on this invoice?

8. What is the net price for this invoice?

9. Assuming the invoice was paid on October 15, what is the total that the customer should pay?

Tran Lin works in the accounts receivable department for Bradford Plastic Manufacturing. She must complete the following invoice:

INVOICE

FROM _Bradford Plastics Manufacturing_

1300 Texas Drive, Richardson, TX 75081

_2-15_   19 _XX_

TO _Thomas Sprinkler Company_

ADDRESS _3502 Willow St._

CITY _New York, N.Y. 10003_

TERMS _2/10, N/30_   ORDER NO. _10320_

Quantity	Description	Unit Cost		Total	
250	Plastic Joints		69		
300	Plastic Elbows		59		
72	Anti-Syphon Valve	9	30		
			List Price		
	Less	25%, 15%, 10%			
		Net Price			
	Less	2/10, N/30			
		Amount Due			

Use this invoice to answer problems 10–14.

10. What is the list price for this invoice?

11. What is the net price for this invoice?

12. What is the total of the trade discounts for this invoice?

13. Assuming the invoice will be paid on February 23, what is the amount of the cash discount?

14. Assuming the invoice will be paid on February 23, what is the total that the customer should pay?

INSTANT REPLAY

UNIT	IMPORTANT POINTS TO REMEMBER	EXAMPLES
UNIT 40 Calculating Single Trade Discounts	The trade discount amount (the portion) is found by multiplying the list price (the base) by the trade discount rate (the rate). To calculate the net price, subtract the trade discount amount from the list price. It is also possible to find the net price by multiplying the complement of the trade discount rate by the list price.	$400 list price ×0.30 trade discount rate $120 trade discount amount $400 list price −120 trade discount amount $280 net price $400 list price × 0.70 complement of 30% discount $280 net price

continued from page 228

UNIT	IMPORTANT POINTS TO REMEMBER	EXAMPLES
UNIT 41 Calculating Series (Chain) Discounts	The complement method is the quickest way to figure the net price when working with a series discount. The following steps can be used: (a) multiply the list price by the complement of the first discount, (b) multiply the answer from step a by the complement of the second discount, and (c) multiply the answer from step b by the complement of the third discount. The result is the net price. The total amount of trade discounts can be determined by subtracting the net price from the list price. It is also possible to find the net price by using the multiple discount method or the single-equivalent discount method.	$3,000 list price × 0.90 complement of 10 percent $2,700 first intermediate answer $2,700 first intermediate answer × 0.80 complement of 20 percent $2,160 second intermediate answer $2,160 second intermediate answer × 0.95 complement of 5 percent $2,052 net price $3,000 list price −2,052 net price $ 948 total of all trade discounts
UNIT 42 Calculating Cash Discounts	To figure cash discounts, use the portion formula (Unit 26). The cash discount amount (the portion) is found by multiplying the net price (the base) by the cash discount rate (the rate). The total amount due is the net price minus the cash discount amount. It is also possible to find the total amount due by multiplying the net price by the complement of the cash discount rate. *Warning: To take a cash discount, the invoice must be paid within the discount period.*	$750 wholesale price × 0.02 cash discount terms $(\frac{2}{10}, \frac{N}{30})$ $ 15 cash discount amount $750 wholesale price − 15 cash discount amount $735 net price $750 wholesale price × 0.98 complement of the cash discount $(\frac{2}{10}, \frac{N}{30})$ $735 net price
UNIT 43 Completing Invoices	Writing up an invoice is a combination of three separate sets of calculations: (a) Calculate the total for each type of merchandise, and then add each individual extension to determine the list price for all merchandise purchased; (b) apply the trade discount (single or series) to the list price; (c) if the invoice is paid within the discount period, apply the cash discount terms to the net price.	See Figure 43–1

INVOICE

FROM _All-Star Sports Equipment_

411 W. 39th St., Bethany, OK 73008

Nov. 14   19 XX

TO _Central College_

ADDRESS _1410 Main Street_

CITY _McPherson, KS 67460_

TERMS _2/10, N/30_   ORDER NO. _13565_

Quantity	Description	Unit Cost		Total	
10	Tennis Rackets	39	50		
5	Soccer Balls	24	35		
18	Volleyballs	19	95		
		List Price			
	Less	25%, 10%			
		Net Price			
	Less	2/10, N/30			
		Amount Due			

| CHAPTER 7 | MASTERY QUIZ |

DIRECTIONS *Round each answer to two places or hundredths. (Each answer counts 5 points.)*

1. T   F   The net price is sometimes referred to as the wholesale price.

2. T   F   When completing an invoice, cash discounts are calculated before trade discounts.

3. What is the single-equivalent discount for an invoice that has series discounts of 30, 15, and 5 percent?

4. What is the single-equivalent discount for an invoice that has series discounts of 20, 10, and 4 percent?

List Price	Trade Discount Rate	Trade Discount Amount	Net (Wholesale) Price
$3,220.00	20%	5. _____	6. _____

List Price	Trade Discounts	Net Price	Total of All Trade Discounts
$1,220.00	35%, 5%, 2%	7. _____	8. _____

Net Price	Cash Discount Terms	Cash Discount Amount (if paid within the discount period)	Total Amount Due
$3,250.10	$\frac{2}{10}, \frac{N}{30}$	9. _____	10. _____
$5,670.40	$\frac{3}{20}, \frac{N}{60}$	11. _____	12. _____

13. On October 10, All-Med Equipment received an invoice for medical supplies. The invoice total was $11,500 with cash discount terms of $\frac{2}{10}$, $\frac{N}{30}$ ROG. The medical supplies were delivered on November 5. If the invoice is paid on November 10, what is the total amount due?

14. On May 4, Landmark Electric Supply received an invoice for electrical fixtures. The invoice total was $7,170 with cash discount terms of $\frac{3}{10}$, $\frac{N}{120}$ EOM. If the invoice is paid on June 12, how much should Landmark pay the supplier?

15. On December 28, Hess Consulting Service received an invoice for office supplies. The invoice total was $12,100. The supplier offers cash discount terms of $\frac{2}{10}$, $\frac{N}{50}$ EOM. If the invoice is paid on February 8, how much should Hess pay the supplier?

Use the information on the following invoice (p. 230) to answer questions 16 through 20.

16. What is the list price for the merchandise included on this invoice?

17. What is the net price for this invoice?

18. What is the total of the trade discounts for this invoice?

19. Assuming the invoice will be paid on November 20, what is the amount of the cash discount?

20. Assuming the invoice will be paid on November 20, what is the total the customer should pay?

Markup, Inventory, and Markdown

CHAPTER PREVIEW

In this chapter, we discuss markup, inventory, and markdown—three applications that are especially important for retailers. The chapter begins by examining the differences between markup based on cost and markup based on the selling price. Then, it describes the steps required to calculate the percent of markup. Next, we look at methods used to determine the dollar value of a firm's inventory. The chapter concludes with a discussion of markdown methods that retailers use to reduce the price of seasonal merchandise or excess inventory.

MATH TODAY

IS THE MARKUP TOO HIGH FOR PRESCRIPTION DRUGS?

Over the last twelve years, prescription drug prices have risen an astonishing 128 percent. That's six times faster than the rate of inflation during the same time period. And while the drug companies may be the darling of investors on Wall Street, both consumer advocates and government officials are irate over what they consider a serious case of price gouging that has resulted in outrageous prices.

Both consumer advocates and government officials argue that the markup on drugs is tied to what they call the desperation index. Simply put, the greater the patient's need and the fewer the treatment options, the higher the price of prescription drugs. And while most drug manufacturers refuse to discuss their pricing policies, it is common knowledge that markups on many prescription drugs exceed 300 percent. For example, Schering-Plough Corporation uses a 1,200 percent markup to price K-DUR—a potassium supplement. The drug Calan—a heart medication manufactured by Searle Corporation—is marked up almost 500 percent. And American Home Products marks up Inderal—another heart medication—more than 1000 percent. As a result of this type of pricing, both consumer advocates and government officials are now calling for controls that would reduce the markups manufacturers use to price their products.

The manufacturers argue that there are at least three reasons why the markup on prescription drugs is higher than the markups on other consumer products. First, drug companies spend an average of 16 percent of sales on research and development. The industry invests more in new product development than any other industry in the United States. This research and development has produced miraculous new treatments for heart disease, high blood pressure, ulcers, and other life-threatening diseases, and it is expensive. Second, there must be some way for drug companies to recoup their research and development expenses before their patents expire at the end of 17 years. After patents expire, other manufacturers are free to offer competing drugs or cheaper generics. Finally, even though research and development efforts have produced miracle drugs, there is still a high risk of failure. Drug companies have spent millions of dollars trying to find a cure for diseases such as cancer, arthritis, and diabetes, and their efforts to date have been fruitless. Industry insiders admit that without the lure of large profits, there will be less and less research and development and ultimately the stream of much-needed, innovative prescription medications will stop.

Source: For more information, see Shawn Tully, "The Plots to Keep Drug Prices High," *Fortune,* December 27, 1993, pp. 120–24; Joseph Weber, "Can a 1,245% Markup on Drugs Really Be Legal?," *Business Week,* November 1, 1993, p. 34; Andrea Rock, "Cut Your Spiraling Drug Costs 70%," *Money,* June 1993, pp. 131+; and Shawn Tully, "Why Drug Prices Will Go Lower," *Fortune,* May 3, 1993, p. 56.

UNIT 44   CALCULATING MARKUP BASED ON COST

LEARNING OBJECTIVES

After completing this unit, you will be able to:

1. calculate the markup amount based on cost.
2. determine the selling price for a cost markup problem.

Before you jump to the wrong conclusion, not all firms or industries use the large markups that are common in the prescription drug industry. And it should also be understood that most businesspeople—manufacturers, wholesalers, and retailers—*are* in business to make a profit, so there may be some justification for above-average markups. To make a profit, businesspeople need to charge customers more for an item than they paid for it—that is they "mark it up." The **markup** (sometimes called the gross profit) is the dollar amount that a retailer adds to the original cost of merchandise. When the markup is added to the original cost, the result is the **selling price.**

Many employees don't realize that there are *two* methods—one based on the original cost and the other based on the selling price—used to calculate the markup amount. The method used to calculate the markup amount also affects the selling price that retailers charge for merchandise. Therefore, it is important to clarify which method to use. Markup based on cost is covered in this unit; markup based on selling price is covered in unit 45.

While any business can use either method of markup, usually manufacturers, wholesalers, and small businesses prefer the cost method. Cost markup problems are a practical application of the portion formula presented in Unit 26. The dollar amount of markup (the portion) is found by multiplying the original cost (the base) by the percent of markup (the rate). The original cost for an electric can opener is $14.00. The markup rate is 25 percent. What is the dollar amount of the markup?

Many people begin their search for the right answer by "plugging in" the facts about a specific problem into the following formula.

$$\begin{array}{ccccc} 100\% & & 25\% \ (R) & & 125\% \\ \text{Original Cost} & + & \text{Markup} & = & \text{Selling Price} \\ \$14.00 \ (B) & & ? \ (P) & & \end{array}$$

Notice that all numbers with a percent (%) sign are written across the top of the formula. Since the markup problems in this unit are based on cost, the original cost represents the base *and* 100% is placed above the original cost in the formula. The 25 percent of markup (rate) is placed above the markup in the formula. Finally, it is possible to add the original cost and markup (stated as percents) and place the answer (125%) over the selling price.

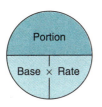

All dollar amounts are written across the bottom of the formula. The original cost ($14) represents the base stated as a dollar amount. With this information, it is now possible to use the portion formula (P = B × R) to solve for the markup amount as illustrated in Example 1.

EXAMPLE 1 Calculating Markup Based on Cost

$$\begin{array}{ll} \$14.00 & \text{original cost (base)} \\ \times \ 0.25 & \text{percent of markup (rate)} \\ \hline \$ \ 3.50 & \text{markup amount (portion)} \end{array}$$

In this example, note that the percent of markup has been changed to the decimal 0.25.

To calculate the selling price, one additional step is necessary: adding the markup amount to the original cost.

EXAMPLE 2 Determining the Selling Price

Find the selling price of the can opener in Example 1.

$$\begin{array}{ccccc} 100\% & + & 25\% & = & 125\% \\ \text{Original Cost} & + & \text{Markup} & = & \text{Selling Price} \\ \$14.00 & + & 3.50 & = & \$17.50 \end{array}$$

$$\begin{array}{ll} \$14.00 & \text{original cost} \\ + \ 3.50 & \text{markup amount} \\ \hline \$17.50 & \text{selling price} \end{array}$$

ALTERNATE METHOD: PERCENT OF MARKUP PLUS 100 PERCENT

Often retailers don't need to know the dollar amount of markup—they need only the selling price. When working a cost markup problem, you can find the selling price directly by multiplying the original cost by the percent of markup *plus* 100 percent.

EXAMPLE 3　Using the Percent of Markup plus 100 Percent

Find the selling price of the can opener in Example 1.

100%	+	25%	=	125% (R)
Original Cost	+	Markup	=	Selling Price
$14.00 (B)			=	? (P)

100　percent	**STEP** ☐1	Determine the adjusted percent of markup by adding 100 percent to the percent of markup (25 percent).
+25　percent of markup		
125　percent of markup (adjusted)		
$14.00 × 1.25 = $17.50	**STEP** ☐2	Multiply the original cost ($14.00) by the adjusted percent of markup (125 percent).

CALCULATOR TIP

Depress clear key.
Enter 100　Depress +
Enter 25　Depress =
Read the answer 125
Enter 14　Depress ×
Enter 1.25　Depress =
Read the answer $17.50

In this example, note that 125 percent has been converted to the decimal 1.25.

HINT: You may want to review the material on portion in Unit 26 before completing this unit.

Using the Cost Method of Markup, Complete the Following Problems

For all problems that do not work out evenly, carry your answers to two decimal places.

Original Cost	Percent of Markup Based on Cost		Markup Amount
$35.00	20%	1.	$35 × .20 = $7
$14.20	12%	2.	
$5.10	32%	3.	

Original Cost	Percent of Markup Based on Cost	Markup Amount
$124.00	$38\frac{3}{4}\%$	**4.** _____
$32.90	45.5%	**5.** _____

Original Cost	Percent of Markup Based on Cost	Selling Price
$8.50	15%	**6.** *$8.50 × 1.15 = $9.78*
$13.70	14%	**7.** _____
$122.00	10%	**8.** _____
$8.80	$24\frac{1}{2}\%$	**9.** _____
$225.00	30.6%	**10.** _____

Original Cost	Percent of Markup Based on Cost	Amount of Markup	Selling Price
$12.00	24%	**11.** *$12.00 × .24 = $2.88*	$14.88
$25.00	60%	**12.** _____	$40.00
$31.00	30%	$9.30	**13.** _____

Original Cost	Percent of Markup Based on Cost	Amount of Markup	Selling Price
$27.38	42%	$11.50	14. _____
$20.00	8%	15. _____	16. _____

17. Northside Mazda sold a new Miata convertible for $18,750. If the car cost the dealership $16,590, what is the markup amount?

$18,750 – $16,590 = $2,160

18. The Strawberry Patch, a children's clothing store, pays $12.50 for infants' T-shirts. If the markup is 21 percent on cost, what is the markup amount?

19. In problem 18, what is the selling price?

20. National Health Foods pays $3.75 for a jar of vitamin C. If the markup is $21\frac{3}{4}$ percent, what is the selling price?

UNIT 45   CALCULATING MARKUP BASED ON SELLING PRICE

LEARNING OBJECTIVES   *After completing this unit, you will be able to:*

1. determine the selling price when markup is based on the selling price.
2. calculate the markup amount based on the selling price method.

In Unit 44, you learned to use the cost method to calculate the markup amount and the selling price. In this unit, we examine how to calculate markup based on the selling price. Although there are exceptions, large retailers prefer the selling price method of markup.

There are at least two major differences between the cost method and the selling price method of calculating markup: With the selling price method, (a) the selling price is the *base* and (b) the percent of markup is based on the unknown selling price. Although the selling price (the base) is unknown, it is possible to find it if you use the complement of the percent of markup.

EXAMPLE 1 Finding the Complement If the Percent of Markup Is 30 Percent

100%	selling price is the base
− 30%	percent of markup
70%	complement of the percent of markup

Once you have determined the complement, it is possible to use the base formula (Base = Portion ÷ Rate) presented in Unit 28 to find the selling price. The selling price (base) is found by dividing the original cost (portion) by the complement of the percent of markup (rate). The original cost for a three-ring notebook is $4. The markup rate based on the selling price is 30 percent. Again, let's begin the search for the right answer with the following formula:

$$70\% \text{ (R)} \quad + \quad 30\% \quad = \quad 100\%$$
$$\text{Original Cost} \quad + \quad \text{Markup} \quad = \quad \text{Selling Price}$$
$$\$4.00 \text{ (P)} \qquad\qquad\qquad\qquad ? \text{ (B)}$$

Notice that all numbers with a percent (%) sign are written across the top of the formula. Since the markup problems in this unit are based on selling price, the selling price represents the base *and* 100% is placed above the selling price in the formula. Also, the 70% complement and 30% percent of markup are placed across the top of the formula.

All dollar amounts are written across the bottom of the formula. The original cost ($4) is the portion. With this information, it is now possible to use the base formula (B = P ÷ R) to solve for the selling price as illustrated in Example 2.

EXAMPLE 2 Determining the Selling Price Using the Selling Price Method

Find the selling price of a three-ring notebook that originally cost $4, when the percent of markup is 30 percent on selling price.

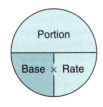

100% −30% 70% complement	**STEP** 1	Determine the complement of the percent of markup.
$4.00 ÷ 0.70 = $5.714	**STEP** 2	Using the base formula, divide the original cost ($4) by the complement of the percent of markup (70 percent).
$5.714 = $5.71	**STEP** 3	If necessary, round your answer to two decimal places.

CALCULATOR TIP

Depress clear key.
Enter 100   Depress −
Enter 30   Depress =
Read the answer 70
Enter 4   Depress ÷
Enter 0.70 Depress =
Read the answer $5.71

In this example, the complement (70 percent) has been converted to the decimal 0.70.

To calculate the dollar amount of markup, one additional step is necessary—subtracting the original cost from the selling price.

EXAMPLE 3   Calculating the Markup Amount

Find the dollar amount of markup for the notebook in Example 2.

70%   +   30%   =   100%

Original Cost +   Markup = Selling Price

$4.00   ? $5.71

> $5.71   selling price
> −4.00   original cost
> $1.71   markup amount

> **CALCULATOR TIP**
>
> Depress clear key.
> Enter 5.71   Depress −
> Enter 4   Depress =
> Read the answer $1.71

Note that when you calculate markup based on the selling price, the **first** answer is the selling price. The **second** answer is the markup amount.

HINT: You may want to review the material on base in Unit 28 before completing this unit.

Using the Selling Price Method of Markup, Complete the Following Problems

For all problems that do not work out evenly, carry your answers to two decimal places.

1. A new water pump cost White Motor Parts $24.20. This retailer sells the pump for $37.50. What is the markup amount on this item?

   $37.50 − $24.20 = $13.30

Percent of Markup		Complement
25%	2.	*100% − 25% = 75%*
15%	3.	
22%	4.	
$10\frac{1}{2}$%	5.	

Original Cost	Percent of Markup Based on Selling Price	Selling Price
$8.00	15%	6. *100% − 15% = 85%* *= .85* *$8.00 ÷ .85 = $9.41*
$12.00	20%	7. _____
$16.00	25%	8. _____
$24.50	27.5%	9. _____

Original Cost	Percent of Markup Based on Selling Price	Selling Price	Markup Amount
$36.00	30%	10. *100% − 30% = 70%* *= .70* *$36.00 ÷ .70 = $51.43*	11. *$51.43 − $36.00* *= $15.43*
$5.40	10%	12. _____	13. _____
$49.00	40%	14. _____	15. _____

16. Atlanta Office Supply pays $275 for an electric typewriter. If the markup amount is $124, what is the selling price?

$275 + $124 = $399

17. If Best Merchandise pays $49 for a patio table that was later sold at a markup of 40 percent on the selling price, what is the selling price?

18. In problem 17, what is the markup amount?

Critical Thinking Problem

Your boss, Mr. Franklin, asks that you calculate the markup amount and the selling price for an eight-speed blender that cost $13.50 by using the cost method of markup. Next, he wants you to calculate the markup amount and the selling price based on the selling price method. Assume the percent of markup in both cases is 20 percent.

19. What is the markup amount if the cost method is used?

20. What is the selling price if the markup amount is based on the cost method?

21. What is the selling price if markup is based on the selling price method?

22. What is the markup amount if the selling price method is used?

23. Assuming the original cost and percent of markup are the same, why does the method used to calculate the markup amount make a difference in the final answers? _____

_____

_____

_____

UNIT 46	CALCULATING PERCENT OF MARKUP

LEARNING OBJECTIVES   *After completing this unit, you will be able to:*

1. find the percent of markup when markup is based on cost.
2. find the percent of markup when markup is based on the selling price.

Generally, retailers know the original cost of merchandise that is offered for sale in their stores. In many cases, they sell this merchandise at retail prices that are suggested by the manufacturer. With this information, it is possible to find the percent of markup. For this type of problem, the first step is to find the markup amount by subtracting the original cost from the selling price.

EXAMPLE 1   Calculating the Markup Amount

A travel clock sells for $29.95; the original cost was $22.10. Find the dollar amount of markup.

CALCULATOR TIP

Depress clear key.
Enter 29.95   Depress   –
Enter 22.10   Depress   =
Read the answer $7.85

$29.95	selling price
– 22.10	original cost
$ 7.85	markup amount

PERCENT OF MARKUP WHEN MARKUP IS BASED ON COST

Percent of markup problems are a practical application of the rate formula presented in Unit 27. The percent of markup (the rate) is found by dividing the markup amount (the portion) by the original cost (the base).

EXAMPLE 2   Find the Percent of Markup (Cost Method)

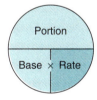

The original cost of a calculator is $7.50 (base); the markup amount is $1.50 (portion) based on cost. Find the percent of markup.

$$
\begin{array}{ccc}
100\% & + & ?\ (R) \\
\text{Original Cost} & + & \text{Markup} \quad = \quad \text{Selling Price} \\
\$7.50\ (B) & + & \$1.50\ (P)
\end{array}
$$

$1.50 ÷ $7.50 = 0.20   **STEP** ☐1   Divide the markup amount ($1.50) by the original cost ($7.50).

0.20 = 20%   **STEP** ☐2   Convert the decimal answer (0.20) to a percent.

CALCULATOR TIP

Depress clear key.
Enter 1.50   Depress   ÷
Enter 7.50   Depress   =
Read the answer 0.20

PERCENT OF MARKUP WHEN MARKUP IS BASED ON THE SELLING PRICE

To find the percent of markup when markup is based on the selling price, you can again use the rate formula: Divide the markup amount (the portion) by the selling price (the base).

| EXAMPLE 3 | Finding the Percent of Markup (Selling Price Method) |

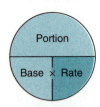

Portion

Base × Rate

The selling price for a desk lamp is $20 (base), and the markup amount is $4.50 (portion) based on the selling price. Find the percent of markup.

$$
\begin{array}{ccc}
& ?\ (R) & 100\% \\
\text{Original Cost} + & \text{Markup} = & \text{Selling Price} \\
& \$4.50\ (P) & \$20.00\ (B)
\end{array}
$$

$4.50 ÷ $20.00 = 0.225   **STEP** [1]   Divide the markup amount ($4.50) by the selling price ($20.00).

0.225 = 22.5%   **STEP** [2]   Convert the decimal answer (0.225) to a percent.

CALCULATOR TIP

Depress clear key.
Enter 4.50   Depress ÷
Enter 20   Depress =
Read the answer 0.225

HINT: You may want to review the material on rate in Unit 27 before completing this unit.

Complete the Following Problems

For all problems that do not work out evenly, carry your answers to two decimal places.

For problems 1 through 8, assume that markup is based on cost.

Original Cost	Selling Price		Markup Amount		Percent of Markup Based on Cost
$ 2.00	$ 3.00	1.	$3.00 – $2.00 = $1.00	2.	$1 ÷ $2 = .50 = 50%
$ 3.50	$ 6.00	3.	_____	4.	_____
$ 7.00	$10.00	5.	_____	6.	_____
$12.50	$15.00	7.	_____	8.	_____

For problems 9 through 16, assume that markup is based on the selling price.

Original Cost	Selling Price	Markup Amount	Percent of Markup Based on Selling Price
$ 1.50	$ 2.29	9. *$2.29 – $1.50 = $.79*	10. *$.79 ÷ $2.29 = .34497 = 34.497% = 34.50%*
$ 5.75	$ 8.00	11. _____	12. _____
$16.70	$21.50	13. _____	14. _____
$30.00	$45.00	15. _____	16. _____

17. Mount Vernon Carpets bought carpeting for $5.20 a yard and sold it for $8.95 a yard. What is the markup amount for a yard of this carpet?

    $8.95 – $5.20 = $3.75

18. R&B Furniture paid $125 for a velvet recliner. The store sells the same recliner for $199. Assuming that R&B uses the cost method of markup, what is the percent of markup?

19. Team Sports, Inc., purchased a shipment of tennis rackets. Each racket cost the store $89. The store advertised the rackets for $119. Assuming that management calculates markup on the selling price of an item, what is the percent of markup?

20. You receive a shipment of 144 ballpoint pens. The invoice amount is $216. You decide to sell the pens for $1.99 each. What is the amount of markup for each ballpoint pen?

21. In problem 20, what is the percent of markup if the markup amount is based on the original cost?

22. In problem 20, what is the percent of markup if the markup amount is based on the selling price?

UNIT 47	DETERMINING THE DOLLAR VALUE OF INVENTORY

LEARNING OBJECTIVES

After completing this unit, you will be able to:

1. describe the difference between the periodic and perpetual inventory systems.
2. find the dollar value of inventory when the specific identification, FIFO, LIFO, and average cost methods are used.

Manufacturers, wholesalers, and retailers must maintain a reasonable amount of inventory that can be sold to their customers. And regardless of how hard they try, some inventory will not be sold at the end of an accounting period. Two methods can be used to determine the number of items that remain unsold. First, some firms physically count all merchandise in stock at least once a year. This method is called the **periodic inventory system.** Other companies maintain what is commonly called a **perpetual inventory system,** which provides up-to-date inventory information at any time. When the perpetual system is used, small firms generally use inventory cards, which are manually updated each time the amount of inventory changes. Normally, larger firms use an electronic cash register, which is coupled to a computer to maintain perpetual inventory records in addition to complete sales records.

Regardless of the method used to maintain inventory records, the accountant must then take this information and implement one of four widely used methods of determining the dollar value of inventory: (1) specific identification; (2) first in, first out; (3) last in, first out; and (4) average cost. The information presented in Figure 47–1 will be used to illustrate each method.

FIGURE 47–1 Inventory Information for Personal Computer Stock No. 700-CD

Date of Original Purchase	Quantity	Unit Cost	Total Cost
Begin. Inventory	5	$780	$ 3,900
April 25, 19XX	12	730	8,760
June 24, 19XX	9	800	7,200
September 15, 19XX	6	820	4,920
November 20, 19XX	4	760	3,040
	36		$27,820

SPECIFIC IDENTIFICATION METHOD

The specific identification method is normally used by companies that maintain perpetual inventory records at actual cost and the number of individual sales are small. Merchandise in this category is usually referred to as big ticket items and comprises items such as computers, appliances, jewelry, or automobiles. Identification is based on a model number, serial number, date of purchase, or purchase price. Using the information presented in Figure 47–1 and the specific identification method, the dollar value for the remaining seven computers is $5,440, as illustrated in Example 1.

EXAMPLE 1 Using the Specific Identification Method

Assume at the end of the accounting period, there were seven personal computers that remain unsold. Also, assume that the seven unsold computers were originally purchased by the seller on the following dates: two units on April 25; three units on September 15; and two units on November 20.

CALCULATOR TIP

Depress clear key.
Enter 730   Depress ×
Enter 2   Depress =
Read the answer $1,460
Enter 820   Depress ×
Enter 3   Depress =
Read the answer $2,460
Enter 760   Depress ×
Enter 2   Depress =
Read the answer $1,520
Enter 1460   Depress +
Enter 2460   Depress +
Enter 1520   Depress =
Read the answer $5,440

$730 × 2 = $1,460   **STEP** 1   Multiply the price for each unit originally purchased on April 25 ($730) by the number of items remaining unsold that were purchased on this date (2).

$820 × 3 = $2,460   **STEP** 2   Multiply the price for each unit originally purchased on September 15 ($820) by the number of items remaining unsold that were purchased on this date (3).

$760 × 2 = $1,520   **STEP** 3   Multiply the price for each unit originally purchased on November 20 ($760) by the number of items remaining unsold that were purchased on this date (2).

$1,460
  2,460
+1,520
$5,440   **STEP** 4   To determine the dollar value of the unsold inventory, add the answers found in steps 1, 2, and 3.

FIRST IN, FIRST OUT METHOD

A lot of firms use the first in, first out (FIFO) method because it matches the typical flow of goods in an ongoing business. When the FIFO method is used, the oldest merchandise purchased is the first merchandise sold. *It is assumed that the unsold inventory is the merchandise purchased at the later dates.* Using the information presented in Figure 47–1 and the FIFO method, the total cost for the remaining seven computers is $5,500, as illustrated in Example 2.

EXAMPLE 2   Using the FIFO Method (the computers left in stock are the most recently purchased computers)

Assume that there are seven unsold computers at the end of the accounting period.

CALCULATOR TIP

Depress clear key.
Enter 760   Depress ×
Enter 4   Depress =
Read the answer $3,040
Enter 7   Depress −
Enter 4   Depress =
Read the answer 3
Enter 820   Depress ×
Enter 3   Depress =
Read the answer $2,460
Enter 3,040 Depress +
Enter 2,460 Depress =
Read the answer $5,500

$760 × 4 = $3,040   **STEP** 1   Multiply the cost of each computer purchased on November 20 ($760) by the number of computers purchased on that date (4).

$$\begin{array}{r} 7 \\ -4 \\ \hline 3 \end{array}$$ **STEP** 2   Subtract the number of computers accounted for in step 1 (4) from the number of unsold computers remaining at the end of the accounting period (7).

$820 × 3 = $2,460   **STEP** 3   Multiply the cost of each computer purchased on September 15 ($820) by the answer from step 2 (3).

$$\begin{array}{r} \$3,040 \\ +2,460 \\ \hline \$5,500 \end{array}$$ **STEP** 4   To determine the dollar value of inventory, add the answers found in steps 1 and 3.

LAST IN, FIRST OUT METHOD

During periods of inflation, a business may use the last in, first out (LIFO) method to reduce the dollar amount of profits and ultimately the firm's tax bill. When the LIFO method is used, the most recently purchased merchandise is the first to be sold. *It is assumed that the unsold inventory is the merchandise purchased at the earlier dates.* Using the information in Figure 47–1 and the LIFO method, the total cost for the remaining seven computers is $5,360, as illustrated in Example 3.

| EXAMPLE 3 | Using the LIFO Method (the computers left in stock are the first computers purchased) |

Assume that there are seven unsold computers at the end of the accounting period.

$780 × 5 = $3,900   **STEP** 1 — Multiply the number of computers included in the beginning inventory (5) by the cost of each computer ($780).

7
−5
2   **STEP** 2 — Subtract the number of computers accounted for in step 1 (5) from the number of unsold computers remaining at the end of the accounting period (7).

$730 × 2 = $1,460   **STEP** 3 — Multiply the cost of each computer purchased on April 25 ($730) by the answer from step 2 (2).

$3,900
+1,460
$5,360   **STEP** 4 — To determine the dollar value of inventory, add the answers found in steps 1 and 3.

CALCULATOR TIP

Depress clear key.
Enter 780   Depress ×
Enter 5   Depress =
Read the answer $3,900
Enter 7   Depress −
Enter 5   Depress =
Read the answer 2
Enter 730   Depress ×
Enter 2   Depress =
Read the answer $1,460
Enter 3,900   Depress +
Enter 1,460   Depress =
Read the answer $5,360

AVERAGE COST METHOD

Unlike other methods used to find a dollar value on inventory, the average cost method does not attempt to identify any specific units that remain unsold. This method does use a "weighted dollar average" for all remaining merchandise. This dollar average is calculated by dividing the total cost of all merchandise available for sale by the total number of available units during the accounting period. Next, multiply the dollar average by the number of unsold units remaining at the end of the accounting period. Using the information presented in Figure 47–1 and the average cost method, the total cost for the remaining seven computers is $5,409.46, as illustrated in Example 4.

| EXAMPLE 4 | Using the Average Cost Method |

Assume that there are seven unsold computers at the end of the accounting period.

$27,820 ÷ 36 = $772.78   **STEP** 1 — Divide the total cost of all merchandise by the total number of available units.

$772.78 × 7 = $5,409.46   **STEP** 2 — Multiply the average cost per computer obtained in step 1 ($722.78) by the number of unsold computers (7).

CALCULATOR TIP

Depress clear key.
Enter 27820   Depress ÷
Enter 36   Depress =
Read the answer $772.78
Enter 772.78   Depress ×
Enter 7   Depress =
Read the answer $5,409.46

COMPARISON OF INVENTORY METHODS

An accountant must choose the method used to value ending inventory. Some factors to consider are listed below. The specific identification method is the most accurate, but is normally used only with big ticket items. The average cost method is also accurate, since each individual unit is valued at a weighted average. The FIFO method reflects the fact that an ongoing business normally sells merchandise in the order that it is received. In inflationary periods LIFO typically results in lower profits and ultimately lower taxes. You should realize that each inventory method results in a different dollar value for ending inventory and can change the amount of profit or loss that business firms report to the Internal Revenue Service. As a result, the Internal Revenue Service requires that business firms must obtain IRS approval before changing the way they report the value of their inventory.

Complete the Following Problems

For all problems that do not work out evenly, carry your answers to two decimal places.

1. T (F) When business firms physically count all merchandise at least once a year, they are using a perpetual inventory system.

2. T F When larger firms use a computer system to update inventory records each time merchandise is sold, they are using a periodic inventory system.

3. T F The FIFO inventory method matches the typical flow of goods in an ongoing business.

4. T F During periods of inflation, business firms can use the average cost inventory method to reduce their tax bill.

5. T F Of the four methods used to determine the dollar value of inventory at the end of an accounting period, the specific identification inventory method is the most accurate.

6. During 1995, Jan Morgan, the inventory manager for Dayton Office Equipment, made four purchases of photocopy machines. On January 31, 48 copiers were purchased. On March 10, 30 copiers were purchased. On June 20, 36 copiers were purchased. On October 30, 50 copiers were purchased. How many copiers did Ms. Morgan purchase during 1995?

Use the following information to work problems 7–10.
The following information for trash compactors was recorded on the inventory records for Richardson Appliances:

15 trash compactors purchased at $140 each
24 trash compactors purchased at $165 each
36 trash compactors purchased at $124 each
40 trash compactors purchased at $152 each

7. How many trash compactors were purchased during this inventory period?

8. What is the total cost for *all* trash compactors purchased during this inventory period?

9. What is the average cost for each trash compactor purchased during this inventory period?

10. What is the dollar value of inventory for trash compactors if 22 trash compactors remain unsold at the end of the accounting period and the average cost method is used?

Use the following information to work problems 11–16.

As the accountant for National Video and Sound, you are trying to determine the value for blank video tapes purchased during the last 12 months. Assume that at the end of the accounting period that there are 210 blank video tapes that remain unsold. Use the form below to find the total cost for each inventory method discussed in this unit.

Video Blank Tapes Model No. Beta—3 Hr.			
Date of Purchase by the Seller	**Quantity**	**Unit Cost**	**Total Cost**
Beginning Inventory	125	$1.50	$187.50
January 2, 199X	250	$1.50	$375.00
April 20, 199X	150	$1.75	$262.50
June 18, 199X	100	$2.00	$200.00
September 30, 199X	150	$1.50	$225.00
November 25, 199X	130	$2.25	$292.50
	11. _____		12. _____

13. _____ Specific Identification—100 units were originally purchased on April 20; 35 units were originally purchased on June 18; 75 units were originally purchased on September 30.

14. _____ Ending Inventory—First In, First Out Method

15. _____ Ending Inventory—Last In, First Out Method

16. _____ Ending Inventory—Average Cost Method

Use the following information to work problems 17–22.

Your supervisor was impressed with the inventory comparisons that you just completed (problems 11–16) and asks that you complete a similar comparison for video disc players. Assume that at the end of the accounting period there are 22 video disc players that remain unsold. Use the form below to find the total cost for each inventory method discussed in this unit.

<table>
<tr><td colspan="4" align="center">Video Disc Player
Model No. SY–40</td></tr>
<tr><td>Date of Purchase
by the Seller</td><td>Quantity</td><td>Unit Cost</td><td>Total Cost</td></tr>
<tr><td>Beginning Inventory</td><td>15</td><td>$465.00</td><td>$6,975.00</td></tr>
<tr><td>February 12, 199X</td><td>5</td><td>$485.50</td><td>$2,427.50</td></tr>
<tr><td>May 20, 199X</td><td>5</td><td>$510.30</td><td>$2,551.50</td></tr>
<tr><td>July 2, 199X</td><td>10</td><td>$520.00</td><td>$5,200.00</td></tr>
<tr><td>August 15, 199X</td><td>5</td><td>$535.75</td><td>$2,678.75</td></tr>
<tr><td>October 28, 199X</td><td>10</td><td>$650.00</td><td>$6,500.00</td></tr>
</table>

17. _____   18. _____

19. _____   Specific Identification—2 units were originally purchased February 12; 4 units were originally purchased May 20; 4 units were purchased July 2; 5 units were purchased August 15; 7 units were originally purchased October 28.

20. _____   Ending Inventory—First In, First Out Method

21. _____   Ending Inventory—Last In, First Out Method

22. _____   Ending Inventory—Average Cost Method

UNIT 48	CALCULATING MARKDOWNS

LEARNING OBJECTIVES   *After completing this unit, you will be able to:*

1. determine the dollar amount of a markdown.
2. calculate the percent of markdown.

A **markdown** is a price reduction that is subtracted from the original selling price. Retailers use markdowns to lower prices on certain merchandise because of excess inventory, seasonal changes, lower prices offered by competitors, or special promotional activities designed to increase sales. Although markdowns may be necessary, retailers who lower prices on merchandise must be willing to accept lower profits on merchandise that has been marked down.

Markdown problems are a practical application of the portion formula presented in Unit 26. The dollar amount of markdown (the portion) is found by multiplying the original selling price (the base) by the percent of markdown (the rate).

EXAMPLE 1 | Calculate the Markdown Amount

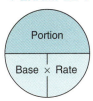

The original selling price of an electric blanket is $80. The retailer offers a 25 percent markdown. Find the dollar amount of the markdown.

$80.00	original selling price (base)
× 0.25	percent of markdown (rate)
$20.00	dollar amount of markdown (portion)

CALCULATOR TIP

Depress clear key.
Enter 80   Depress ×
Enter 0.25 Depress =
Read the answer $20

In this example, note that the percent of markdown (25 percent) has been changed to the decimal 0.25.

To calculate the reduced selling price, one additional step is necessary: Subtract the dollar amount of markdown from the original selling price.

EXAMPLE 2 | Determining the Reduced Selling Price

CALCULATOR TIP

Depress clear key.
Enter 80 Depress –
Enter 20 Depress =
Read the answer $60

Find the reduced selling price for the electric blanket in Example 1.

$80.00	original selling price
–20.00	dollar amount of markdown
$60.00	reduced selling price

ALTERNATE METHOD: USING COMPLEMENTS

You can find the reduced selling price directly by multiplying the original selling price by the complement of the percent of markdown. To find the complement, mentally subtract the percent of markdown from 100 percent. The reduced selling price is the same regardless of the method used.

EXAMPLE 3 | Using the Complement Method

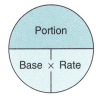

Find the reduced selling price for the electric blanket in Example 1 by the complement method.

100%		STEP 1	Find the complement of the percent of markdown (25 percent).
– 25%			
75%	complement		

$ 80		STEP 2	Multiply the original selling price ($80) by the complement (75 percent).
×0.75			
$ 60			

CALCULATOR TIP

Depress clear key.
Enter 100   Depress –
Enter 25   Depress =
Read the answer 75
Enter 80   Depress ×
Enter 0.75 Depress =
Read the answer $60

In this example, the complement (75 percent) has been converted to the decimal 0.75.

CHAPTER EIGHT

CALCULATING THE PERCENT OF MARKDOWN

If the original selling price and the reduced selling price are known, it is possible to calculate the percent of markdown by using the rate formula presented in Unit 27. First find the dollar amount of markdown by subtracting the reduced selling price from the original selling price. Then, divide the dollar amount of markdown (the portion) by the original selling price (the base).

EXAMPLE 4 Calculate the Percent of Markdown

The original selling price for a wrist watch, $50, is reduced to $44. Find the percent of markdown.

$50 −44 $ 6	STEP 1	Subtract the reduced selling price ($44) from the original selling price ($50) to find the dollar amount of markdown.
$6 ÷ $50 = 0.12	STEP 2	Divide the dollar amount of markdown ($6) by the original selling price ($50).
0.12 = 12%	STEP 3	Convert the decimal answer (0.12) to a percent.

CALCULATOR TIP

Depress clear key.
Enter 50   Depress −
Enter 44   Depress =
Read the answer $6
Enter 6   Depress ÷
Enter 50   Depress =
Read the answer 0.12

Note that when you calculate the percent of markdown, you always divide by the *original* selling price—*not* the reduced selling price.

HINT: You may want to review the material on portion in Unit 26 and the material on rate in Unit 27 before completing this unit.

Complete the Following Problems

For all problems that do not work out evenly, carry your answers to two decimal places.

Original Selling Price	Percent of Markdown		Markdown Dollar Amount		Reduced Selling Price
$ 50.00	10%	1.	*.10 × $50.00* *= $5.00*	2.	*$50.00 − $5.00* *= $45.00*
$ 40.00	30%	3.		4.	
$ 62.00	20%	5.		6.	

254

$ 75.00	15%	**7.** _____	**8.** _____

$110.00	25%	**9.** _____	**10.** _____

Original Selling Price	Reduced Selling Price	Dollar Amount of Markdown	Percent of Markdown
$ 25.00	$ 15.00	**11.** $25.00 − $15.00 = $10.00	**12.** $10.00 ÷ $25.00 = .40 = 40%
$120.00	$ 90.00	**13.** _____	**14.** _____
$ 6.95	$ 4.50	**15.** _____	**16.** _____
$149.95	$ 99.95	**17.** _____	**18.** _____

19. At the Toy Store, a new electronic game was marked down to $39.50. If the original selling price had been $59.95, what is the markdown amount?

$59.95 − $39.50 = $20.45

20. In problem 19, what is the percent of markdown?

21. The Stereo Shop purchased a custom-made cabinet for $375. The retailer marks up this type of merchandise 40 percent on cost. What is the markup amount?

22. In problem 21, what is the selling price of the custom cabinet?

23. In problem 21, what is the dollar amount of markdown if The Stereo Shop reduces the original selling price to $400?

24. In problem 23, what is the percent of markdown?

Personal Math Problem

Buy-Rite Drugs—a local drug store—charges $3.99 for a 20-ounce bottle of shampoo. At a nationally recognized discount store, in a town 10 miles away, the regular price for a 15-ounce bottle of the same shampoo is $3.49.

25. What is the cost per ounce for the 20-ounce bottle of shampoo?

26. If the discount store lowers the price of a 15-ounce bottle of shampoo by 25 percent, what is the cost per ounce for the 15-ounce bottle of shampoo?

27. If you were the consumer in this problem, would you purchase shampoo from the local retailer or the discount store? Why? _____

_____

_____

_____

_____

_____

UNIT	IMPORTANT POINTS TO REMEMBER	EXAMPLES
UNIT 44 Calculating Markup Based on Cost	The markup amount (the portion) is found by multiplying the original cost (the base) by the percent of markup (the rate). To calculate the selling price, add the markup amount to the original cost.	$16.50 original cost × 0.20 percent of markup (on cost) $ 3.30 markup amount $16.50 original cost + 3.30 markup amount $19.80 selling price
UNIT 45 Calculating Markup Based on Selling Price	When calculating markup based on the selling price, the selling price (the base) is found by dividing the original cost (the portion) by the *complement* of the percent of markup (the rate). To calculate the markup amount, one additional step is necessary—subtract the original cost from the selling price.	100% – 20% 80% complement of the percent of markup $5.00 ÷ 0.80 = $6.25 original complement selling cost price $6.25 selling price –5.00 original cost $1.25 markup amount
UNIT 46 Calculating Percent of Markup	To calculate the percent of markup use the following two steps: First calculate the dollar amount of markup. Then, to calculate percent of markup when markup is based on cost, divide the dollar amount of markup by the original cost. To calculate percent of markup when markup is based on the selling price, divide the dollar amount of markup by the selling price.	*Cost Method* $2.00 ÷ $8.00 = 0.25 = 25% markup original percent of amount cost markup *Selling Price Method* $6.00 ÷ $12.00 = 0.50 = 50% markup selling percent of amount price markup

continued from page 257		
UNIT	**IMPORTANT POINTS TO REMEMBER**	**EXAMPLES**

UNIT 47
Determining the Dollar Value of Inventory

The following information will be used to illustrate the four methods of determining the dollar value of unsold inventory.

```
Begin. Inv.   6 @ $34 = $   204
Jan. 23      20 @   30 =     600
May 14       15 @   35 =     525
Nov. 25       5 @   28 =     140
             46            $1,469
```

Assume that there are 10 items unsold at the end of the accounting period.

Specific Identification Method

Assume that three items were purchased Jan. 23 and seven items were purchased May 14.

```
3  ×  $30  =  $ 90
7  ×  $35  =   245
              $335
```

First In, First Out Method

```
5  ×  $28  =  $140
5  ×  $35  =   175
              $315
```

Last In, First Out Method

```
6  ×  $34  =  $204
4  ×  $30  =   120
              $324
```

Average Cost Method

```
$1,469  ÷  46  =  $31.93
$31.93  ×  10  =  $319.30
```

UNIT 48
Calculating Markdowns

To find the markdown amount, multiply the original selling price (the base) by the percent of markdown (the rate). To calculate the reduced selling price, subtract the markdown amount from the original selling price. To calculate the percent of markdown, divide the markdown amount by the original selling price.

```
$30.00   original selling price
× 0.10   percent of markdown
$ 3.00   markdown amount
```

```
$30.00   original selling price
– 3.00   markdown amount
$27.00   reduced selling price
```

```
$3   ÷   $30   =   0.10   =   10%

mark-    original         percent
down     selling            of
amount   price           markdown
```

CHAPTER 8	MASTERY QUIZ

DIRECTIONS *Round each answer to two decimal places or hundredths. (Each answer) counts 5 points.)*

Original Cost	Percent of Markup Based on Cost	Amount of Markup	Selling Price
$35.00	16%	1. _____	2. _____
$ 5.50	25%	3. _____	4. _____

Original Cost	Percent of Markup Based on Selling Price	Amount of Markup	Selling Price
$ 2.20	40%	5. _____	6. _____
$100.00	30%	7. _____	8. _____

9. Kid's Stuff Clothing paid $18 for a wool sweater and sold the sweater for $29.95. What is the markup amount for this item?

10. The High Tech Store paid $120 for a 19-inch portable television. If the markup amount is $69.95, what is the selling price?

11. A pair of denim jeans cost Hip Clothing $22. If this store marks up this item 45 percent on the selling price, what is the selling price?

12. Beethoven Music bought a classical cassette tape for $6.20 and sold it for $9. Assuming that Beethoven uses the selling price method of markup, what is the percent of markup?

Use the following information to work problems 13–16.
Assume that at the end of the accounting period there are 12 items that remain unsold.

Date of Purchase	Quantity	Unit Cost	Total Cost
Begin. Inventory	8	$200	$1,600
June 21, 19XX	10	175	1,750
October 15, 19XX	6	220	1,320
December 18, 19XX	8	210	1,680
	32		$6,350

13. What is the dollar value of unsold inventory if the specification identification method is used? (Assume that seven items were originally purchased on June 21 and that five items were originally purchased on October 15.

14. What is the dollar value of unsold inventory if the first in, first out method is used?

15. What is the dollar value of unsold inventory if the last in, first out method is used?

16. What is the dollar value of unsold inventory if the average cost method is used?

17. Great Wallpapers, Inc., purchased a roll of vinyl wallpaper for $12.50. This retailer marks up this type of merchandise 30 percent on cost. What is the markup amount?

18. In problem 17, what is the selling price of the wallpaper?

19. In problem 17, what is the markdown amount if Great Wallpapers reduces the original selling price to $15?

20. In problem 19, what is the percent of markdown?

| CHAPTERS 5–8 | CUMULATIVE REVIEW |

SPECIAL NOTE TO STUDENTS: To help you build a foundation for success in math, we have included a cumulative review at the end of every three or four chapters in the text. If you do not score at least 80 on this cumulative review, you may want to review the type of problems that you missed.

DIRECTIONS: *If necessary, round each answer to two decimal places or hundredths. (Each answer counts 5 points.)*

1. "Pay to the order of Barbara Cain, Mark Lamplighter," is what type of endorsement?

2. On May 4, Jane Martinez wrote check number 1204 for $56.70. If her previous balance was $2,311.78, what is her new balance after she records this check?

3. On June 10, Sandy Miguel deposited $1,560 in her checking account. If her previous balance was $1,925.40, what is her new balance after she records this deposit?

4. At the end of last month, Bart Carlton had a $1,080 balance on his bank card. If the bank charges a monthly rate of $1\frac{1}{2}$ percent based on the previous month's balance, what is the finance charge?

5. During the first week of November, Latavia Brown worked $43\frac{1}{2}$ hours. Assuming that Ms. Brown earns $8 an hour and is paid time-and-a-half for all hours worked in excess of 40, what is her gross pay amount?

6. Bill Umbaugh is paid a monthly salary of $1,500 plus $4\frac{1}{2}$ percent commission on sales. If Bill sold merchandise valued at $52,310 during the month of April, what was his gross pay for the month?

7. Anne Carlisle earns $3,450 a month. Assuming a 6.2 percent tax rate for Social Security and a 1.45 percent tax rate for Medicare, what is Anne's monthly Social Security tax deduction?

8. In problem 7, what is Anne's monthly Medicare tax deduction?

9. Mark Masters has annual wages of $22,100. Assuming a 5.4 percent tax rate and a wage ceiling of $7,000 for SUTA, what is the amount of SUTA tax that his employer must pay?

10. A water pump produced by Precision Manufacturing has a list price of $99. If Precision Manufacturing offers retailers a trade discount of 24 percent, what is the trade discount amount?

Net Price	Cash Discount Terms	Cash Discount Amount	Total Amount Due
$3,240	$\frac{2}{10}$, $\frac{N}{30}$	11. _____	12. _____
$1,900	$\frac{1}{10}$, $\frac{N}{60}$	13. _____	14. _____

15. An electric coffee maker has a list price of $25. The manufacturer offers retailers series discounts of 15 percent and 10 percent. What is the net price for one of the coffee makers?

16. A cotton western shirt cost Cotler's Western Wear $21. If this store marks up this item 40 percent based on cost, what is the selling price?

17. A 12-volt battery cost Chief Auto Parts $22.05. If this store marks up this item 30 percent based on selling price, what is the selling price?

18. Shakespeare Music Store bought a classical CD for $10.08 and sold it for $14. Assuming that Shakespeare uses the selling price method of markup, what is the percent of markup?

19. During 1995, June Cook, the inventory manager for Battle Creek Office Equipment, made three purchases of laser computer printers. On February 1, 25 printers were purchased. On June 30, 36 printers were purchased. On November 1, 48 printers were purchased. How many printers did Ms. Cook purchase during 1995?

20. "Jeans R Us" originally priced a pair of denim jeans for $30. In order to increase sales, the store's manager reduced the price to $24. What is the percent of markdown?

CHAPTER 9

Interest Calculations and Promissory Notes

CHAPTER PREVIEW

We begin this chapter with simple interest calculations. We also describe how to determine the principal, rate, and time factors when working interest problems. Then, our focus shifts to discounting noninterest-bearing and interest-bearing promissory notes.

MATH TODAY

INTEREST RATES: IT PAYS TO SHOP AROUND

Is there a relationship between low interest rates that banks pay depositors and the interest rates you pay for home, car, and other installment loans? You bet there is!

Banks and other financial institutions—like most businesses—are in business to make a profit. Simply put, they pay people to deposit their money. Then, they turn around and use the depositors' money to fund consumer loans. The dollar difference between what banks pay depositors and what they charge borrowers can be used to pay typical business expenses and provide a profit for the bank's stockholders.

With these basic facts in mind, how can you ensure that you get the best deal when you borrow money? According to financial experts, the best way to make sure you get the lowest interest rates when you purchase something and pay for it over an extended time is to shop around. Mike and Veronica Hess, a young married couple in San Diego, found that interest rates on a new-car loan were as low as 7 percent and as high as 19 percent. And the dollar amount of savings—over $4,200—for a 48-month, $15,000 car loan at 7 percent is substantial when compared to a 19 percent loan.

The experts also warn that it pays to think about the length of time required to repay a loan. Generally speaking, the shorter the repayment period, the better. Although monthly payments for a short-term loan will be higher, the total interest paid to the lender will be less because you have the borrowed money for a shorter period of time. Betty and George San Miguel thought their loan officer was crazy when she suggested that the Orlando couple consider a 5-year mortgage instead of a 10-year mortgage when the coupled wanted to borrow $80,000 to purchase a piece of real estate for investment purposes. In their mind, they thought their monthly payments would double if the time for repayment was cut in half. But the loan officer quickly pointed out that the monthly payment for an $80,000 loan repaid over 10 years with an interest rate of 8 percent would be $971. If the same loan were repaid over five years, the monthly payment would increase $651 to $1,622. And while the monthly payment does increase, the San Miguels will save almost $20,000 if they choose the five-year loan.

Source: For more information, see Phillip Zweig, "Mortgages: Now, You Can Get the Best of Both Worlds," *Business Week,* December 20, 1993, p. 114; Beth Kobliner, "Don't Let These Obstacles Bar the Door to a Mortgage," *Money,* November 1993, pp. 42–43+; Gene Koretz, "Home Buyers Don't Want to Bet on Low Inflation," *Business Week,* May 24, 1993, p. 22; and "Looking for a Mortgage? It Can Pay to Shop Around," *Nation's Business,* May 1993, p. 78.

UNIT 49   CALCULATING SIMPLE INTEREST

LEARNING OBJECTIVES

After completing this unit, you will be able to:

1. calculate simple interest.
2. calculate interest for periods of less than one year.
3. calculate the repayment amount for a loan.

Mike and Veronica Hess—one of the couples in the opening case—saved over $4,200 by just comparing different interest rates on car loans. The other couple—Betty and George San Miguel—will save over $20,000 by paying off their real estate loan in 5 years instead of 10 years. Are these chance happenings? No, not at all. In both cases, these couples made wise financial decisions that resulted in substantial dollar savings for the cost of borrowing money. You can obtain the same type of dollar savings if you understand the basics of interest calculations.

Interest is the amount paid for the use of borrowed money. To find the dollar amount of interest, multiply the principal by the rate and then by the time. The **principal** (P) is the amount of money borrowed. The **rate** (R) is the percent of the principal that is charged as interest by the lender. The **time** (T) is the period for which the money is borrowed; it may be expressed in days, months, or years. To calculate the interest amount, use the simple interest formula.

$$\text{Interest} = \text{Principal} \times \text{Rate} \times \text{Time, or } I = P \times R \times T.$$

EXAMPLE 1 | Find the Interest Amount for a $6,000 Loan at 10 Percent for Three Years

CALCULATOR TIP

Depress clear key
Enter 6,000   Depress   ×
Enter 0.10   Depress   ×
Enter 3   Depress   =
Read the answer $1,800

Interest = P × R × T
Interest = $6,000 × 0.10 × 3

$600 = $6,000 × 0.10

STEP 1   Multiply the principal ($6,000) by the rate (10 percent). Be sure to convert the percent to a decimal (0.10).

$1,800 = $600 × 3

STEP 2   Multiply the answer from step 1 ($600) by the time factor (three years).

You can use the simple interest formula for loans that are made for less than one year, but you must adjust the time factor first. If the time factor for a loan is expressed in months, you must divide the number of months by 12 (the number of months in one year).

EXAMPLE 2 | Find the Interest for a $2,000 Loan at 8 Percent for Five Months

CALCULATOR TIP

Depress clear key
Enter 2,000   Depress   ×
Enter 0.08   Depress   ×
Enter 5   Depress   ÷
Enter 12   Depress   =
Read the answer $66.67

Interest = P × R × T
Interest = $2,000 × 0.08 × 5 ÷ 12

$160 = $2,000 × 0.08

STEP 1   Multiply the principal ($2,000) by the rate (0.08). Note that 8 percent has been changed to the decimal 0.08.

$800 = $160 × 5

STEP 2   Multiply the answer from step 1 ($160) by the time factor (five months).

$66.67 = $800 ÷ 12

STEP 3   Divide the answer from step 2 ($800) by 12. Round the final answer to two decimal places.

Before working a problem where the time factor is stated in days, you must know the difference between an exact year and a banker's year (sometimes called a commercial year). An **exact year** contains 365 days. A **banker's year**, sometimes called *ordinary interest*, contains 360 days. A number of bankers and other lenders will use a banker's year when calculating interest because 360 days is easier to work with than 365 days. When the time factor is expressed in days, divide the number of days by either 360 or 365.

| | EXAMPLE 3 | Find the Interest for a $3,000 Loan at 9 Percent for 45 Days (Use an Exact Year) |

CALCULATOR TIP

Depress clear key
Enter 3,000   Depress ×
Enter 0.09   Depress ×
Enter 45   Depress ÷
Enter 365   Depress =
Read the answer $33.29

Interest = P × R × T
Interest = $3,000 × 0.09 × 45 ÷ 365

$270 = $3,000 × 0.09

STEP 1 Multiply the principal ($3,000) by the rate (0.09). Note that 9 percent has been changed to the decimal 0.09.

$12,150 = $270 × 45

STEP 2 Multiply the answer from step 1 ($270) by the time factor (45 days).

$33.29 = $12,150 ÷ 365

STEP 3 Since this loan is based on the exact year, divide the answer from step 2 ($12,150) by 365. Round the final answer to two decimal places.

| EXAMPLE 4 | Find the Interest for a $7,500 Loan at $7\frac{1}{2}$ Percent for 93 Days (Use a Banker's Year) |

Interest = P × R × T
Interest = $7,500 × 0.075 × 93 ÷ 360

$562.50 = $7,500 × 0.075

STEP 1 Multiply the principal ($7,500) by the rate (0.075). Note that $7\frac{1}{2}$ percent has been changed to the decimal 0.075.

$52,312.50 = $562.50 × 93

STEP 2 Multiply the answer from step 1 ($562.50) by the time factor (93 days).

$145.31 = $52,312.50 ÷ 360

STEP 3 Since this loan is based on a banker's year, divide the answer from step 2 ($52,312.50) by 360. Round the final answer to two decimal places.

CALCULATING REPAYMENT AMOUNT

The *repayment amount* for a loan is the original loan amount *plus* the interest. Notice in Example 5 that the interest calculated in Example 4 ($145.31) has been added to the original loan amount ($7,500).

| EXAMPLE 5 | Find the Repayment Amount for the $7,500 Loan in Example 4 |

$7,500.00 original loan amount
+   145.31 interest
$7,645.31 repayment amount

Complete the Following Problems

If necessary, round your answer to two decimal places.

1. How many days are there in a banker's year?
360

2. How many days are there in an exact year?

Find the Simple Interest for the Following Problems

3. $2,000 at 12% for 4 years

4. $3,500 at 9% for 2 years

5. $1,500 at 10% for 5 months

6. $2,500 at 7% for 8 months

7. $2,000 at 11% for 180 days (based on an exact year)

8. $4,500 at 10% for 75 days (based on an exact year)

9. $21,200 at 9% for 152 days (based on an exact year)

10. $5,000 at 8% for 120 days (based on a banker's year)

11. $15,600 at 9.2% for 30 days (based on a banker's year)

12. $24,900 at $8\frac{1}{4}$% for 61 days (based on a banker's year)

13. Jeff Washington wants to borrow $9,400 for six months from the Richardson Credit Union. If the credit union charges Jeff $9\frac{1}{2}$ percent interest on this loan, how much interest will the credit union collect?

14. In problem 13, what is the amount that Jeff will repay the credit union at the end of six months?

15. Commercial Bank uses both the exact year and the banker's year when determining simple interest on loans. If the bank makes a loan for $42,500 at 8 percent interest for 50 days, what is the interest amount if the exact year is used?

16. In problem 15, what is the interest amount if the banker's year is used?

Personal Math Problem

Theresa Evans opened the envelope and read the advertisement for a $5,000 loan. According to the ad, all she had to do was complete the application form and the consumer finance company would lend her $5,000. There were no strings attached, and she could use the money to pay bills, to take a vacation, to make a down payment on a new car, or for anything else that she thought was important. And the best part, according to the advertisement, was that the company charged only 18 percent annual interest.

17. If Ms. Evans borrows $5,000 and repays the loan at the end of one year, what is the dollar amount of interest for one year?

18. What is the dollar amount of interest each month?

19. If you were Ms. Evans, would you borrow $5,000 from this finance company?

Why or why not? _____

_____

_____

_____

UNIT 50	FINDING PRINCIPAL, RATE, AND TIME

LEARNING OBJECTIVES

After completing this unit, you will be able to:

1. find the principal in an interest problem.
2. find the rate in an interest problem.
3. find the time factor in an interest problem.

As presented in Unit 49, the basic formula for calculating interest is Interest = Principal × Rate × Time. In this unit, we use three similar formulas to find the principal, the rate, and the time factor in an interest problem. To help remember each formula, you may want to use the model shown here.

$$\frac{\text{Interest}}{\text{Principal} \times \text{Rate} \times \text{Time}}$$

The line dividing interest from principal, rate, and time indicates division. To find the specific formula you need, simply cover the unknown part with your finger. Then perform the indicated mathematical operation. For example, if the principal portion of the model is covered, the formula is

$$\text{Principal} = \frac{\text{Interest}}{\text{Rate} \times \text{Time}}$$

To calculate the principal, the interest amount is divided by the result of multiplying the rate by the time.

EXAMPLE 1

Find the Principal when the Interest Is $60, the Interest Rate Is 9 Percent, and the Time Factor Is 120 Days (Use a Banker's Year)

CALCULATOR TIP

Depress clear key
Enter 0.09 Depress ×
Enter 120   Depress ÷
Enter 360   Depress =
Read the answer 0.03
Enter 60   Depress ÷
Enter 0.03 Depress =
Read the answer $2,000

$$\text{Principal} = \frac{\text{Interest}}{\text{Rate} \times \text{Time}}$$

STEP 1 — Use the model to determine the correct formula for principal.

$$\text{Principal} = \frac{\text{Interest}}{0.09 \times 120 \div 360}$$

STEP 2 — Multiply the rate (9 percent) by the time factor. Note that the time factor for this problem is stated as $\frac{120}{360}$.

$$\text{Principal} = \frac{\$60}{0.03}$$

STEP 3 — Divide the interest amount ($60) by the answer from step 2 (0.03).

Principal = $2,000

To determine the rate formula in an interest problem, cover the rate portion of the model with your finger. The formula for rate is

$$\text{Rate} = \frac{\text{Interest}}{\text{Principal} \times \text{Time}}$$

To calculate the rate, the interest amount is divided by the result of multiplying the principal by the time factor.

EXAMPLE 2   Find the Rate when the Interest Is $80, the Principal Is $3,200, and the Time Factor Is 90 Days (Use a Banker's Year)

CALCULATOR TIP

Depress clear key
Enter 3,200 Depress ×
Enter 90   Depress ÷
Enter 360   Depress =
Read the answer 800
Enter 80   Depress ÷
Enter 800   Depress =
Read the answer 0.10

$$\text{Rate} = \frac{\text{Interest}}{\text{Principal} \times \text{Time}}$$

STEP |1| Use the model to determine the correct formula for rate.

$$\text{Rate} = \frac{\text{Interest}}{\$3,200 \times 90 \div 360}$$

STEP |2| Multiply the principal ($3,200) by the time factor. Note the time factor for this problem is stated as $\frac{90}{360}$.

$$\text{Rate} = \frac{\$80}{\$800}$$

STEP |3| Divide the interest amount ($80) by the answer from step 2 ($800).

Rate = 0.10 = 10%

To determine the time formula when working an interest problem, cover the time portion of the model with your finger. The formula for time is

$$\text{Time} = \frac{\text{Interest}}{\text{Principal} \times \text{Rate}}$$

To calculate the time factor, the interest amount is divided by the result of multiplying the principal by the rate. When working a time problem, one additional step is required—you must *multiply* your answer by the number of days in a banker's year or exact year.

EXAMPLE 3   Find the Time when the Interest Is $200, the Principal Is $5,000, and the Rate Is 8 Percent (Use a Banker's Year)

CALCULATOR TIP

Depress clear key
Enter 5,000 Depress ×
Enter 0.08   Depress =
Read the answer 400
Enter 200   Depress ÷
Enter 400   Depress =
Read the answer 0.5
Enter 0.5   Depress ×
Enter 360   Depress =
Read the answer 180 days

$$\text{Time} = \frac{\text{Interest}}{\text{Principal} \times \text{Rate}}$$

STEP |1| Use the model to determine the correct formula for time.

$$\text{Time} = \frac{\text{Interest}}{\$5000 \times 0.08}$$

STEP |2| Multiply the principal ($5,000) by the rate (8 percent).

$$\text{Time} = \frac{\$200}{\$400}$$

STEP |3| Divide the interest amount ($200) by the answer from step 2 ($400).

Time = 0.5 × 360

STEP |4| Multiply the answer from step 3 (0.5) by 360 days.

Time = 180 days

HINT: To improve the accuracy of your final answer, you may want to carry out all intermediate answers to at least five decimal places.

EXAMPLE 4   Find the Time when the Interest Is $220, the Principal Is $8,000, and the Rate Is 6 Percent (Use a Banker's Year)

$$\text{Time} = \frac{\text{Interest}}{\text{Principal} \times \text{Rate}}$$

STEP	1	Use the model to determine the correct formula for time.

$$\text{Time} = \frac{\text{Interest}}{\$8,000 \times 0.06}$$

STEP	2	Multiply the principal ($8,000) by the rate (6 percent).

$$\text{Time} = \frac{\$220}{480}$$

STEP	3	Divide the interest amount ($220) by the answer from step 2 ($480). Use at least 5 decimals.

$$\text{Time} = 0.45833 \times 360$$

STEP	4	Multiply the answer from step 3 (0.45833) by 360 days.

Time = 165 days (rounded)

Had the answer in Step 3 (0.45833) been rounded off to 0.5, the final answer would have been 180 days instead of 165 days. While 180 days is a "correct" answer, it is not as accurate as it should be. By using at least five decimals, the final answer (165 days) is much more accurate.

Complete the Following Problems

Round your final answer to two decimal places. Use a banker's year unless otherwise indicated.

	Principal	Rate	Time	Interest
1.	$14,500	9%	1 year	*$1,305* $I = P \times R \times T$ *$14,500 × .09 = $1,305* *$1,305 × 1 = $1,305*
2.	$20,560	_____	80 days	$822.40
3.	$7,000	11%	90 days	_____

	Principal	Rate	Time	Interest
4.	_____	10%	45 days	$350
5.	$52,000	7%	_____	$910
6.	$2,500	12%	262 days	_____
7.	$15,400	_____	72 days	$277.20
8.	$35,000	8.5%	_____	$1,190

	Principal	Rate	Time	Interest
9.	_____	10%	216 days	$3,780
10.	$2,950	$7\frac{3}{4}$%	126 days	_____

11. Michelle Visconti invested $21,070 in a savings account that paid 6 percent interest. How much did she earn if the money is left on deposit for 150 days?

12. The local savings and loan association made a $35,000 loan for 180 days. The interest was $1,400. What was the interest rate?

13. If the First Federal Bank charges 9 percent interest for a 120-day loan, the interest amount is $264. What is the principal amount?

14. The Federated Employees' Union charges 11 percent interest for all short-term loans made for less than one year. If Patrick Marks borrows $18,000, he must pay $1,584 interest. What is the length of the loan?

UNIT 51	DETERMINING MATURITY DATES AND THE NUMBER OF DAYS BETWEEN TWO DATES

LEARNING OBJECTIVES   *After completing this unit, you will be able to:*

1. calculate the maturity date for a loan.
2. find the number of days between two dates.

Two special problems develop when calculating interest. First, both the lender and the borrower must know the maturity date for the loan. The **maturity date** is the date on which a loan becomes payable. Second, to calculate the interest amount, it is often necessary to find the number of days between the date a loan was originated and the maturity date. To solve both of these problems, you must know the number of days in each month.

There are two methods commonly used to determine the number of days in each month of the year. You may remember the following rule:

Thirty days has September, April, June, and November; all the rest have 31 days, except February, which has 28 days (or 29 days in a leap year).

It is also possible to determine the number of days in each month by using the knuckle approach. To use this approach, you begin with the first knuckle next to the thumb. Each knuckle represents a month with 31 days. Each "valley" between knuckles represents a month with less than 31 days.

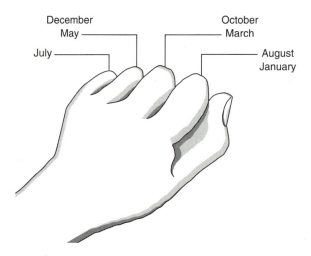

The month of February presents a special problem. Any year evenly divisible by 4 with no remainder is a **leap year.** During a leap year, February has 29 days. In all other years, February has 28 days.

When the time factor is expressed in days, the maturity date is found by counting the number of days from the date the loan was originated. Note that the first day is not counted, but the last day is counted.

EXAMPLE 1   Find the Maturity Date for a 120-Day Loan Originated on March 5

	31	March
Loan origination	− 5	March
	26	March
	+ 30	April
	+ 31	May
	+ 30	June
Maturity date	+ 3	July
Total days	120	

STEP ☐1   Subtract the date of the loan origination (5) from the total number of days in the month (31 in March).

STEP ☐2   To the difference found in step 1 (26), add the total number of days in the next month, then the next, and keep going until the total reaches 120 days.

As you learned in Unit 49, you must know the number of days when calculating the interest for a loan. It is possible to find the number of days between two dates by counting the number of days between the loan origination date and the maturity date. As in Example 1, you do not count the first day, but the last day is counted.

EXAMPLE 2   Find the Number of Days between April 15 and October 10

	30	April
Loan origination	− 15	April
	15	April
	+ 31	May
	+ 30	June
	+ 31	July
	+ 31	August
	+ 30	September
Maturity date	+ 10	October
Total days	178	

STEP ☐1   Subtract the date of the loan origination (15) from the number of days in the month (30 in April).

STEP ☐2   To the difference found in step 1 (15), add the number of days in each month up to the maturity date.

Complete the Following Problems

If necessary, round your answer to two decimal places.

1. In your own words, define the maturity date. _____

_____

2. How many days are there in the month of November 1995?

3. How many days are there in the month of January 1995?

4. How many days are there in the month of August 1996?

5. How many days are there in the month of February 1994?

6. How many days are there in the month of February 1996?

Find the maturity date for each of the following problems.

Loan Origination Date	Number of Days	Maturity Date
July 14, 1995	120	**7.** _____
January 8,1995	80	**8.** _____
June 25, 1995	150	**9.** _____
October 30, 1995	210	**10.** _____
February 20, 1995	105	**11.** _____
March 5, 1995	130	**12.** _____

Find the number of days between the loan origination date and the maturity date for each of the following problems.

Loan Origination Date	Maturity Date	Number of Days
July 21, 1995	August 10, 1995	**13.** _____
January 5, 1996	July 12, 1996	**14.** _____
September 29, 1995	February 4, 1996	**15.** _____
October 6, 1995	November 25, 1995	**16.** _____
March 29, 1995	September 21, 1995	**17.** _____
June 11, 1995	December 10, 1995	**18** _____

Use the following information to solve problems 19–21.

Assume that you borrow $5,200 from the local credit union on December 11, 1995. The credit union charges $9\frac{1}{2}$ percent interest on this loan. On April 28, 1996, you decide to pay off this loan.

19. How many days are there between the loan origination date and the maturity date?

20. How much interest will the credit union collect on this loan? (Assume that a banker's year is used).

21. What is the repayment amount for this loan?

Critical Thinking Problem

On June 10, Judith Macauley, president of Thermal Windows, Inc., realizes that she must borrow $12,000 to cover necessary business expenses. If the company does well over the next three months, she could pay the money back at the end of 90 days. If the company doesn't do so well, it may take 150 days to pay the money back. In either case, the money is available from Thermal Window's bank at $9\frac{1}{2}$ percent.

22. If the money is paid back in 90 days, what is the maturity date?

23. What is the amount of interest for the 90-day loan? (Assume that a banker's year is used.)

24. If the money is paid back in 150 days, what is the maturity date?

25. What is the amount of interest for the 150-day loan? (Assume that a banker's year is used.)

26. If you were Ms. Macauley, which loan would you choose? _____

_____

_____

UNIT 52	PROMISSORY NOTES, DISCOUNT BANK NOTES, AND DISCOUNTING NONINTEREST-BEARING NOTES

LEARNING OBJECTIVES   *After completing this unit, you will be able to:*

1. describe the parts of a promissory note.
2. calculate the bank discount and proceeds for a simple discount bank note.
3. calculate the bank discount and proceeds when a business firm discounts a noninterest-bearing promissory note.

A **promissory note** is a written pledge by a borrower to pay a specific sum of money to a lender on a future date. Generally, a promissory note is legally binding and enforceable through the court system. The individual or business borrowing the money is called the **maker.** The party lending the money and to whom repayment is made is called the **payee.** Both the maker and the payee along with other requirements for a promissory note are illustrated in Figure 52–1.

SIMPLE DISCOUNT BANK NOTES

With a simple discount bank note, the bank deducts the dollar amount of interest from the original value of the promissory note before any money is given to the borrower. Because the bank subtracts the money when the loan is originated, the dollar amount of interest is referred to as a bank discount. To calculate the discount amount, use the simple interest formula ($I = P \times R \times T$) described in Unit 49. Although the simple interest formula is used, there are important differences when working with simple discount bank notes. With simple discount bank notes, the bank discount charge is subtracted when the loan is originated. The actual dollar amount that the borrower receives is called the **proceeds.** To calculate proceeds, subtract the bank discount amount from the face value of the promissory note. At maturity, the borrower repays the face value of the note. With a simple interest note described in Unit 49, the interest is added to the face value of the note to determine the amount that must be repaid.

The steps illustrated in Example 1 can be used to determine the bank discount amount and the proceeds for Jason Peterson, owner of Peterson Fabrics Shops, who signs a simple discount bank note at Phoenix National Bank for $15,000 at 8 percent for 90 days.

$ _____ $18,000 (1) _____ Abilene _____, Texas, _____ June 6 (4) _____ A.D. 19 _____ 95

_____ Sixty days (3) _____ after date, without grace, for value received, I, we, or either of us, promise to

pay to the order of _____ The Barton Company (7) _____

_____ (1) Eighteen thousand and no/100 -------------------- Dollars

at _____ First Bank _____ with interest from _____ June 6 _____ to maturity at the rate of _____ 12 (2) _____ per cent, per annum

AND FROM MATURITY AT THE RATE OF FIFTEEN PERCENT, PER ANNUM, WE THE MAKERS, SURETIES, ENDORSERS AND GUARANTORS OF THIS NOTE HEREBY SEVERALLY WAIVE PRESENTATION FOR PAYMENT, NOTICE OF NON-PAYMENT, PROTEST, AND NOTICE OF PROTEST AND DILIGENCE IN BRINGING SUIT AGAINST ANY PARTY HERETO, AND CONSENT THAT THE TIME OF PAYMENT MAY BE EXTENDED BY RENEWAL NOTE OR OTHERWISE ONE OR MORE TIMES FOR PERIODS DISCRETIOARY WITH THE HOLDER WITHOUT NOTICE THEREOF TO ANY OF THE SURETIES, ENDORSERS AND/OR GUARANTORS ON THIS NOTE. IT IS FURTHER EXPRESSLY AGREED THAT IF THIS NOTE IS PLACED IN THE HANDS OF AN ATTORNEY FOR COLLECTION, OR IS COLLECTED THROUGH THE PROBATE OF BANKRUPTCY COURT, OR THROUGH OTHER LEGAL PROCEEDINGS, THEN IN ANY OF SAID EVENTS, A REASONABLE AMOUNT SHALL BE ADDED AND COLLECTED AS ATTORNEY AND COLLECTION FEES

Due _____ August 5, 1995 (5) _____   *Paul Robertson* (6) _____

Address _____ 326 East Elm Street _____   Financial Vice-President

Phone _____ 555-1732 _____   Chicago Manufacturing

1. The principal ($18,000) is the amount of the debt. It is the amount of the credit transaction.
2. The rate (12 percent) expresses the value paid for use of the borrowed money. It is usually stated in annual or yearly terms.
3. The time (60 days) is the period for which the money is borrowed.
4. The date (June 6) is the date the note was issued.
5. The maturity date (August 5) is the day the principal and interest are due. It is often called the due date.
6. The maker (Chicago Manufacturing) is the individual or company issuing the note and borrowing the money.
7. The payee (The Barton Company) is the individual or company extending the credit.

FIGURE 52–1

| EXAMPLE 1 | Find the Bank Discount and Proceeds for a $15,000 Loan at 8 Percent for 90 Days (Use a Banker's Year) |

Bank discount = P × R × T

Bank discount = $15,000 × 0.08 × 90 ÷ 360

CALCULATOR TIP

Depress clear key
Enter 15,000 Depress ×
Enter 0.08 Depress ×
Enter 90 Depress ÷
Enter 360 Depress =
Read the answer   300
Enter 15,000 Depress −
Enter 300 Depress =
Read the answer $14,700

$1,200 = $15,000 × 0.08   STEP [1]   Multiply the loan amount $15,000 by the rate (8 percent).

$108,000 = $1,200 × 90   STEP [2]   Multiply the answer from step 1 ($1,200) by the number of days (90).

$300 = $108,000 ÷ 360   STEP [3]   Since this loan is based on a banker's year, divide the answer (bank discount) from step 2 ($108,000) by 360. Round the final answer to two decimal places.

$15,000	Bank note
− 300	Bank discount
$14,700	Proceeds

STEP 4 — To determine the proceeds, subtract the bank discount determined in step 3 ($300) from the face value of the note ($15,000).

At maturity, Peterson Fabric Shops must repay $15,000—the original face value of the promissory note. Since the borrower has use of only $14,700, the actual or effective interest rate is higher than the stated rate of 8 percent for the loan in Example 1. The formula for calculating rate presented in Unit 50 can be used to determine the true or effective rate of interest. When determining the effective rate of interest for a simple discount bank note, the p in the formula stands for proceeds and not the original loan amount.

EXAMPLE 2 — Find the Effective Rate of Interest for the Loan in Example 1 (Use a Banker's Year)

CALCULATOR TIP

Depress clear key
Enter 14,700 Depress ×
Enter 90 Depress ÷
Enter 360 Depress =
Read the answer 3,675
Enter 300 Depress ÷
Enter 3,675 Depress =
Read the answer 0.0816

$$R = \frac{I}{P \times T}$$

STEP 1 — State the formula for rate.

$$R = \frac{Interest}{\$14,700 \times 90 \div 360}$$

STEP 2 — Multiply the proceeds ($14,700) for the promissory note by the time factor. Note the time factor for this problem is stated as $\frac{90}{360}$.

$$R = \frac{300}{3675}$$

STEP 3 — Divide the interest amount ($300) by the answer from step 2 ($3,675). This answer should be carried out to at least 4 decimal places.

$$R = 0.0816 = 8.16\%$$

STEP 4 — Convert the decimal answer to a percent.

DISCOUNTING A NONINTEREST-BEARING NOTE[1]

Business firms that sell merchandise or services on a credit basis often accept promissory notes. And when these business firms need cash, they may sell their promissory notes to banks or other financial institutions.

As illustrated in Example 3, the procedure used to discount a noninterest-bearing note is similar to the procedure used to determine the bank discount and proceeds for a simple discount bank note, with one additional step. Before you calculate the bank discount amount and the proceeds, you must first determine the discount period. The **discount period** is the number of days the bank will hold the promissory note until the maturity date.

[1]Promissory notes can be either interest-bearing notes or noninterest-bearing notes. The discount procedure for a noninterest-bearing note is discussed in this unit. The discount procedure for an interest-bearing note is described in Unit 53.

EXAMPLE 3   Determine the Bank Discount and Proceeds for an $18,000 Noninterest-Bearing Promissory Note with a Maturity Date of June 30 (Use a Banker's Year)

On March 10, Sam Johnson, president of SJ Cosmetics, realizes that the company must raise cash immediately. He decides to sell an $18,000 noninterest-bearing promissory note with a maturity date of June 30 that the firm received from one of its customers. The banker for SJ Cosmetics agrees to buy the promissory note but charges a 12 percent discount rate. What is the amount of the bank discount and the amount of proceeds?

31	March	**STEP**	**1**	Determine the discount period between March 10 and June 30.
<u>10</u>				
21	March			
30	April			
31	May			
<u>30</u>	June			
112	days			

CALCULATOR TIP

Depress clear key
Enter 18,000   Depress ×
Enter 0.12   Depress ×
Enter 112   Depress ÷
Enter 360   Depress =
Read the answer   672
Enter 18,000   Depress −
Enter 672   Depress =
Read the answer $17,328

$BD = \$18,000 \times 0.12 \times 112 \div 360$

STEP **2**   Multiply the face value of the promissory note ($18,000) by the bank discount rate (12 percent).

$\$2,160 = \$18,000 \times 0.12$

$\$241,920 = \$2,160 \times 112$ **STEP** **3**   Multiply the answer from step 2 ($2,160) by the number of days in the discount period (112 days).

$\$672 = \$241,920 \div 360$ **STEP** **4**   Divide the answer from step 3 ($241,920) by 360. Round the final answer to two decimal places.
(bank discount)

$18,000	bank note
− 672	bank discount
$17,328	proceeds

STEP **5**   To determine the proceeds, subtract the bank discount determined in step 4 ($672) from the face value of the note ($18,000).

Complete the Following Problems

If necessary, round your answer to two decimal places. For all problems in this unit, use a banker's year.

 1. In your own words, define promissory note. _____

_____

_____

 2. In your own words, describe the difference between a simple discount bank note and a simple interest note. _____

_____

_____

_____

For the following problems, determine the bank discount and proceeds.

Dollar Value of Simple Discount Note	Bank Discount Rate	Time Factor	Bank Discount Amount		Proceeds
$ 8,500	10%	60 days	**3.** _____$141.67_____ $8,500 \times .10 \times \dfrac{60}{360}$ $8,500 \times .10 = \$850$ $\$850 \times 60 = \$51,000$ $\$51,000 \div 360 = \$141.67$	**4.**	_$8,500 - \$141.67$_ $= \$8,358.33$
$11,000	8%	90 days	**5.** _____	**6.**	_____
$5,200	7%	150 days	**7.** _____	**8.**	_____
$22,400	11.2%	72 days	**9.** _____	**10.**	_____
$33,150	9.4%	164 days	**11.** _____	**12.**	_____

For problems 13 through 18, determine the effective rate of interest. Use four decimal places for these problems.

Interest	Proceeds (principal)	Time Factor	Effective Interest Rate
$1,000	$23,000	180 days	13. _____
$ 275	$15,000	60 days	14. _____
$ 525	$34,975	90 days	15. _____
$ 116	$6,884	100 days	16. _____
$ 110	$10,890	45 days	17. _____
$ 660	$ 8,840	250 days	18. _____

19. On August 14, Reynolds Plumbing Supply decides to discount a promissory note that has a maturity date of December 10. How many days are there in the discount period?

20. On March 5, Contemporary Artworks accepted a 90-day promissory note from the Dutch Art Gallery. What is the maturity date for this note?

21. In problem 20, how many days are in the discount period if Contemporary Artworks discounts the note on April 2?

22. Grand Saline Chemical Supply accepted a $3,600 noninterest-bearing note from one of its customers. The note has a maturity date of October 22. If the note is discounted at the bank for a period of 41 days and the bank charges 11 percent, how much is the bank discount?

23. In problem 22, what is the amount of the proceeds that Grand Saline Chemical Supply will receive after the discount?

24. On May 3, Cowboy Clothing accepted a 60-day, $22,400 noninterest-bearing note from one of its customers. If the note is discounted on May 20 and the bank charges 12 percent, how much is the bank discount?

25. In problem 24, what is the amount of the proceeds that Cowboy Clothing will receive after the discount?

Critical Thinking Problem

Mary Torino, one of the owners of Torino and Masters Plastics, Inc., is trying to decide which loan she should use to solve her firm's financial problems. She has obtained commitments from two banks for a $20,000 loan for a period of six months. One bank has offered a simple interest loan with an interest rate of 9 percent. The second bank offered a simple discount bank note with an interest rate of 8.75 percent.

Hint: You may want to review the material on simple interest loans presented in Unit 49 before answering the following problems.

26. What is interest amount for the simple interest loan at 9 percent?

27. What is the effective interest rate for the simple interest loan?

28. What is the discount amount for the simple discount bank note at 8.75 percent?

29. What is the effective interest rate for the simple discount bank note?

30. If you were Ms. Torino, which loan would you choose? Why? _____

_____

_____

_____

UNIT 53   DISCOUNTING INTEREST-BEARING NOTES

LEARNING OBJECTIVES

After completing this unit, you will be able to:

1. determine the maturity date for a promissory note.
2. determine the number of days in the discount period.
3. calculate the maturity value for an interest-bearing note.
4. calculate the bank discount amount and the proceeds when a business firm discounts an interest-bearing promissory note.

In the last unit, we examined the steps required to find the proceeds when a business firm discounts a noninterest-bearing promissory note before the maturity date. In this unit, we examine the steps required to discount an interest-bearing promissory note. In both cases, business firms often accept promissory notes when merchandise or services are sold on credit.

For example, Boston Manufacturing accepts a 90-day, $18,500 promissory note from Orlando Computer Sales on July 8. The customer, Orlando Computer Sales, agrees to pay 10 percent interest until the maturity date October 6. In this case, the seller—Boston Manufacturing—has two options. First, Boston Manufacturing can hold the note until the maturity date and receive the face value of the note plus interest. Second, Boston Manufacturing can discount Orlando's promissory note in order to obtain cash before the maturity date.

A QUICK REVIEW OF DATING

Before discussing the steps required to discount an interest-bearing note, you may find it helpful to review the procedure for determining the maturity date and the number of days in the discount period. The diagram below illustrates both the maturity date and the number of days in a discount period for the promissory note described above.

Notice in Example 1, the steps used to determine the maturity date for the promissory note between Boston Manufacturing and Orlando Computer Sales.

EXAMPLE 1   Find the Maturity Date for a 90-Day Promissory Note Originated on July 8

31 July – 8 ‾‾‾‾ 23	**STEP** 1 Subtract the origination date of the promissory note (8) from the total number of days in the month (31 in July).

23 July
       31 August
       30 September
Maturity Date ---- 6 October
       90 total days
	STEP 2 To the difference found in step 1 (23), add the total number of days in the next month, then the next, and keep going until the total reaches 90 days.

As defined in the last unit, the discount period is the number of days the bank will hold the promissory note until the maturity date. Assume that Boston Manufacturing decides to discount the Orlando Computer Sales promissory note on August 10. The steps required to determine the discount period are illustrated in Example 2.

EXAMPLE 2   Find the Number of Days in the Discount Period If a Promissory Note is Discounted on August 10

31 August – 10 ‾‾‾‾ 21	**STEP** 1 Subtract the date of the loan discount (10) from the number of days in the month (31 in August).

21 August 30 September 6 October ‾‾‾‾ 57 days in discount period	**STEP** 2 To the difference found in step 1 (21), add the number of days in each month up to the maturity date.

STEPS REQUIRED TO DISCOUNT AN INTEREST-BEARING NOTE

Once the maturity date and the discount period are known, it is possible to determine both the bank discount and the proceeds from a discounted promissory note.

EXAMPLE 3   Find the Bank Discount and Proceeds for a Discounted Promissory Note.

On March 3, Carlton Manufacturing receives a 9 percent, 120-day, $9,000 promissory note from Pittsburgh Appliances, Inc. Carlton discounts the note on May 22. The bank charges 11 percent for discounting this type of note. (Use a banker's year.)

CALCULATOR TIP

Depress clear key
Enter 9,000   Depress   ×
Enter 0.09   Depress   ×
Enter 120   Depress   ÷
Enter 360   Depress   =
Read the answer   270
Enter 9,000   Depress   +
Enter 270   Depress   =
Read the answer $9,270

$270 = \$9,000 \times 0.09 \times 120 \div 360$	**STEP** 1	Find the amount of interest for the original note. Multiply the face value of the original note ($9,000) by the interest rate (9 percent) by the time factor ($\frac{120}{360}$).
$\begin{array}{r} \$9,000 \\ +\ \ 270 \\ \hline \$9,270 \end{array}$	**STEP** 2	Calculate the maturity value. Add the interest amount from step 1 ($270) to the face value of the original note ($9,000).
$\begin{array}{r} 31\ \text{March} \\ -\ \ 3 \\ \hline 28\ \text{March} \\ +30\ \text{April} \\ +31\ \text{May} \\ +30\ \text{June} \\ +\ 1\ \text{July (maturity date)} \\ \hline 120\ \text{days} \end{array}$	**STEP** 3	Determine the maturity date.
$\begin{array}{r} 31\ \text{May} \\ -22 \\ \hline 9\ \text{May} \\ +30\ \text{June} \\ +\ 1\ \text{July} \\ \hline 40\ \text{days} \end{array}$	**STEP** 4	Find the discount period.

CALCULATOR TIP

Depress clear key
Enter 9,270   Depress   ×
Enter 0.11   Depress   ×
Enter 40   Depress   ÷
Enter 360   Depress   =
Read the answer   113.30
Enter 9,270   Depress   −
Enter 113.30   Depress   =
Read the answer $9,156.70

$\$113.30 = 9,270 \times 0.11 \times 40 \div 360$	**STEP** 5	Calculate the bank discount. Multiply the *maturity value* ($9,270) by the bank discount rate (11 percent) by the time factor for the discount period ($\frac{40}{360}$).
$\begin{array}{r} \$9,270.00 \\ -\ \ 113.30 \\ \hline \$9,156.70 \end{array}$	**STEP** 6	Calculate the proceeds. Subtract the bank discount ($113.30) from the maturity value ($9,270).

Assuming that Carlton Manufacturing discounts the interest-bearing promissory note on May 22, a bank discount of $113.30 will be subtracted from the maturity value. Carlton Manufacturing will receive $9,156.70 on May 22. And the bank will receive the $9,270 maturity value on the July 1 maturity date.

Complete the Following Problems

If necessary, round your answer to two decimal places. For all problems in this unit, use a banker's year.

1. What is the maturity date for a 105-day promissory note that originates on April 3?

2. What is the maturity date for a 62-day promissory note that originates on October 12?

3. What is the maturity date for a 150-day promissory note that originates on March 5?

4. On November 5, Panhandle Art Gallery accepted a 90-day promissory note from a customer. What is the maturity date for this note?

5. In problem 4, how many days are in the discount period if Panhandle discounts the note on December 20?

6. On June 14, Justin's Construction Supply accepted a 180-day promissory note from a customer. What is the maturity date for this note?

7. In problem 6, how many days are in the discount period if Justin's Construction Supply discounts the note on September 2?

For problems 8–11, determine the dollar amount of interest.

Face Value of Promissory Note	Interest Rate	Time Factor	Interest Amount
$12,000	8%	30 days	8. _____
$21,000	9%	140 days	9. _____
$ 7,500	11%	190 days	10. _____
$16,700	$7\frac{1}{2}$%	100 days	11. _____

For problems 12–15, determine the maturity value.

Face Value of Promissory Note	Interest Rate	Time Factor	Maturity Value
$ 5,000	10%	120 days	12. _____
$34,000	9%	85 days	13. _____
$ 4,200	11.2%	70 days	14. _____

Face Value of Promissory Note	Interest Rate	Time Factor	Maturity Value
$14,350	7.9%	105 days	15. _____

For problems 16–19, determine the bank discount.

Maturity Value	Discount Rate	Discount Period	Bank Discount
$ 1,900	10%	60 days	16. _____
$16,400	11%	180 days	17. _____
$25,000	12%	90 days	18. _____
$36,500	11.5%	54 days	19. _____

For problems 20–23, determine the proceeds.

Maturity Value	Discount Rate	Discount Period	Proceeds
$ 8,000	9%	120 days	20. _____

Maturity Value	Discount Rate	Discount Period	Proceeds
$44,000	12%	80 days	21. _____
$21,700	$11\frac{1}{4}\%$	90 days	22. _____
$13,250	12.5%	135 days	23. _____

24. On October 24, Jasperware Manufacturing accepted a 45-day, $12,000 noninterest-bearing note from one of its customers. How much is the bank discount if the note is discounted on November 1 and the bank charges 12 percent?

25. In problem 24, what is the amount of the proceeds that Jasperware will receive after the discount?

On March 18, Jackson Bros. Plumbing Supply accepted an interest-bearing note for $24,580 from Thomas and Martin Plumbing Company. The note was for 110 days and Thomas and Martin agreed to pay 10 percent interest.

26. What is the amount of interest that Thomas and Martin must pay at maturity?

27. What is the maturity value of this note?

28. What is the maturity date for this note?

29. If Jackson Bros. discounts this note on May 4, how many days are there in the discount period?

30. If the bank charges Jackson Bros. 13 percent to discount the note, what is the dollar amount of the bank discount?

31. What is the amount of the proceeds that Jackson Bros. will receive if the note is discounted on May 4?

Kapoor Photographic Supply agreed to accept an interest-bearing $4,800 note for new photographic equipment purchased by Bell Photo Processing, Inc. The 75-day note was dated December 2. As part of the financial arrangement, Bell Photo agreed to pay $8\frac{1}{2}$ percent interest until maturity.

32. What is the amount of interest that Bell must pay at maturity?

33. What is the maturity value for this note?

34. What is the maturity date for this note?

35. If Kapoor Photographic discounts this note on December 15, how many days are there in the discount period?

36. If the bank charges Kapoor Photographic 12 percent to discount the note, what is the dollar amount of the bank discount?

37. What is the amount of the proceeds that Kapoor Photographic will receive if the note is discounted on December 15?

38. Assuming that Kapoor Photographic does discount the note, who will receive the maturity value on the maturity date?

INSTANT REPLAY

UNIT	IMPORTANT POINTS TO REMEMBER	EXAMPLES
UNIT 49 Calculating Simple Interest	The dollar amount of interest is found by multiplying the principal (the amount of money borrowed) by the rate by the time. This formula is usually abbreviated as follows: $$I = P \times R \times T$$ The *repayment amount* is the original loan amount plus the interest.	Find the interest amount for an $8,000 loan at 12 percent interest for 240 days (assume that the loan is based on a banker's year). $I = P \times R \times T$ $I = \$8,000 \times 0.12 \times 240 \div 360$ $I = \$640$ The repayment amount is **$8,000 + 640 = $8,640**
UNIT 50 Finding Principal, Rate, and Time	To help remember the formulas for P, R, and T, use the model shown here. $$\frac{Interest}{Principal \times Rate \times Time}$$ To find the specific formula you need, simply cover the unknown part with your finger. Then, perform the indicated mathematical operation.	$I = P \times R \times T$ $P = \dfrac{I}{R \times T}$ $R = \dfrac{I}{P \times T}$ $T = \dfrac{I}{P \times R}$ Then, Answer $\times$ 360 days

continued from page 294 UNIT	IMPORTANT POINTS TO REMEMBER	EXAMPLES

UNIT 51 Determining Maturity Dates and the Number of Days between Two Dates	The *maturity date* is the date on which a loan becomes payable. When the time factor is expressed in days, the maturity date is found by counting the number of days from the date the loan was originated. It is also possible to find the number of days between two dates by counting the number of days between the loan origination date and the maturity date.	Find the maturity date for an 80-day loan originated on April 3:
		
		                   30 April
		 Loan origination **− 3**
		                   27 April
		                +31 May
		 Maturity date **+22** June
		    Total days **80**
		
		 Find the number of days between July 10 and November 15:
		
		                   31 July
		 Loan origination **−10**
		                   21 July
		               +31 August
		               +30 September
		               +31 October
		 Maturity date **+15** November
		    Total days **128**

UNIT 52 Promissory Notes, Discount Bank Notes, and Discounting Noninterest-Bearing Notes	With a simple discount bank note, the bank deducts the dollar amount of interest when the loan is originated. The actual dollar amount that the borrower receives is called the proceeds. The same discount procedure can be used to determine the proceeds that a business firm receives when a note from a customer is discounted in order to raise immediate cash.	Find the bank discount amount and the proceeds for a $5,000 note if the bank charges 12 percent for 30 days. **$50 = $5,000 × 0.12 × 30 ÷ 360** (bank discount) **$5000 − 50 = $4,950** (proceeds)
UNIT 53 Discounting Interest-Bearing Notes	To discount an interest-bearing note, follow these steps: (*a*) find the interest; (*b*) calculate the maturity value; (*c*) find the maturity date; (*d*) find the discount period; (*e*) calculate the bank discount; and (*f*) calculate the proceeds.	ABC accepts a 60-day, 8% note for $3,000 on May 10. ABC discounts the note on May 20. The bank charges 10%. (a) **$40 = $3,000 × 0.08 × 60 ÷ 360** (b) **$3,040 = $3,000 + 40** (c) maturity date = **July 9** (d) discount period = **50 days** (e) **$42.22 = $3,040 × .10 × 50 ÷ 360** (f) **$2,997.78 = $3,040 − 42.22**

CHAPTER 9	MASTERY QUIZ

DIRECTIONS   *Round each answer to two decimal places or hundredths. (Each answer counts 5 points.)*

1. What is the formula for calculating interest?   _____

2. How many days are there in an exact year?   _____

3. What is the formula for calculating rate?   _____

Complete the Following Problems

(Assume that each problem is based on a banker's year)

Principal	Rate	Time	Interest
$24,500	10%	1 year	4. _____
5. _____	9%	120 days	$570
$ 5,100	8%	120 days	6. _____
$4,400	10%	7. _____	$77
$12,800	8. _____	180 days	$448

9. Find the maturity date for a 90-day loan originating on September 19.

10. Find the maturity date for a 135-day loan originating on June 13.

11. Find the number of days between October 3 and January 25.

12. The owners of Irving Electronics Repair borrowed $3,000 from Tulsa State Bank. Irving Electronics signed a 70-day simple discount bank note with a rate of 9.6 percent. What is the dollar amount of the bank discount? (Use an exact year.)

13. In problem 12, what is the amount of the proceeds that Irving Electronics Repair will receive?

Brown Cow Dairies agreed to accept a $3,100 note from one of its customers. The 75-day note was dated April 11. As part of the financial arrangement, the customer agreed to pay 9 percent interest until maturity. (Use a banker's year.)

14. What is the amount of interest that the customer must pay?

15. What is the maturity value for this note?

16. What is the maturity date for this note?

17. If Brown Cow discounts this note on May 4, how many days are there in the discount period?

18. If the bank charges Brown Cow 11.5 percent to discount the note, what is the dollar amount of the bank discount?

19. What is the amount of the proceeds that Brown Cow will receive if the note is discounted on May 4?

20. Assuming that Brown Cow does discount the note, who will receive the maturity value on the maturity date?

CHAPTER 10

Compound Interest, Present Value, and Home Mortgages

CHAPTER PREVIEW

In this chapter, we expand the simple interest calculations presented in Chapter 9 to solve some special interest applications that are used by both individuals and businesses. We begin this chapter with compound interest calculations. Then, we discuss the steps required to work present value problems. Finally, we examine the topics of home mortgages and loan amortization.

MATH TODAY

THE TIME VALUE OF MONEY CONCEPT

Many Americans dream of retiring while they are still young enough to enjoy their wealth. In fact, most won't accumulate enough wealth to turn the dream into reality. Over half of all Americans believe they can live on Social Security and their company pension after they quit work. Actually, Social Security and pensions cover only about one-third to one-half of retirement expenses, and if you retire early, when you are in your 40s or 50s, you may not be able to collect benefits until you reach 65.

Yet even a person with an average salary can retire early. It just takes planning, discipline, and a thorough understanding of the time value of money. The time value of money is a concept that recognizes that money can be invested and can earn interest over a period of time. Take the case of Marty and John French. This young couple in their 30s decided it was time to start an investment program when a lawyer notified them that Marty's uncle had died and left them $90,000. To accomplish their financial goals, they began to analyze their current financial situation. They immediately realized that one of their "best" investments would be to pay off their high-interest credit card debts and two car loans. Total cost for debt reduction equaled $15,000. Next, they decided to purchase a $75,000 certificate of deposit at a local bank.

While many of their friends suggested that they should purchase corporate bonds, stocks, or mutual funds, they admit that they are conservative people and don't like to take risks. Since the certificate of deposit was guaranteed by the government, they considered this investment virtually "risk-free." Naturally, they wanted the highest return possible, so they compared different rates at local banks, savings and loan associations, and credit unions. After a lengthy search, they found one bank that offered a 5 percent certificate of deposit, and they could lock this rate in for three years. At the end of three years, their $75,000 investment would be worth $86,850. They will have earned $11,850 by letting their interest compound and grow. At the end of three years, they can use the money they accumulate for anything they value—another certificate of deposit, another investment option, or financing a major purchase like a car or a home.

Source: For more information, see Justin Martin, "Generation X: Save, Baby, Save," *Fortune/1994 Investor's Guide,* Fall 1994, pp. 127–28; "How to Handle a Windfall," *Consumer Reports,* June 1993, p. 361; Brian Dumaine, "Leaving the Rat Race Early," *Fortune/1993 Investor's Guide,* Fall 1993, pp. 111–112; and John Manners, "Retire Wealthy," *Money,* September 1992, p. 68.

UNIT 54	WORKING COMPOUND INTEREST PROBLEMS

LEARNING OBJECTIVES   *After completing this unit, you will be able to:*

1. determine the compounded principal amount.
2. use compound interest tables.
3. determine the effective interest rate.

The old saying goes: "I've been rich and I've been poor, but believe me, rich is better." And while being rich doesn't guarantee happiness, the accumulation of money does provide financial security and is a goal worthy of pursuit. Regardless of how much money you want or what you want to use the money for, the principles of compound interest can help you obtain your financial goals.

With *compound interest,* the interest earned for one period is added to the principal before interest for the next period is calculated. For example, money deposited in a savings account earns interest. If the interest is left on deposit, the new balance is the original principal *plus* the interest for the period. To calculate the compounded principal in an interest problem, first figure the interest for the first period (say, a year). Then, add the interest to the original principal to find the new compounded principal amount. Next, multiply the compounded principal amount by the interest rate to obtain the interest for the second year. Finally, add the second-year interest to the first-year compounded principal amount to obtain the second-year compounded principal amount.

EXAMPLE 1 Find the Compounded Principal Amount for a $15,000 Investment that Pays 6 Percent Interest Compounded Annually for Two Years

15,000 × 0.06 = $900   **STEP** [1] Multiply the original principal amount ($15,000) by the interest rate (6 percent). Note that 6 percent has been converted to the decimal 0.06.

$15,000
+   900   **STEP** [2] Add the interest for the first year to the original principal amount.
$15,900

$15,900 × 0.06 = $954   **STEP** [3] Multiply the first-year principal amount ($15,900) from step 2 by the interest rate (6 percent or 0.06).

$15,900
+   954   **STEP** [4] Add the interest for the second year to the first-year principal amount.
$16,854

CALCULATOR TIP

Depress clear key
Enter 15,000   Depress   ×
Enter 0.06   Depress   =
Read the answer $900
Enter $15,000 Depress   +
Enter $900   Depress   =
Read the answer $15,900
Enter $15,900 Depress   ×
Enter 0.06   Depress   =
Read the answer $954
Enter $15,900 Depress   +
Enter 954   Depress   =
Read the answer $16,854

To find the interest amount, subtract the original principal amount from the compounded principal.

EXAMPLE 2 Find the Interest Amount when the Original Principal is $15,000 and the Compounded Principal is $16,854

$16,854   compounded principal
−15,000   original principal amount
$ 1,854   interest amount

The procedure in Example 1 works well when the original principal must be compounded for two or three periods. When interest must be calculated for many periods, you may find that it is easier to use a compound interest table like the one shown in Table 54–1 (p. 302). The table shows the compound value for $1 for different interest rates and time periods.*

To use a compound interest table, read down the left side of the table to locate the correct number of periods. Then move across that row to find the interest rate. Finally, multiply this table factor by the original principal amount in the problem to find the compounded principal.

---

*It is also possible to find the compounded principal by the following formula:

$$\text{Compounded Principal} = PV(1 + i)^n$$

where pv is the original principal, i is the interest rate, and n is the number of time periods. Note that for the problem in Example 3, the answer is the same when the formula is used.

$$\text{Compounded Principal} = PV(1 + i)^n$$
$$\text{Compounded Principal} = \$20{,}000(1 + 0.05)^{10}$$
$$\text{Compounded Principal} = \$20{,}000(1.629)$$
$$\text{Compounded Principal} = \$32{,}580$$

TABLE 54-1   Interest Table for $1 Compounded on an Annual Basis

Period	1%	2%	3%	4%	5%	6%	7%	8%	9%	10%	11%	12%
1	1.010	1.020	1.030	1.040	1.050	1.060	1.070	1.080	1.090	1.100	1.110	1.120
2	1.020	1.040	1.061	1.082	1.103	1.124	1.145	1.166	1.188	1.210	1.232	1.254
3	1.030	1.061	1.093	1.125	1.158	1.191	1.225	1.260	1.295	1.331	1.368	1.405
4	1.041	1.082	1.126	1.170	1.216	1.262	1.311	1.360	1.412	1.464	1.518	1.574
5	1.051	1.104	1.159	1.217	1.276	1.338	1.403	1.469	1.539	1.611	1.685	1.762
6	1.062	1.126	1.194	1.265	1.340	1.419	1.501	1.587	1.677	1.772	1.870	1.974
7	1.072	1.149	1.230	1.316	1.407	1.504	1.606	1.714	1.828	1.949	2.076	2.211
8	1.083	1.172	1.267	1.369	1.477	1.594	1.718	1.851	1.993	2.144	2.305	2.476
9	1.094	1.195	1.305	1.423	1.551	1.689	1.838	1.999	2.172	2.358	2.558	2.773
10	1.105	1.219	1.344	1.480	1.629	1.791	1.967	2.159	2.367	2.594	2.839	3.106
11	1.116	1.243	1.384	1.539	1.710	1.898	2.105	2.332	2.580	2.853	3.152	3.479
12	1.127	1.268	1.426	1.601	1.796	2.012	2.252	2.518	2.813	3.138	3.498	3.896
13	1.138	1.294	1.469	1.665	1.886	2.133	2.410	2.720	3.066	3.452	3.883	4.363
14	1.149	1.319	1.513	1.732	1.980	2.261	2.579	2.937	3.342	3.797	4.310	4.887
15	1.161	1.346	1.558	1.801	2.079	2.397	2.759	3.172	3.642	4.177	4.785	5.474
16	1.173	1.373	1.605	1.873	2.183	2.540	2.952	3.426	3.970	4.595	5.311	6.130
17	1.184	1.400	1.653	1.948	2.292	2.693	3.159	3.700	4.328	5.054	5.895	6.866
18	1.196	1.428	1.702	2.206	2.407	2.854	3.380	3.996	4.717	5.560	6.544	7.690
19	1.208	1.457	1.754	2.107	2.527	3.026	3.617	4.316	5.142	6.116	7.263	8.613
20	1.220	1.486	1.806	2.191	2.653	3.207	3.870	4.661	5.604	6.727	8.062	9.646
25	1.282	1.641	2.094	2.666	3.386	4.292	5.427	6.848	8.623	10.835	13.585	17.000
30	1.348	1.811	2.427	3.243	4.322	5.743	7.612	10.063	13.268	17.449	22.892	29.960
40	1.489	2.208	3.262	4.801	7.040	10.286	14.974	21.725	31.409	45.259	65.001	93.051
50	1.645	2.692	4.384	7.107	11.467	18.420	29.457	46.902	74.358	117.390	184.570	289.000

EXAMPLE 3   Find the Compounded Principal for a $20,000 Investment that Pays 5 Percent Interest Compounded Annually for 10 Years

The table factor is 1.629 — **STEP 1** — In the compound interest table, locate the number of periods (10) and move across that row to the column for the interest rate (5 percent).

$20,000 × 1.629 = $32,580 — **STEP 2** — Multiply the principal ($20,000) by the table factor (1.629).

WORKING INTEREST PROBLEMS WHEN COMPOUNDING OCCURS ON A SEMIANNUAL, QUARTERLY, OR MONTHLY BASIS

It is also possible to modify the above procedure to work a problem when interest is compounded semiannually, quarterly, or monthly. The information that follows will help you determine the number of times that compounding occurs in each of the above situations.

Basis for Compounding	Number of Periods Each Year
Annual	Once a year (1)
Semiannual	Two times a year (2)
Quarterly	Four times a year (4)
Monthly	Twelve times a year (12)

Before completing problems where compounding occurs on a semiannual, quarterly, or monthly basis, it is necessary to complete two additional steps. First, *multiply* the *time* factor by the number of times compounding occurs. That is, multiply by 2 if compounding occurs twice a year; multiply by 4 if compounding occurs 4 times a year; and multiply by 12 if compounding occurs 12 times a year. Second, *divide* the *interest rate* by the number of times compounding occurs. Now it is possible to use the answers found in steps 1 and 2 to find an interest factor from the compound interest table.

EXAMPLE 4 Find the Compounded Principal for a $16,000 Investment that Pays 6 Percent Interest Compounded Semiannually for 7 Years

CALCULATOR TIP

Depress clear key
Enter 16,000   Depress   ×
Enter 1.513   Depress   =
Read the answer $24,208

7 years × 2 = 14 periods **STEP** 1 Multiply the time factor (7) by the number of times compounding occurs (2).

6% ÷ 2 = 3% **STEP** 2 Divide the interest rate (6 percent) by the number of times compounding occurs (2).

The table factor is 1.513 **STEP** 3 In the compound interest table, locate the number of periods (14) and move across that row to the column for the interest rate (3 percent).

$16,000 × 1.513 = $24,208 **STEP** 4 Multiply the principal ($16,000) by the table factor (1.513).

For the problem in Example 4, the depositor will receive 3 percent interest (6 percent ÷ 2 = 3 percent) at the end of the first 6-month period. Then the interest amount for the first six months is added to the original principal before the interest amount for the second 6-month period is calculated. In fact, an additional 3 percent interest is added to the previous balance every 6 months for a total of 14 periods (7 years × 2 = 14 periods). And at the end of 14 periods (7 years), the original principal is now worth $24,208.

WORKING INTEREST PROBLEMS WHEN COMPOUNDING OCCURS ON A DAILY BASIS

Many banks, savings and loan associations, and credit unions compound interest on a daily basis. Basically, the same procedures can be used to determine the compounded principal amount, *but* you must use a special table—like the one illustrated in Table 54–2—when interest is compounded daily.

TABLE 54–2   Interest Table for $1 Compounded Daily (based on a 360-day year)

Number of Years	Percent					
	5%	6%	7%	8%	9%	10%
1	1.051	1.061	1.073	1.083	1.094	1.105
2	1.107	1.128	1.150	1.174	1.197	1.221
3	1.164	1.197	1.234	1.271	1.310	1.350
4	1.225	1.271	1.323	1.377	1.433	1.492
5	1.284	1.345	1.420	1.492	1.568	1.649
6	1.356	1.433	1.522	1.616	1.716	1.822
7	1.426	1.522	1.632	1.751	1.878	2.014
8	1.500	1.616	1.751	1.896	2.054	2.225
9	1.578	1.716	1.878	2.054	2.248	2.459
10	1.649	1.822	2.014	2.225	2.459	2.718
15	2.117	2.459	2.857	3.320	3.857	4.481
20	2.718	3.320	4.055	4.952	6.048	7.387
25	3.551	4.481	5.754	7.387	9.485	12.178
30	4.576	6.049	8.165	11.020	14.875	20.077

EXAMPLE 5   Find the Compounded Principal for an $8,000 Investment that Pays 7 Percent Interest Compounded Daily for 4 Years

The table factor is 1.323   **STEP** 1   In the daily compound interest table, locate the number of periods (4) and move across that row to the column for the interest rate (7 percent).

$8,000 × 1.323 = $10,584   **STEP** 2   Multiply the principal ($8,000) by the table factor (1.323)

DETERMINING THE EFFECTIVE INTEREST RATE

When interest is calculated on a semiannual, quarterly, monthly, or daily basis, the depositor receives more than the stated rate of interest because of the effect of compounding. Thus, the stated interest rate is lower than the actual or effective interest rate. The **effective interest rate** is determined by dividing the interest amount received in one year by the original principal. This problem is an application of the rate formula presented in Unit 50 of Chapter 9.

EXAMPLE 6   Determine the Effective Interest Rate for a $12,000 Investment that Pays 6 Percent Compounded Semiannually for 1 Year

$12,000 × 1.061 = $12,732   **STEP** 1   Using Table 54–1, determine the compounded principal for $12,000 invested at 6 percent compounded semiannually for 1 year.

$12,732
−12,000
$ 732

$$R = \frac{I}{P \times T}$$

$$R = \frac{I}{\$12,000 \times 1}$$

$$R = \frac{\$732}{\$12,000}$$

R = 0.061 = 6.1 percent

STEP 2 To determine the interest amount, subtract the original principal ($12,000) from the compounded principal ($12,732).

STEP 3 State the formula for rate.

STEP 4 Multiply the original principal ($12,000) by the time factor (1 year).

STEP 5 To determine the effective interest rate, divide the interest amount ($732) by the answer from step 4 ($12,000).

As shown in Example 6, the effective rate of interest is 6.1 percent, which is slightly higher than the stated interest rate of 6 percent. The difference between the effective rate and the stated rate is the result of compounding the original principal on a semiannual basis instead of an annual basis.

Complete the Following Problems

Annual Interest Rate	Time	Compounded		Table Factor
7%	4 years	Annually	1.	1.311

7% ÷ 1 = 7%   4 years × 1 = 4   7%, 4 periods

6%	5 years	Semiannually	2.	
8%	3 years	Quarterly	3.	
12%	1 year	Monthly	4.	
6%	15 years	Semiannually	5.	

NOTE: for the following problems, round your answer to two decimal places.

Principal	Annual Interest Rate	Time	Compounded	Compounded Principal Value
$ 2,500	10%	3 years	Annually	6. _____
$14,250	8%	4 years	Quarterly	7. _____
$37,500	6%	15 years	Semiannually	8. _____
$ 6,800	12%	1 year	Monthly	9. _____
$11,275	6%	8 years	Semiannually	10. _____
$44,900	7%	16 years	Annually	11. _____
$16,400	8%	5 years	Quarterly	12. _____
$51,640	12%	1 year	Monthly	13. _____

Annual Interest Rate	Time	Compounded	Table Factor
8%	3 years	daily	14. _____
5%	5 years	daily	15. _____
9%	10 years	daily	16. _____

Principal	Annual Interest Rate	Time	Compounded	Compounded Principal Value
$ 7,000	5%	3 years	daily	17. _____
$12,000	10%	4 years	daily	18. _____
$14,500	6%	20 years	daily	19. _____
$17,100	7%	15 years	daily	20. _____

21. George Begg invested $22,000 at Team Bank. The interest rate was 8 percent compounded quarterly. At the end of one year, what is the value of this investment? (Use Table 54–1.)

$$\left.\begin{array}{l} 8\% \div 4 = 2\% \\ 1\ yr \times 4 = 4\ periods \end{array}\right\} = 1.082 \qquad \$22,000 \times 1.082 = \$23,804$$

22. In problem 21, how much interest did Mr. Begg earn?

23. In problem 21, what is the effective rate of interest?

24. Charise Bacon invested $30,000 at 8 percent compounded semiannually for seven years. Using Table 54–1, what is the table factor for this problem?

25. In problem 24, what is the compounded principal at the end of seven years?

26. In problem 24, what is the interest amount?

27. On January 15, J.M. Toby invested $10,000 at 5 percent compounded daily for six years. Using Table 54–2, what is the table factor for this problem?

28. In problem 27, what is the compounded principal at the end of six years?

29. Daniel Wataumba purchased a 36-month, $25,000 certificate of deposit at a Nashville bank. The bank agreed to pay Daniel 8 percent compounded on a daily basis. Assuming that Daniel does not withdraw the interest, how much is the compounded principal at the end of 36 months?

Personal Math Problem

Janice Garfield is trying to decide where to purchase a $50,000 certificate of deposit. Greater Phoenix Bank offers a 24-month certificate of deposit that pays 4 percent compounded annually. Another Phoenix bank, Fidelity National, has a 24-month certificate of deposit that pays 4 percent compounded quarterly.

30. What is the compounded principal for the certificate of deposit offered by Greater Phoenix Bank at the end of 24 months?

31. What is the compounded principal for the certificate of deposit offered by Fidelity National at the end of 24 months?

32. If you were Ms. Garfield, which certificate of deposit would you choose? Why? _____
_____
_____
_____

33. In your own words, explain why one certificate of deposit earns more money than the other when both advertise a 4 percent interest rate. _____
_____
_____

UNIT 55   WORKING PRESENT VALUE PROBLEMS

LEARNING OBJECTIVES   *After completing this unit, you will be able to:*

1. describe the difference between present value problems and compound interest problems.
2. use tables to determine present value amounts.

TABLE 55–1 Present Value for $1

Period	1%	2%	3%	4%	5%	6%	7%	8%	9%	10%	11%	12%
1	0.990	0.980	0.971	0.962	0.952	0.943	0.935	0.926	0.917	0.909	0.901	0.893
2	0.980	0.961	0.943	0.925	0.907	0.890	0.873	0.857	0.842	0.826	0.812	0.797
3	0.971	0.942	0.915	0.889	0.864	0.840	0.816	0.794	0.772	0.751	0.731	0.712
4	0.961	0.924	0.885	0.855	0.823	0.792	0.763	0.735	0.708	0.683	0.659	0.636
5	0.951	0.906	0.863	0.822	0.784	0.747	0.713	0.681	0.650	0.621	0.593	0.567
6	0.942	0.888	0.837	0.790	0.746	0.705	0.666	0.630	0.596	0.564	0.535	0.507
7	0.933	0.871	0.813	0.760	0.711	0.665	0.623	0.583	0.547	0.513	0.482	0.452
8	0.923	0.853	0.789	0.731	0.677	0.627	0.582	0.540	0.502	0.467	0.434	0.404
9	0.914	0.837	0.766	0.703	0.645	0.592	0.544	0.500	0.460	0.424	0.391	0.361
10	0.905	0.820	0.744	0.676	0.614	0.558	0.508	0.463	0.422	0.386	0.352	0.322
11	0.896	0.804	0.722	0.650	0.585	0.527	0.475	0.429	0.388	0.350	0.317	0.287
12	0.887	0.788	0.701	0.625	0.557	0.497	0.444	0.397	0.356	0.319	0.286	0.257
13	0.879	0.773	0.681	0.601	0.530	0.469	0.415	0.368	0.326	0.290	0.258	0.229
14	0.870	0.758	0.661	0.577	0.505	0.442	0.388	0.340	0.299	0.263	0.232	0.205
15	0.861	0.743	0.642	0.555	0.481	0.417	0.362	0.315	0.275	0.239	0.209	0.183
16	0.853	0.728	0.623	0.534	0.458	0.394	0.339	0.292	0.252	0.218	0.188	0.163
17	0.844	0.714	0.605	0.513	0.436	0.371	0.317	0.270	0.231	0.198	0.170	0.146
18	0.836	0.700	0.587	0.494	0.416	0.350	0.296	0.250	0.212	0.180	0.153	0.130
19	0.828	0.686	0.570	0.475	0.396	0.331	0.277	0.232	0.194	0.164	0.138	0.116
20	0.820	0.673	0.554	0.456	0.377	0.312	0.258	0.215	0.178	0.149	0.124	0.104
25	0.780	0.610	0.478	0.375	0.295	0.233	0.184	0.146	0.116	0.092	0.074	0.059
30	0.742	0.552	0.412	0.308	0.231	0.174	0.131	0.099	0.075	0.057	0.044	0.033
40	0.672	0.453	0.307	0.208	0.142	0.097	0.067	0.046	0.032	0.022	0.015	0.011
50	0.608	0.372	0.228	0.141	0.087	0.054	0.034	0.021	0.013	0.009	0.005	0.003

In the last unit, we discussed the steps required to complete a compound interest problem. The purpose of compound interest problems is to determine the dollar amount, sometimes referred to as future value, of the original investment *plus* interest after a specific number of investment periods. For example, if $10,000 is invested at 6 percent compounded annually for five years, the investor will receive $13,380 at the end of five years.

In this unit, we discuss present value. The purpose of **present value** is to determine how much money must be invested today at current interest rates in order to obtain a specific dollar amount at a future date. For example, Mike Williams wants to accumulate $40,000 during the next 7 years in order to pay for his daughter's college education. How much should Mr. Williams invest today to reach his goal?

The easiest way to solve this type of problem is to use a present value table like the one illustrated in Table 55–1.* First, read down the left side of

---

*It is also possible to find the present value by the following formula:

$$\text{Present value} = \frac{MV}{(1 + i)^n}$$

where *mv* is the maturity value, *i* is the interest rate, and *n* is the number of time periods. Note that for the problem in Example 1, the answer is approximately the same. The difference ($4.26) is due to round off.

$$\text{Present value} = \frac{\$40,000}{(1 + 0.06)^7}$$

$$\text{Present value} = \frac{\$40,000}{1.504}$$

$$\text{Present value} = \$26,595.74$$

the table to locate the correct number of periods. Then, move across that row to find the interest rate. Finally, multiply this table factor by the dollar amount that the investor wants to obtain by the future date.

EXAMPLE 1   Find the Present Value Amount that Can Be Invested at 6 Percent for 7 Years in order to Obtain $40,000

CALCULATOR TIP

Depress clear key
Enter 40,000   Depress   ×
Enter 0.665   Depress   =
Read the answer $26,600

The table factor is 0.665

STEP [1]   In the present value table, locate the number of periods (7) and move across that row to the column for the interest rate (6 percent).

$40,000 × 0.665 = $26,600

STEP [2]   Multiply the dollar amount that the investor wants to obtain ($40,000) by the table factor (0.665).

WORKING PRESENT VALUE PROBLEMS WHEN COMPOUNDING OCCURS ON A SEMIANNUAL, QUARTERLY, OR MONTHLY BASIS

Like compound interest problems, it is also possible to modify the above procedure to work a problem when the investment is compounded on a semiannual, quarterly, or monthly basis. First, *multiply* the *time* factor by the number of times compounding occurs. That is, multiply by 2 if compounding occurs twice a year; multiply by 4 if compounding occurs 4 times a year; and multiply by 12 if compounding occurs 12 times a year. Second, *divide* the *interest rate* by the number of times compounding occurs within a year. Third, use the answers found in steps 1 and 2 to find a table factor from the present value table.

EXAMPLE 2   Find the Present Value that Can Be Invested at 4 Percent Interest Compounded Quarterly for Four Years in order to Obtain $25,000

---

4 years × 4 = 16 periods

STEP [1]   Multiply the time factor (4) by the number of times compounding occurs (4).

4% ÷ 4 = 1%

STEP [2]   Divide the interest rate (4 percent) by the number of times compounding occurs (4).

The table factor is 0.853

STEP [3]   In the present value table, locate the number of periods (16) and move across that row to the column for the interest rate (1 percent).

$25,000 × 0.853 = $21,325

STEP [4]   Multiply the dollar amount that the investor wants to obtain ($25,000) by the table factor (0.853).

Complete the Following Problems

If necessary, round your answer to two decimal places.

1. In your own words, describe the purpose of compound interest problems.

_____

_____

_____

2. In your own words, describe the purpose of present value problems.

_____

_____

_____

Annual Interest Rate	Time	Compounded		Present Value Table Factor
5%	12 years	Annually	**3.**	_0.557_

$5\% \div 1 = 5\%$ $12 \text{ years} \times 1 = 12 \text{ periods} = 0.557$

Annual Interest Rate	Time	Compounded		Present Value Table Factor
6%	8 years	Semiannually	**4.**	_____
4%	5 years	Quarterly	**5.**	_____
10%	20 years	Annually	**6.**	_____
8%	3 years	Quarterly	**7.**	_____

Dollar Amount To Be Obtained	Annual Interest Rate	Time	Compounded		Present Value Amount
$10,000	8%	10 years	Annually	**8.**	_____

Dollar Amount To Be Obtained	Annual Interest Rate	Time	Compounded	Present Value Amount
$ 5,000	4%	9 years	Semiannually	**9.** _____
$75,000	8%	4 years	Quarterly	**10.** _____
$55,000	10%	8 years	Annually	**11.** _____
$32,000	7%	2 years	Annually	**12.** _____
$11,000	6%	15 years	Semiannually	**13.** _____
$ 6,200	3%	20 years	Annually	**14.** _____
$19,400	10%	6 years	Semiannually	**15.** _____
$44,100	8%	3 years	Quarterly	**16.** _____
$33,700	12%	8 years	Semiannually	**17.** _____

18. In five years, Pronto Press, Inc., will need $16,200 to expand their printing operations. How much should Pronto invest today if the investment will earn 6 percent compounded semiannually in order to accumulate $16,200 in five years?

$$\left. \begin{array}{l} 6\% \div 2 = 3\% \\ 5 \text{ years} \times 2 = 10 \text{ periods} \end{array} \right\} = 0.744 \quad \$16,200 \times 0.744 = \$12,052.80$$

19. Kathy and Manuel Garcia would like to purchase a home in three years when Manuel is discharged from the U.S. Army. They estimate that they will need $8,000 for the down payment and will finance the balance of the purchase price. Assuming that they establish a savings account that pays 4 percent compounded quarterly, how much should they deposit to insure that they accumulate $8,000 in 3 years?

20. Keisha Turley estimates that she needs an additional $500,000 in order to retire when she is 65. Assuming that her investment will earn 6 percent compounded semiannually, how much should she deposit today if she wants to retire in 20 years?

21. In problem 20, how much interest will Ms. Turley earn during the 20-year period?

22. Jim Warner wants to purchase a new Pontiac Trans Am in four years. He estimates that the car will cost $28,000. If Jim invests $22,000 today in an investment that earns 4 percent compounded semiannually, will he have enough money to pay for the car? (Use Table 54–1.)

Critical Thinking Problems

Jane Ford, president of Ford Manufacturing, is trying to decide which way to pay for a new, automated packaging machine. The machine will cost $26,000. She also estimates that the old machine will last another three years.

Option 1: Ms. Ford could make an investment that will accumulate enough money to make the purchase at the end of three years. The potential investment will earn 5 percent each year compounded annually.

Option 2: Ms. Ford could use an installment purchase plan offered by the manufacturer and pay for the machine over 12 months. The manufacturer does charge the buyer 10 percent interest when the installment plan is used.

23. Assuming that Ms. Ford chooses option 1, how much money should she invest in order to accumulate $26,000 in three years?

24. If option 1 is used, what is the total purchase price for the machine?

25. If Ms. Ford chooses option 2 and pays for the machine over 12 months, how much is the finance charge?

26. If you were Ms. Ford, which of the two options would you choose? Why? __

_____

_____

_____

UNIT 56	FINANCING THE HOME PURCHASE

LEARNING OBJECTIVES   *After completing this unit, you will be able to:*

1. explain the difference between fixed rate and adjustable rate mortgages.
2. calculate the down payment amount for a home mortgage.
3. calculate the monthly payment amount for a home mortgage.

For most people, buying a home is the largest single purchase they make during their lifetime. While there are exceptions to the following guidelines, many loan officers suggest that you should spend no more than 25 to 30 percent of your take-home pay on housing. Other lenders suggest that your home should cost about $2\frac{1}{2}$ times your annual income.

Unless you pay cash for a home, you must finance your home purchase over a number of years. A **home mortgage** is a long-term loan used to pay for a specific piece of real estate. Banks, savings and loan associations, and credit unions are the most common sources for home financing. Generally, these financial institutions offer two types of loans. A **fixed-rate mortgage** is a loan with a fixed interest rate and fixed monthly payments during the life of the loan. An **adjustable-rate mortgage** (**ARM**) is a loan that has an interest rate that increases or decreases during the life of the loan. Interest rates for this type of loan are usually tied to some index such as the interest rates on U.S. Treasury bills or the average national mortgage rate. Interest adjustments and changes in payment amounts are made regularly, usually at intervals of one, three, or five years. A homeowner who thinks interest rates will increase over the life of the loan will probably choose a fixed-rate mortgage. On the other hand, a homeowner who thinks interest rates will rise modestly, stay stable, or decline should choose an adjustable-rate mortgage.

DOWN PAYMENTS AND CLOSING COSTS

Regardless of the type of mortgage chosen, lenders expect that homeowners will make a down payment when financing a home purchase. In fact, most banks, savings and loan associations, credit unions, and mortgage companies expect borrowers to make a down payment of at least 5 to 20 percent of the purchase price. Down payment calculations are a practical application of the portion formula presented in Unit 26. The dollar amount of the down payment (the portion) is found by multiplying the purchase price (the base) by the percent of down payment (the rate).

EXAMPLE 1 Find the Dollar Amount of Down Payment for a Home with a Purchase Price of $75,000 that Requires a 10 Percent Down Payment

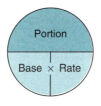

Portion

Base × Rate

$75,000 (Base)
× 0.10 (Rate)
$7,500.00 (Portion)

In this example, note the down payment percent (10 percent) has been changed to the decimal 0.10. To calculate the amount of the loan, one additional step is necessary: Subtract the down payment amount from the purchase price.

CALCULATOR TIP

Depress clear key
Enter 75,000 Depress ×
Enter 0.10 Depress =
Read the answer $7,500

EXAMPLE 2 Find the Loan Amount for the Home Purchase in Example 1

$75,000 home purchase price
− 7,500 down payment amount
$67,500 loan amount

CALCULATOR TIP

Depress clear key
Enter 75,000 Depress −
Enter 7,500 Depress =
Read the answer $67,500

In addition to the down payment, both the buyer and seller of a home may have to pay closing costs. **Closing costs** are the fees and charges paid when a real estate transaction is completed. Although determining the actual costs that buyer and seller pay at closing is beyond the scope of this text, typical closing costs include title search fees, title insurance, attorney's fees, survey costs, appraisal fees, recording fees, and prepaid interest.

CALCULATING MONTHLY PAYMENT AMOUNTS

Generally, home mortgages are repaid over a 15- to 30-year period. In order to determine the monthly payment amount required to repay the principal and interest—often referred to as P&I, most lenders use a home amortization table like the one illustrated in Table 56–1. To use this amortization table, read down the left side of the table to locate the number of years for a specific home mortgage. Then, move across that row to find the interest rate for the home mortgage. Finally, multiply this table factor by the loan amount. **NOTE:** the factors in Table 56–1 are based on $1,000 units of the loan amount.

TABLE 56–1 Home Mortgage Amortization Chart (Principal and Interest Factors per $1,000 of Loan Amount)

No. of Years	6%	$6\frac{1}{2}$	7%	$7\frac{1}{2}$	8%	$8\frac{1}{2}$	9%	$9\frac{1}{2}$	10%	$10\frac{1}{2}$	11%	$11\frac{1}{2}$	12%
5	19.33	19.57	19.80	20.04	20.28	20.52	20.76	21.00	21.25	21.49	21.74	21.99	22.24
10	11.10	11.35	11.61	11.87	12.13	12.40	12.67	12.94	13.22	13.49	13.78	14.06	14.35
15	8.44	8.71	8.99	9.27	9.56	9.85	10.14	10.44	10.75	11.05	11.37	11.68	12.00
20	7.16	7.46	7.75	8.06	8.36	8.68	9.00	9.32	9.65	9.98	10.32	10.66	11.01
25	6.44	6.75	7.07	7.39	7.72	8.05	8.39	8.74	9.09	9.44	9.80	10.16	10.53
30	6.00	6.32	6.65	6.99	7.34	7.69	8.05	8.41	8.78	9.15	9.52	9.90	10.29

EXAMPLE 3   Find the Monthly Payment Amount for a $60,000 Home Mortgage with an Interest Rate of $7\frac{1}{2}$ Percent Repaid over 30 Years

The table factor is $6.99	**STEP** 1	In the amortization table, locate the number of years for this home mortgage (30) and move across that row to the column for the interest rate ($7\frac{1}{2}$ percent).
$60,000 ÷ $1,000 = 60	**STEP** 2	To determine the number of $1,000 units in this loan, divide the loan amount ($60,000) by $1,000.
$6.99 × 60 = $419.40	**STEP** 3	Multiply the table factor from step 1 ($6.99) by the answer from step 2 (60).

CALCULATOR TIP

Depress clear key
Enter 60,000   Depress   ÷
Enter 1,000   Depress   =
Read the answer 60
Enter 6.99   Depress   ×
Enter 60   Depress   =
Read the answer $419.40

In addition to the amount required to pay principal and interest, most lenders also require that borrowers make monthly deposits into an escrow account for payment of taxes and insurance. The money deposited in an *escrow account* protects the lender from financial loss due to unpaid real estate taxes or damage from fire or other hazards. For the $60,000 home purchased in Example 3, real estate taxes total $1,800 a year and insurance premiums total $900 a year. The steps required to calculate the monthly escrow deposits for real estate and insurance are illustrated in Example 4.

EXAMPLE 4   Find the Monthly Escrow Deposit when Real Estate Taxes Total $1,800 a Year and Insurance Premiums Total $900 a Year.

$1,800 ÷ 12 = $150	**STEP** 1	Divide the annual real estate tax amount ($1,800) by 12.
$900 ÷ 12 = $75	**STEP** 2	Divide the annual insurance premium ($900) by 12.
$150 + 75 $225	**STEP** 3	To determine the total amount that should be deposited in the escrow account, add the answers from step 1 and step 2.

CALCULATOR TIP

Depress clear key
Enter 1,800   Depress   ÷
Enter 12   Depress   =
Read the answer $150
Enter 900   Depress   ÷
Enter 12   Depress   =
Read the answer $75
Enter 150   Depress   +
Enter 75   Depress   =
Read the answer $225

To calculate the total monthly payment amount required to repay principal and interest, real estate taxes, and insurance, one final step is necessary. You must add the amounts for principal and interest and escrow payments.

EXAMPLE 5   Find the Total Monthly Payment Amount for the Home Purchased in Example 3

$419.40 (P&I) +225.00 (escrow) $644.40 total payment		Add the monthly payment amounts or P&I from Example 3 ($419.40) and the total monthly amount for real estate taxes and insurance from Example 4 ($225).

CALCULATOR TIP

Depress clear key
Enter 419.40   Depress   +
Enter 225   Depress   =
Read the answer $644.40

Complete the Following Problems

If necessary, round your answer to two decimal places. Also, use Table 56–1 for all home mortgage problems.

 1. In your own words, describe the difference between a fixed-rate mortgage and an adjustable-rate mortgage. _____

_____

_____

_____

_____

2. All Country Mortgage, Inc., requires that all home buyers make a 20 percent down payment in order to obtain their lowest interest loan. What is the down payment amount for an $88,000 home mortgage?

$88,000 × .20 = $17,600

3. LaTeisha and Andrew Brown just talked to a loan officer at Tuscon Savings and Loan. The loan officer told them that they must make a $4,500 down payment in order to obtain a home with a purchase price of $90,000. How much is the loan amount?

4. Linda and Bobby Goodman have just made the decision to purchase a new home. The purchase price for their new home is $120,400. If the mortgage company requires that borrowers make a 15 percent down payment, what is the dollar amount of their down payment?

5. In problem 4, what is the dollar amount of the loan for the Goodman's home mortgage?

Interest Rate	Number of Years	Payment Factor from Table 56–1
8%	15 years	**6.** _____
6%	20 years	**7.** _____
9%	30 years	**8.** _____
$7\frac{1}{2}$%	25 years	**9.** _____
$9\frac{1}{2}$%	15 years	**10.** _____
10%	30 years	**11.** _____

Home Mortgage Amount	Interest Rate	Number of Years	Payment Amount for Principal and Interest
$ 70,000	8%	30 years	**12.** _____
$110,000	9%	20 years	**13.** _____
$125,000	$7\frac{1}{2}$%	15 years	**14.** _____
$150,000	$8\frac{1}{2}$%	30 years	**15.** _____
$ 55,100	$9\frac{1}{2}$%	20 years	**16.** _____
$105,300	7%	30 years	**17.** _____
$155,250	8%	15 years	**18.** _____
$171,300	$8\frac{1}{2}$%	30 years	**19.** _____
$144,350	11%	25 years	**20.** _____

21. You are the loan officer for a local savings and loan association. As part of your job, you are required to determine monthly payment amounts for prospective home buyers. What is the monthly payment for principal and interest for a 30-year, $110,000 mortgage if the interest rate is $8\frac{1}{2}$ percent?

22. Patsy Maricopa wants to purchase a new three bedroom condominium. In order to purchase this home, she will obtain a 15-year, $98,500 home mortgage with an interest rate of $10\frac{1}{2}$ percent. What is the monthly payment amount for principal and interest?

23. What is the monthly escrow payment for real estate taxes if annual taxes for a home are $2,100?

24. What is the monthly escrow payment for insurance if the yearly premium for insurance coverage is $850?

Use the Following Information to Answer Questions 25–30

Mike and Marty Nations are trying to decide if they can afford to purchase a $125,000 home, and they ask for your help. The bank requires a 5 percent down payment on all 30-year home mortgages. The current interest rate for this type of mortgage is 9 percent. The Nationses estimate that real estate taxes will total $1,600 a year and insurance coverage will cost $1,000 a year.

25. How much down payment will the Nationses have to pay to purchase this home?

26. What is the loan amount required to purchase this home?

27. What is the monthly payment required for principal and interest?

28. What is the monthly escrow payment required to pay real estate taxes?

29. What is the monthly escrow payment required to pay for insurance coverage?

30. What is the total monthly payment for principal and interest, real estate taxes, and insurance?

UNIT 57	REPAYING A HOME MORTGAGE AND PREPARING AN AMORTIZATION CHART

LEARNING OBJECTIVES

After completing this unit, you will be able to:

1. find the total dollar amounts of monthly payments and interest charges for a home mortgage.
2. determine the amount of interest and the amount of principal reduction for a monthly home mortgage payment.
3. construct an amortization chart.

As pointed out in the last unit, a home purchase is the largest single purchase that most people make during their lifetime. The calculations in this unit only underscore the importance of making wise decisions when financing a home purchase.

TOTAL REPAYMENT COSTS

For most home buyers, calculating the total amount required to repay a home mortgage can be startling. Also, calculating the amount of interest that the home buyer pays over a 15- to 30-year period can be revealing. Take the case of Julio and Tina Santana. Thirty years ago they borrowed $85,000 to purchase a home. The interest rate for their home mortgage was 7 percent. Monthly payments for principal and interest were $565.25. As shown in Example 1, the Santanas paid $203,490 to repay their $85,000 home mortgage over a 30-year period of time.

EXAMPLE 1 Find the Total Repayment Amount for a 30-Year Mortgage when Monthly Payments Are $565.25

$565.25 × 12 = $6,783	**STEP** 1	Multiply the monthly payment amount ($565.25) by the number of months in a year (12).
$6,783 × 30 = $203,490	**STEP** 2	Multiply the answer from step 1 ($6,783) by the number of years for the home mortgage (30).

CALCULATOR TIP

Depress clear key
Enter 565.25   Depress   ×
Enter 12   Depress   =
Read the answer $6,783
Enter 6,783   Depress   ×
Enter 30   Depress   =
Read the answer $203,490

Using the information obtained in Example 1, it is now possible to determine the amount of interest that the Santanas paid over the 30-year life of their $85,000 home mortgage by completing one additional step. As shown in Example 2, the Santanas paid $118,490 in interest over a 30-year period of time.

EXAMPLE 2 Find the Total Interest Charged for the $85,000 Loan in Example 1

CALCULATOR TIP

Depress clear key
Enter 203,490   Depress   –
Enter 85,000   Depress   =
Read the answer $118,490

$203,490	total repayment amount
– 85,000	original home mortgage
$118,490	total interest

If you are thinking that the Santanas are paying a lot of money to repay an $85,000 loan, you're right. There are ways to reduce the amount of interest that is repaid over the term of loan. First, *always* shop around for the lowest interest rates available. Interest rates at banks, savings and loan associations, credit unions, and mortgage companies are competitive, and you will never know if you can get a lower rate unless you ask. Second, consider applying for a 15- or 20-year mortgage instead of the typical 30-year mortgage. While the monthly payments will be higher, you will pay less interest over a shorter period of time.

ANALYSIS OF MONTHLY PAYMENT AMOUNTS

A part of each monthly payment is used to pay the interest on the loan for the month. The remainder is used to reduce the principal balance of the loan. By using the simple interest formula presented in Unit 49, we can determine the amount of interest that must be paid each month. Then, it is possible to determine the amount that can be used to reduce the principal balance.

EXAMPLE 3   Find the Monthly Interest and the Amount that Can Be Used to Reduce the Principal Balance for an 8 Percent Loan with a Principal Balance of $54,000 and Monthly Payments of $396

I = $54,000 × 0.08 × 1 ÷ 12 = $4,320

$54,000 × 0.08 = $4,320   **STEP** [1]   Multiply the loan balance ($54,000) by the interest rate (8 percent). This answer represents interest for one year.

CALCULATOR TIP

Depress clear key
Enter 54,000   Depress   ×
Enter 0.08   Depress   =
Read the answer $4,320
Enter 4,320   Depress   ÷
Enter 12   Depress   =
Read the answer $360
Enter $396   Depress   −
Enter 360   Depress   =
Read the answer $36

$4,320 ÷ 12 = $360   **STEP** [2]   To determine the amount of interest for one month, divide the answer from step 1 ($4,320) by the number of months in one year (12).

$396
−360
$ 36   **STEP** [3]   Subtract the amount of interest for one month ($360) from the monthly payment amount ($396). This is the amount that can be used to reduce the principal balance.

As illustrated in Example 4, one additional step is necessary to calculate the new principal balance.

EXAMPLE 4   Find the New Principal Balance for the $54,000 Home Mortgage in Example 3

$54,000   original principal balance
−   36   amount used to reduce balance
$53,964   new principal balance

CALCULATOR TIP

Depress clear key
Enter 54,000   Depress   −
Enter 36   Depress   =
Read the answer $53,964

The same calculations illustrated in Examples 3 and 4 can be used to determine the interest amount and the amount that can be used to reduce the principal balance for any month during a loan agreement. In fact, you can use these calculations to build an amortization (repayment) chart like the one illustrated in Table 57–1.

As illustrated in Table 57–1, the amount of interest will decrease and the amount that can be used to reduce the principal balance will increase when each monthly payment is made.

Complete the Following Problems

If necessary, round your answer to two decimal places.

1. How many monthly payments are there in a 15-year home mortgage?

12 × 15 = 180

TABLE 57–1 Partial Amortization Table for a 30-Year, $100,000 Home Mortgage (The Interest Rate for the Loan Is 9 Percent and the Monthly Payment Is $805.)

Payment Number	Principal Balance	Interest Amount	Amount Used to Reduce Principal Balance	New Principal Balance
1	$100,000.00	$750.00	55.00	$99,945.00
2	99,945.00	749.59	55.41	99,889.59
3	99,889.59	749.17	55.83	99,833.76
4	99,833.76	748.75	56.25	99,777.51
5	99,777.51	748.33	56.67	99,720.84
6	99,720.84	747.91	57.09	99,663.75
7	99,663.75	747.48	57.52	99,606.23
8	99,606.23	747.05	57.95	99,548.28
9	99,548.28	746.61	58.39	99,489.89
10	99,489.89	746.17	58.83	99,431.06
11	99,431.06	745.73	59.27	99,371.79
12	99,371.79	745.29	59.71	99,312.08

2. How many monthly payments are there in a 20-year home mortgage?

3. How many monthly payments are there in a 25-year home mortgage?

4. How many monthly payments are there in a 30-year home mortgage?

Monthly Payment Amount	Term of Home Mortgage	Total Repayment Amount
$ 600	30 years	**5.** _____
$ 512	15 years	**6.** _____
$1,000	20 years	**7.** _____
$1,190	25 years	**8.** _____
$1,400	30 years	**9.** _____

10. Find the total interest charged for a 20-year, $115,000 loan with monthly payments of $1,100.

11. Find the total interest charged for a 25-year, $78,000 loan with monthly payments of $709.

12. Find the total interest charged for a 15-year, $52,000 loan with monthly payments of $607.

13. Find the total interest charged for a 20-year, $107,000 loan with monthly payments of $1,178.

HINT: You may want to review the material on simple interest in Unit 49 before completing problems 13–19.

Principal Loan Balance	Interest Rate	Monthly Interest Amount
$112,000	8%	**14.** _____
$ 43,200	10%	**15.** _____
$ 89,500	9%	**16.** _____
$123,400	$8\frac{1}{2}\%$	**17.** _____
$ 39,000	11%	**18.** _____
$ 92,300	$9\frac{1}{2}\%$	**19.** _____
$156,600	9%	**20.** _____

21. Myka Steinberg borrowed $65,000 to purchase a home. The interest rate for this 15-year loan was 9 percent with monthly payments for principal and interest of $659. What is the amount of interest for the first month?

22. In problem 20, what is the principal balance after the first monthly payment?

23. The principal balance for a 30-year loan with an interest rate of $8\frac{1}{2}$ percent is $67,250. If the monthly payment is $710, what is the principal balance after the next monthly payment?

24. Back in 1986, Pete Mills obtained a 20-year, $53,200 loan to purchase a town home. The interest rate for the loan was 10 percent. Now the principal balance is $20,100. The monthly payment for this loan is $513. What is the balance after the next monthly payment?

Use the Following Information to Construct an Amortization Chart (Questions 25–28)

Prescilla Earthman obtained a 20-year, $102,000 loan to purchase a two-bedroom home in Tulsa, Oklahoma. The interest rate for the loan is 8 percent, and the monthly payments are $853.

Payment Number	Principal Balance	Interest Amount	Amount Used to Reduce Principal Balance	New Principal Balance
1st	$102,000	_____	_____	**25.** _____
2nd	_____	_____	_____	**26.** _____
3rd	_____	_____	_____	**27.** _____
4th	_____	_____	_____	**28.** _____

Personal Math Problem

Margaret Pride must borrow $125,700 to pay for her new home. She is qualified for both a 15- and a 30-year loan. The interest rate for both loans is 7 percent.

29. What is the monthly payment for the 15-year loan? (Use Table 56–1)

30. What is the total repayment amount if Ms. Pride chooses the 15-year loan?

31. What is the monthly payment for the 30-year loan? (Use Table 56–1)

32. What is the total repayment amount if Ms. Pride chooses the 30-year loan?

 33. If you were Ms. Pride, which loan would you choose? Why? _____

_____

_____

_____

INSTANT REPLAY		
UNIT	**IMPORTANT POINTS TO REMEMBER**	**EXAMPLES**
UNIT 54 Working Compound Interest Problems	With *compound interest,* the interest earned for one period is added to the principal before interest for the next period is calculated. To find the total interest amount, subtract the original principal from the compounded principal.	Find the compounded principal for a $10,000 investment that pays 8 percent interest compounded annually for five years. (Use Table 54–1.) $10,000 × 1.469 = $14,690
	When interest must be calculated for several time periods, you may find it is easier to use a compound interest table. Interest can be compounded annually, semiannually, quarterly, monthly, or daily.	$14,690 –10,000 $ 4,690 total interest

continued from page 325 **UNIT**	**IMPORTANT POINTS TO REMEMBER**	**EXAMPLES**
UNIT 55 Working Present Value Problems	The purpose of present value is to determine how much money must be invested today at current interest rates in order to obtain a specific dollar amount at a future date. The easiest way to solve this type of problem is to use a present value table like the one illustrated in Table 55–1. When working present value problems, adjustments can be made when compounding occurs on a semiannual, quarterly, or monthly basis.	Find the present value amount to obtain $30,000 in 6 years. The interest rate is 5 percent. (Use Table 55–1.) **$30,000 × 0.746 = $22,380**
UNIT 56 Financing the Home Purchase	Down payment amounts can be determined by using the portion formula. Then it is possible to find the loan amount by subtraction. Monthly payment amounts for principal and interest can be found by using Table 56–1. In addition, most lenders require that home buyers make monthly deposits to an escrow account to cover real estate and insurance costs.	**$50,000 × .10 = $5,000** home mortgage down payment $50,000 home mortgage – 5,000 down payment $45,000 loan amount **45,000 ÷ 1,000 = 45** **45 × $6.32 = $284.40** (Table factor for $6\frac{1}{2}$% loan for 30 years)
UNIT 57 Repaying a Home Mortgage and Preparing an Amortization Chart	Total repayment costs are found by multiplying the monthly payment amounts by the number of payments in the term of the loan. Total interest charged for a loan is determined by subtracting the original loan amount from the total repayment amount. A part of each monthly payment is used to pay interest on the loan; the remainder is used to reduce the principal balance of the loan.	**$300 × 12 × 30 = $108,000** monthly total payment repayment cost $108,000 total repayment cost – 53,000 loan amount $ 55,000 total interest - - - - - - - - - - - - - - - - **$41,000 × 0.08 ÷ 12 = $273.33** principal interest months monthly balance rate in a interest year amount $400.00 monthly payment –273.33 monthly interest 126.67 amount used to reduce principal balance

CHAPTER 10	MASTERY QUIZ

DIRECTIONS *Round each answer to two decimal places, or hundredths (Each answer counts 5 points)*

1. Find the compound principal value for a $12,500 investment that pays 7 percent interest compounded annually for 12 years. (Use Table 54–1.)

2. Find the compound principal value for a $7,300 investment that pays 6 percent interest compounded semiannually for five years. (Use Table 54–1.)

3. Find the compounded principal value for a $26,530 investment that pays 8 percent compounded quarterly for four years. (Use Table 54–1.)

4. In problem 3, what is the interest amount at the end of four years?

5. Find the compounded principal value for a $12,500 investment that pays 5 percent compounded daily for 5 years. (Use Table 54–2.)

Use Table 55–1 to Complete Problems 6–10

Dollar Amount to Be Obtained	Annual Interest Rate	Time	Compounded	Present Value Amount
$12,000	4%	10 years	Annually	**6.** _____
$ 5,600	6%	4 years	Semiannually	**7.** _____
$55,000	8%	5 years	Quarterly	**8.** _____
$90,000	11%	20 years	Annually	**9.** _____

Dollar Amount to Be Obtained	Annual Interest Rate	Time	Compounded	Present Value Amount
$20,000	12%	4 years	Quarterly	**10.** _____

11. What is the down payment for a $142,000 home mortgage if the lender requires that the home buyer make a 15 percent down payment?

12. In problem 11, what is the amount of the loan?

13. What is the monthly payment amount for principal and interest for a $76,500 loan if the interest rate is $7\frac{1}{2}$ percent and the repayment period is 15 years? (Use Table 56–1.)

14. What is the monthly escrow payment for real estate taxes if annual taxes for a home are $1,400?

15. Find the total repayment amount for a 20-year home mortgage if the monthly payment is $825.

16. What is the total interest charged for a 30-year, $79,500 loan with monthly payments of $502?

Use the Following Information to Answer Questions 17–20.

Bill Bronte must borrow $71,000 to pay for his new home. The interest rate for this 20-year loan is 8 percent with monthly payments for principal and interest of $594.

17. What is the amount of interest for the first month?

18. What is the principal balance after the first monthly payment?

19. What is the amount of interest for the second month?

20. What is the principal balance after the second monthly payment?

Depreciation

This chapter of *Business Mathematics* covers depreciation—a topic that is of particular interest to managers and accountants. The chapter begins by examining original cost, salvage value, useful life, and book value. Then, it describes the steps required to calculate five different types of depreciation: the straight-line method, the units-of-production method, the sum-of-the years'-digits method, the declining balance method, and the modified accelerated cost recovery system.

MATH TODAY

HOW DEPRECIATION AFFECTS PROFITS REPORTED IN ANNUAL REPORTS

Question Why do corporations go to so much trouble and expense to send out a glossy, multicolored, attractive annual report filled with pictures of smiling faces?

Answer Most chief executive officers believe that the annual report is their best chance to inspire shareholder confidence in the company and its management. To this end, millions of annual reports are sent out by companies every year. IBM sent out 1.5 million and AT&T sent out 4 million. They can cost large corporations over $1 million, and more than half are never even opened, much less read.

While most people think that the upbeat letter from the chairman of the board or the financial statements are the most important material, the true story behind the numbers contained in an annual report is in the footnotes. According to Kenneth L. Fisher, author and money manager, that's where they bury the bodies where the fewest folks find them—in the fine print.* Even though most firms use generally accepted accounting principles, there is still room to apply aggressive—and, according to some experts, questionable—accounting practices that some firms can use to make their profits look better than they really are.

One item always reported in the footnotes is the method used to calculate depreciation. Depreciation—like other business expenses—is subtracted from sales revenues to determine the amount of the firm's profit or loss. By just changing the way depreciation is calculated, a firm can decrease expenses and increase profits. That's exactly what happened a few years ago with General Motors. The company reported profits of $4.9 billion, and yet, at least $790 million of the auto giant's profits came not from selling automobiles, but from increasing the useful life from 35 to 45 years for some buildings and other assets. Because the useful life was extended, annual depreciation expenses decreased *and* profits increased. Unfortunately, the majority of stockholders never realized where the increase in profits came from because they never examined the footnotes.

*Kenneth L. Fisher, "Thanks, Dad," *Forbes,* August 8, 1988, p. 122. For more information, see Howard M. Schilit, "Financial Reports: Tricks Used to Fool Investors," *Consumers' Research,* December 1993, pp. 10–14; Janet Bodnar, "How to Play Detective with Shareholder Reports," *Kiplinger's Personal Finance Magazine,* October 1993, p. 122+; Jason Zweig and John Chamberlain, "Windbag Theory," *Forbes,* August 3, 1992, pp. 43–44; and Kenneth L. Fisher, "Thanks, Dad," *Forbes,* August 8, 1988, p. 122.

UNIT 58   CALCULATING STRAIGHT-LINE DEPRECIATION

LEARNING OBJECTIVES   *After completing this unit, you will be able to:*

1. determine annual depreciation expense using the straight-line method.
2. calculate book value when the straight-line method of depreciation is used.
3. calculate depreciation for a partial year.

Unlike most books, the best place to start for the real "meat" of the annual report is not the beginning, but in the footnotes buried in the back of the report. Only by examining the footnotes contained in the General Motors annual report would you discover that a change in the method used to calculate depreciation inflated the firm's reported profits by $790 million. And while most firms are neither as large as General Motors nor have as many assets to depreciate, managers of smaller firms have to make the same types of decisions when choosing a method for calculating depreciation expense.

For an asset with a useful life of one year or less, the cost of the asset is deducted as a business expense in the year it is purchased. For an asset that will be used for more than one year (a building, a piece of equipment or machinery, an automobile), the cost of the item is distributed over the life of the item. The process of distributing the cost of an asset over the life of the item is called **depreciation.** Depreciation calculations are especially important because depreciation expense, along with other business expenses like salaries, utilities, rent, and advertising, must be subtracted from income to determine a firm's profit or loss. This unit covers the straight-line method of depreciation. The remaining units of Chapter 11 cover four other methods that can be used to calculate depreciation expense.

Regardless of the method used, there are some terms that you must know before working depreciation problems. The **original cost** is the total amount paid for an asset. The original cost also includes freight charges and any installation costs necessary to make the asset operational. The **useful life** is the estimated length of time an asset will be used. The **salvage value** is the estimated value of an asset at the end of its useful life.

If you know the original cost, useful life, and salvage value, you can use the following formula to determine the straight-line depreciation expense:

$$\text{Annual depreciation expense} = \frac{\text{Original cost} - \text{Salvage value}}{\text{Useful life}}$$

EXAMPLE 1   Calculating Depreciation Using the Straight-Line Method

Apple Tree Children's Care purchased a new Chevrolet van for $14,875. The van has a salvage value of $2,875 and a useful life of five years. What is the annual depreciation expense?

Annual depreciation expense

$= \dfrac{\text{Original cost} - \text{Salvage value}}{\text{Useful life}}$

STEP 1   State the formula for straight-line depreciation.

$= \dfrac{\$14,875 - 2,875}{\text{Useful life}}$

STEP 2   Subtract the salvage value ($2,875) from the original cost ($14,875).

$\$12,000 \div 5 = \$2,400$

STEP 3   Divide the answer from step 2 ($12,000) by the useful life (five years).

CALCULATOR TIP

Depress clear key
Enter 14,875   Depress   –
Enter 2,875   Depress   =
Read the answer $12,000
Enter 12,000   Depress   ÷
Enter 5   Depress   =
Read the answer $2,400

One of the advantages of using the straight-line method is that the annual depreciation expense is the same each year. In Example 1, the annual depreciation expense ($2,400) is the same for the first, second, third, fourth, and fifth years.

The **book value** of an asset is the original cost minus the total depreciation to date. In Figure 58–1, the book value for the first, second, third, fourth, and fifth years is determined by subtracting total depreciation at the end of each year from the original cost. At the end of year five, the book value for the van in Example 1 is equal to the salvage value. The book value can *never* be less than the salvage value of an asset.

After annual depreciation expense and book values have been calculated, it is possible to construct a depreciation schedule like the one illustrated in Figure 58–1.

DEPRECIATION FOR A PARTIAL YEAR

Business firms purchase assets when they are needed. This means that it is often impossible to calculate a full year's depreciation during the first year. To

332

CHAPTER ELEVEN

FIGURE 58–1 Depreciation Schedule for a Van with an Original Cost of $14,875, with a Salvage Value of $2,875, and a Useful Life of Five Years

End of Year	Annual Depreciation	Total Depreciation	Book Value
1	$2,400	$ 2,400	$12,475
2	$2,400	$ 4,800	$10,075
3	$2,400	$ 7,200	$ 7,675
4	$2,400	$ 9,600	$ 5,275
5	$2,400	$12,000	$ 2,875

calculate a partial year's depreciation, divide the annual depreciation by 12—the number of months in a calendar year. Then multiply the answer by the number of months the asset is used in the first year.

EXAMPLE 2 Calculating Depreciation for a Partial Year

Jensen Decorative Arts purchased a new telephone system that cost $2,100. The salvage value is $300, and the system has a useful life of five years. The system was placed in service on June 1. What is the depreciation expense and first-year book value at the end of the first year?

CALCULATOR TIP

Depress clear key
Enter 2,100 Depress −
Enter 300 Depress =
Read the answer $1,800
Enter 1,800 Depress ÷
Enter 5 Depress =
Read the answer $360
Enter 360 Depress ÷
Enter 12 Depress =
Read the answer $30
Enter 30 Depress ×
Enter 7 Depress =
Read the answer $210
Enter 2,100 Depress −
Enter 210 Depress =
Read the answer $1,890

$\frac{\$2,100 - 300}{5} = \$360$ **STEP 1** Determine the straight-line depreciation expense.

$\$360 \div 12 = \$30$ **STEP 2** To determine the depreciation expense for one month, divide the annual depreciation expense from step 1 ($360) by 12—the number of months in a calendar year.

$\$30 \times 7 = \$210$ **STEP 3** Multiply the answer to step 2 ($30) by the number of months the asset is used the first year (7).

$\begin{array}{r} \$2,100 \\ -\ 210 \\ \hline \$1,890 \end{array}$ **STEP 4** To determine the first-year book value, subtract the first-year depreciation expense in step 3 ($210) from the original cost ($2,100).

Use the Straight-Line Method to Complete the Following Problems

For all problems that do not work out evenly, carry your answers to two decimal places.

1. In your own words, define original cost. _____

_____

2. In your own words, define useful life. _____

_____

3. In your own words, define salvage value. _____

_____

4. What is the formula for calculating straight-line depreciation? _____

_____

For problems 5 through 21, use the straight-line method of depreciation to determine the annual depreciation expense.

Original Cost	Useful Life	Salvage Value	Annual Depreciation Expense
$ 4,000	10 years	$500	**5.** $\dfrac{\$4,000 - \$500}{10} = \dfrac{\$3,500}{10} = \$350$
$13,500	8 years	–0–	**6.** _____
$ 1,400	5 years	$200	**7.** _____
$ 5,250	6 years	$800	**8.** _____
$22,454	8 years	$950	**9.** _____
$ 3,900	5 years	$200	**10.** _____

11. Exacta Manufacturing purchased a drill press for $975. The machine has a useful life of eight years and a salvage value of $135. If the firm uses the straight-line method of depreciation, what is the annual depreciation expense?

12. In problem 11, what is the book value for the drill press at the end of the first year?

13. In problem 11, what is the book value for the drill press at the end of the second year?

14. In problem 11, what is the book value for the drill press at the end of the eighth year?

15. Northwestern Development Company purchased a 22-unit apartment complex for $440,000. Northwestern's accountants suggest that the firm use the straight-line method of depreciation to depreciate the apartments over a 20-year period. The accountants also estimate that the apartments will have a salvage value of $62,000. What is the annual depreciation expense?

16. In problem 15, what is the book value at the end of the first year?

17. In problem 15, what is the book value at the end of the fifth year?

18. In problem 15, what is the *total* amount of depreciation that Northwestern Development is entitled to at the end of 20 years?

19. Globe Exports purchased a new computer system that cost $2,150. The salvage value is $250, and the system has a useful life of five years. The system was placed in service on April 1. What is the depreciation expense for the period April 1 through December 31?

20. Contemporary Fashions purchased new store display equipment that cost $12,400. The salvage value is $1,000 and the equipment has a useful life of eight years. The equipment was placed in service on August 1. What is the depreciation expense for the period August 1 through December 31?

21. In problem 20, what is the book value at the end of the first year?

UNIT 59	COMPUTING UNITS-OF-PRODUCTION DEPRECIATION

LEARNING OBJECTIVES   *After completing this unit, you will be able to:*

1. determine annual depreciation expense using the units-of-production method.
2. calculate book value when the units-of-production method is used.

When an asset has a useful life expressed in number of units produced, number of hours used, or number of miles driven, the units-of-production method can be used to calculate depreciation expense. This method is similar to the straight-line method of depreciation discussed in Unit 58, but there are two differences. First, with the units-of-production method, useful life is expressed in number of units produced (or driven) instead of number of years. Second, the amount of depreciation per unit is multiplied by the units produced in a 12-month period.

The depreciation per unit can be determined by using the following formula:

$$\text{Depreciation per unit} = \frac{\text{Original cost} - \text{Salvage value}}{\text{Useful life (in units produced)}}$$

EXAMPLE 1   Calculating Depreciation Using the Units-of-Production Method

Mikaska Manufacturing, Inc., purchased a commercial metal press for $54,900. The press has a salvage value of $2,900. It is expected that this machine will produce 200,000 units during its normal life. What is the depreciation amount if the machine produces 34,000 units in 1995?

$$\frac{\$54,900 - \$2,900}{\text{Useful life (in units produced)}}$$

STEP 1   Subtract the salvage value from the original cost.

$$\$52,000 \div 200,000 = \$0.26$$

STEP 2   To determine the depreciation per unit, divide the answer from step 1 ($52,000) by the useful life expressed in units (200,000).

Annual depreciation expense $= \$0.26 \times 34,000$

$= \$8,840$

STEP 3   To determine the annual depreciation expense, multiply the depreciation per unit from step 2 ($0.26) by the actual number of units produced in 1995 (34,000).

CALCULATOR TIP

Depress clear key
Enter 54,900   Depress −
Enter 2,900   Depress =
Read the answer 52,000
Enter 52,000   Depress ÷
Enter 200,000   Depress =
Read the answer $0.26
Enter 0.26   Depress ×
Enter 34,000   Depress =
Read the answer 8,840

It is now possible to construct a depreciation schedule like the one illustrated in Figure 59–1. Note that the schedule shows the number of units

---

FIGURE 59–1   Depreciation Schedule for a Commercial Metal Press with an Original Cost of $54,900, with a Salvage Value of $2,900, and a Useful Life of 200,000 Units

End of Year	Units Produced Each Year	Depreciation Expense for the Year	Total Depreciation	Book Value
1	34,000	$ 8,840	$ 8,840	$46,060
2	79,300	$20,618	$29,458	$25,442
3	41,200	$10,712	$40,170	$14,730
4	45,500	$11,830	$52,000	$ 2,900

---

produced each year, depreciation expense, and the book value at the end of each year. The book value for the first, second, third, and fourth years is determined by subtracting total depreciation at the end of each year from the original cost. Like straight-line depreciation, the book value can *never* be less than the salvage value of the asset.

Use the Units-of-Production Method to Complete the Following Problems

For all problems that do not work out evenly, carry your answer to two decimal places.

1. A machine cost $7,000. It has no salvage value and has a useful life of 350,000 units. What is the depreciation amount for each unit produced?

$$\frac{\$7,000}{350,000} = \$.02$$

2. In problem 1, what is the depreciation amount if the machine produces 31,300 units during the first year?

3. A machine that costs $3,520 has a useful life of 32,000 hours. It has no salvage value. What is the depreciation amount for each hour the machine is used?

4. In problem 3, what is the depreciation amount if the machine is used 4,410 hours during the second year?

5. An automobile cost $13,470. It has a salvage value of $700 and a useful life of 100,000 miles. What is the depreciation amount for each mile driven? (Carry your answer to three decimal places.)

6. In problem 5, what is the depreciation amount if the car is driven 21,420 miles during the first year?

Given the following information, prepare a depreciation schedule based on the units-of-production method.

Pace Moving & Storage purchased a new moving van that cost $23,450. The salvage value is $2,450. The useful life for the van is 150,000 miles.

End of Year	Miles Driven	Annual Depreciation Expense	Total Depreciation	Book Value
1	32,734	7. *$4,582.76*	8. *$4,582.76*	9. *$18,867.24*
		32,734 × $.14 = $4,582.76		
2	56,210	10. _____	11. _____	12. _____
3	40,330	13. _____	14. _____	15. _____
4	20,726	16. _____	17. _____	18. _____

Critical Thinking Problem

James Slovak purchased a $\frac{3}{4}$ ton pickup truck for use in his landscaping business. The truck cost $14,450. James estimates that the salvage value for the truck is $2,350. He also estimates that the truck will last five years and will probably be driven 110,000 miles.

19. If James chooses the units-of-production method and drives the truck 21,430 miles the first year, what is the amount of depreciation for the first year?

20. If James chooses the straight-line method, what is the annual depreciation expense?

21. If you were Mr. Slovak, which depreciation method would you choose? Why?

_____

_____

_____

_____

_____

_____

UNIT 60	USING THE SUM-OF-THE-YEARS'-DIGITS METHOD OF DEPRECIATION

LEARNING OBJECTIVES

After completing this unit, you will be able to:

1. determine annual depreciation expense using the sum-of-the-years'-digits method.
2. calculate the book value when the sum-of-the-years'-digits method is used.

Although a number of business owners like the straight-line and the units-of-production methods of depreciation because they are simple, those methods may not be the best methods for determining depreciation expense. Because some assets—like automobiles—lose more value during the early years of their useful life, the sum-of-the-years'-digits method may be a more accurate method of matching depreciation expense with this loss of value.

The sum-of-the-years'-digits method is an accelerated method of depreciation because it assumes a greater amount of depreciation in the first year and a progressively smaller amount of depreciation in each remaining year. Like the straight-line and units-of-production methods, this method is based on the original cost, salvage value, and useful life of an asset. It is called the sum-of-the-years'-digits method because you must find the sum of the digits for the years of the asset's useful life. For example, the sum of the years' digits for an asset with a useful life of five years would be 15 (5 years + 4 years + 3 years + 2 years + 1 year = 15 years).[1]

The first step in calculating annual depreciation expense with this method is to form a depreciation fraction. The denominator is the sum of the years' digits. Taking the previous example of a five-year useful life, the denominator would be 15. The numerator (top number) in the depreciation fraction is the number of years remaining in the asset's useful life. For an asset with a useful life of five years, the following numerators would be used:

---

[1] To determine sum of the years for assets with a longer useful life, it may be easier to use the formula

$$\frac{n\,(n+1)}{2}$$

where *n* is the useful life of the asset. For example, the sum of the years for an asset with a useful life of 20 years is 210:

$$\frac{n\,(n+1)}{2} = \frac{20\,(20+1)}{2} = 210.$$

Year	Years Remaining	Numerator
First	5	5
Second	4	4
Third	3	3
Fourth	2	2
Fifth	1	1

In other words, for an asset with a useful life of five years, the depreciation fractions for each year would be as follows:

Years of Use	Depreciation Fraction
Year 1	$\frac{5}{15}$
Year 2	$\frac{4}{15}$
Year 3	$\frac{3}{15}$
Year 4	$\frac{2}{15}$
Year 5	$\frac{1}{15}$

Now you can determine annual depreciation expense by using the following formula:

$$\text{Annual depreciation expense} = \left(\text{Original cost} - \text{Salvage value} \right) \times \text{Depreciation fraction}$$

EXAMPLE 1 Calculating Depreciation Using the Sum-of-the-Year's-Digits Method

Four Corners Delivery Service purchased a new delivery vehicle that cost $22,410. The vehicle has a salvage value of $1,800 and an estimated life of five years.

$22,410
− 1,800
$20,610

STEP 1 Subtract the salvage value ($1,800) from the original cost ($22,410).

5 + 4 + 3 + 2 + 1 = 15

STEP 2 Determine the denominator (bottom number) for the depreciation fraction.

First year = 5 years remaining

STEP 3 Determine the numerator (top number) for the depreciation fraction.

$\frac{5}{15}$

STEP 4 Form the depreciation fraction for the first year.

First year depreciation expense = $20,610 × $\frac{5}{15}$

= $6,870

STEP 5 Multiply the answer from step 1 ($20,610) by the depreciation fraction ($\frac{5}{15}$).

To determine the annual depreciation expense for the second, third, fourth, and fifth year, multiply the difference between original cost and salvage value by the appropriate depreciation fraction.

FIGURE 60–1   Depreciation Schedule for a Delivery Vehicle That Cost $22,410 with a Salvage Value of $1,800 and an Estimated Life of Five Years

End of Year	Depreciation Fraction	Annual Depreciation	Total Depreciation	Book Value
1	$\frac{5}{15}$	$6,870	$ 6,870	$15,540
2	$\frac{4}{15}$	$5,496	$12,366	$10,044
3	$\frac{3}{15}$	$4,122	$16,488	$ 5,922
4	$\frac{2}{15}$	$2,748	$19,236	$ 3,174
5	$\frac{1}{15}$	$1,374	$20,610	$ 1,800

It is also possible to construct a depreciation schedule like the one illustrated in Figure 60–1. Like the straight-line and units-of-production methods, book value for the first, second, third, fourth, and fifth years is found by subtracting total depreciation at the end of each year from the original cost. As always, the salvage value is a bottom amount; book value cannot fall below it.

Use the Sum-of-the-Years'-Digits Method to Complete the Following Problems

For all problems that do not work out evenly, carry your answer to two decimal places.

1. For an asset with a useful life of four years, what is the depreciation fraction for the first year?

    $\frac{4}{10}$

2. For an asset with a useful life of six years, what is the depreciation fraction for the second year?

3. For an asset with a useful life of 10 years, what is the depreciation fraction for the third year?

4. For an asset with a useful life of 20 years, what is the depreciation fraction for the fifth year?

For problems 5 through 12, determine the depreciation expense for the year indicated.

Original Cost	Useful Life	Salvage Value	Year of Depreciation		Annual Depreciation Expense
$ 1,800	4 years	–0–	1st year	**5.**	$\frac{4}{10} \times \frac{\$1,800}{1} = \$720$
2,500	5 years	$100	2nd year	**6.**	
$14,700	10 years	–0–	1st year	**7.**	
$ 560	4 years	–0–	3rd year	**8.**	
$ 3,900	8 years	$300	4th year	**9.**	
$ 1,100	5 years	$100	3rd year	**10.**	
$22,400	10 years	$400	1st year	**11.**	
$35,600	20 years	–0–	5th year	**12.**	

Use the following information to answer problems 13 through 20.

The Western Corporation purchased a new forklift for $26,500. It has a salvage value of $2,000 and a useful life of four years. The company uses the sum-of-the-years'-digits method for calculating depreciation.

13. What is the first-year depreciation expense?

14. What is the book value at the end of the first year?

15. What is the second-year depreciation expense?

16. What is the second-year book value?

17. What is the third-year depreciation expense?

18. What is the third-year book value?

19. What is the fourth-year depreciation expense?

20. What is the fourth-year book value?

Critical Thinking Problem

Jana Wade purchased a new car that will be used for business purposes. The car cost $16,400. Jana estimates that the salvage value is $2,000 and the useful life is five years or 90,000 miles. During the first year, she drove the car 31,250 miles. Ms. Wade is trying to decide whether to use the straight-line method, units-of-production method, or sum-of-the-years'-digits method.

21. If the goal is to maximize depreciation expense the first year, which method should Ms. Wade use?

22. Of the three depreciation methods listed above, which method would you use if you were Ms. Wade? Why? _____

_____

_____

| UNIT 61 | USING THE DECLINING-BALANCE METHOD OF DEPRECIATION |

LEARNING OBJECTIVES   *After completing this unit, you will be able to:*

1. determine annual depreciation expense using the declining-balance method.
2. calculate the book value when the declining-balance method is used.

Like the sum-of-the-years'-digits method (Unit 60), the declining-balance method of depreciation is an accelerated method of depreciation. That is, more depreciation is assumed in the early years of an asset's life than in the latter years. And like the sum-of-the-years' digits method, this method matches the loss of value for some assets—like automobiles—which lose more value during the early years of useful life. The declining-balance method differs from other depreciation methods in two important ways:

1. Salvage value is *not* subtracted from the original cost when calculating depreciation amounts for each yearly amount. However, the salvage value is still a "bottom amount" that the book value cannot fall below. That is, you cannot depreciate the asset below the salvage value.
2. The declining-balance method uses up to twice (200 percent) the straight-line rate to generate larger amounts of depreciation in the early years of an asset's life. Today, it is also common to use 150 percent or 125 percent of the straight-line rate when using the declining-balance method.

To work a depreciation problem with the declining-balance method, you must use the following steps:

STEP 1　Divide 100 percent by the number of years the asset will be used (useful life) to determine the straight-line rate of depreciation.

STEP 2　If the maximum depreciation amount is desired, multiply the straight-line rate by 2 to obtain the declining-balance rate. If less than the maximum depreciation amount is desired, multiply the straight-line rate by either 1.5 or 1.25.

STEP 3　Multiply the original cost of the asset by the declining-balance rate to obtain the first-year depreciation expense.

STEP 4　Subtract the first-year depreciation amount from the original cost to determine the first-year book value.

EXAMPLE 1　Calculating Depreciation Using the Declining-Balance Method

Fountain Head Resorts purchased a new computer system for $3,400. It has a salvage value of $400 and a useful life of five years. Using the declining-balance method, determine the depreciation amounts and book values for each of the five years the computer system is used.

$$\frac{20\%}{5)\overline{100\%}}$$

STEP 1　Divide 100 percent by the number of years the asset will be used (five) to determine the straight-line rate of depreciation.

$$\begin{array}{r} 20\% \\ \times\ 2 \\ \hline 40\% \end{array}$$

STEP 2　Multiply the straight-line rate (20 percent) by 2 to obtain the declining-balance rate.

$$\begin{array}{r} \$3,400 \\ \times\ .40 \\ \hline \$1,360 \end{array}$$

STEP 3　Multiply the original cost ($3,400) by the declining-balance rate (40 percent or 0.40) to obtain the **first-year depreciation amount.**

$$\begin{array}{r} \$3,400 \\ -1,360 \\ \hline \$2,040 \end{array}$$

STEP 4　Subtract the first-year depreciation ($1,360) from the original cost to determine the **first-year book value.**

NOTE: For the remaining useful life of the asset, steps 3 and 4 are repeated *but* the preceding year's book value is substituted for the original cost.

$$\begin{array}{r} \$2,040 \\ \times\ 0.40 \\ \hline \$\ \ 816 \end{array}$$

STEP 5　Multiply the first-year book value ($2,040) by the declining-balance rate (40 percent) to determine the **second-year depreciation.**

$$\begin{array}{r} \$2,040 \\ -\ \ 816 \\ \hline \$1,224 \end{array}$$

STEP 6　Subtract the second year depreciation ($816) from the first-year book value ($2,040) to determine the **second-year book value.**

$$\begin{array}{r} \$1,224.00 \\ \times\ \ \ 0.40 \\ \hline \$\ \ 489.60 \end{array}$$

STEP 7　Multiply the second-year book value ($1,224) by the declining-balance rate (40 percent) to determine the **third-year depreciation.**

$$\begin{array}{r} \$1,224.00 \\ -\ \ 489.60 \\ \hline \$\ \ 734.40 \end{array}$$

STEP 8　Subtract the third year depreciation ($489.60) from the second-year book value ($1,224) to determine the **third-year book value.**

CALCULATOR TIP

Depress clear key
Enter 100　　Depress ÷
Enter 5　　　Depress =
Read the answer 20%
Enter 20　　Depress ×
Enter 2　　　Depress =
Read the answer 40%
Enter 3,400　Depress ×
Enter 0.40　Depress =
Read the answer $1,360
Enter 3,400　Depress –
Enter 1,360　Depress =
Read the answer $2,040

To complete the problem, the last two steps are repeated, but the preceding year's book value is substituted for the original cost.

$734.40 × 0.40 $293.76	STEP 9	Multiply the third-year book value ($734.40) by the declining-balance rate (40 percent) to determine the **fourth-year depreciation.**
$734.40 −293.76 $440.64	STEP 10	Subtract the fourth-year depreciation ($293.76) from the third-year book value ($734.40) to determine the **fourth-year book value.**
$440.64 −400.00 $ 40.64	STEP 11	**CAUTION: If another year's depreciation is taken, the book value will fall below the salvage value. Therefore, to calculate the fifth-year depreciation amount, the salvage value ($400) is subtracted from the fourth-year book value ($440.64).**
$440.64 − 40.64 $400.00	STEP 12	Subtract the fifth-year depreciation ($40.64) from the fourth-year book value ($440.64) to determine the **fifth-year book value.**

Use the Declining-Balance Method to Complete the Following Problems

For all problems that do not work out evenly, carry your answer to two decimal places. Use two times the straight-line rate.

For problems 1 through 5, find the declining-balance depreciation rate.

Useful life	Declining-Balance Depreciation Rate
10 years	1. _____ *100% ÷ 10 = 10%* _____ *10% × 2 = 20%*
8 years	2. _____
5 years	3. _____
20 years	4. _____
4 years	5. _____

For problems 6 through 9, use the declining balance method to find the first-year depreciation expense.

Original Cost	Useful Life	Depreciation Rate	First-Year Depreciation Expense
$ 3,300	5 years	100% ÷ 5 = 20% 20% × 2 = 40%	6. $3,300 × .40 = $1,320
$ 4,850	10 years	_____	7. _____
$ 2,750	4 years	_____	8. _____
$13,200	20 years	_____	9. _____

For problems 10 through 17, use the declining-balance method to find the first-year depreciation expense and the first-year book value.

Original Cost	Useful Life	Depreciation Rate	First-Year Depreciation Expense	First-Year Book Value
$ 1,500	5 years	40% 100% ÷ 5 = 20% 20% × 2 = 40%	10. $1,500 × .40 = $600	11. $1,500 – $600 = $900
$ 8,250	10 years	_____	12. _____	13. _____
$12,800	8 years	_____	14. _____	15. _____
$ 2,970	4 years	_____	16. _____	17. _____

Given the following information, answer problems 18 through 25.

Julia Scott purchased an automobile for business use. The car cost $14,500 with a salvage value of $1,500 and an estimated life of four years. Use the declining-balance method.

18. What is the first-year depreciation expense?

100% ÷ 4 = 25%   *$14,500 × 0.50 = $7,250*
25% × 2 = 50%

19. What is the first-year book value?

20. What is the second-year depreciation expense?

21. What is the second-year book value?

22. What is the third-year depreciation expense?

23. What is the third-year book value?

24. What is the fourth-year depreciation expense?

25. What is the fourth-year book value?

UNIT 62 **USING THE MODIFIED ACCELERATED COST RECOVERY SYSTEM**

LEARNING OBJECTIVES *After completing this unit, you will be able to:*

1. determine annual depreciation expense using the modified accelerated cost recovery system.
2. calculate book value when the modified accelerated cost recovery system is used.

TABLE 62–1   Recovery Classes for MACRS for Property Placed in Service after December 31, 1986

3-Year Class   Certain horses and specialized manufacturing tools.

5-Year Class   Automobiles and light trucks, typewriters, calculators, copiers, computers and duplication equipment, any semiconducting manufacturing equipment, and certain research and development equipment.

7-Year Class   Office furniture, appliances, carpets, and furniture used in residential rental property, and *any property that does not have a specified class life.*

10-Year Class   Ships, barges, and other water transportation equipment and certain agricultural buildings.

15-Year Class   Municipal wastewater treatment plants and certain telephone equipment.

20-Year Class   Municipal sewers.

27.5-Year Class   Residential rental property.

31.5-Year Class   Nonresidential rental real estate placed in service before May 13, 1993.

39-Year Class   Nonresidential real estate placed in service after May 12, 1993.

The first four depreciation methods—straight-line, units-of-production, sum-of-the-years'-digits, and declining-balance—are generally used for financial accounting purposes. Thus, when a firm reports financial information to stockholders, bankers, suppliers, or prospective investors, they use the methods we have already covered in Chapter 11. *But when a firm prepares its tax return, it must use a different method to calculate depreciation expense.*

The Tax Reform Act of 1986—the last major change in tax law—requires that businesses use the *modified accelerated cost recovery system (MACRS)* to depreciate assets that were placed in service after December 31, 1986.* When using this method—sometimes referred to as the federal income tax method—you must be concerned with the following two factors:

1. the specific recovery classification for each asset that you want to depreciate.
2. the MACRS depreciation rates for each asset you want to depreciate.

MACRS RECOVERY CLASSIFICATIONS

To use the MACRS, business assets are generally placed in one of the nine recovery classifications illustrated in Table 62–1. Note that any property not specifically placed in one of the nine recovery classifications is placed in the *7-year class.*

MACRS DEPRECIATION RATES

Special depreciation tables available from the Internal Revenue Service are used to determine annual depreciation expense—See Table 62–2. Note that

---

*Business firms may make an *irrevocable* decision to use the straight-line method of depreciation instead of the MACRS. Although the straight-line method may be easier to use, the MACRS generates greater depreciation expense in the early years of an asset's life. And since depreciation expense is used to reduce profits and ultimately taxes, most businesses opt to use the modified accelerated cost recovery system.

TABLE 62–2   Depreciation Rates for MACRS for Property Placed in Service after December 31, 1986

Year	| Recovery Class 3-Year	5-Year	7-Year	10-Year	15-Year	20-Year
1	33.33%	20.00%	14.29%	10.00%	5.00%	3.75%
2	44.45	32.00	24.49	18.00	9.50	7.219
3	14.81	19.20	17.49	14.40	8.55	6.677
4	7.41	11.52	12.49	11.52	7.70	6.177
5		11.52	8.93	9.22	6.93	5.713
6		5.76	8.92	7.37	6.23	5.285
7			8.93	6.55	5.90	4.888
8			4.46	6.55	5.90	4.522
9				6.56	5.91	4.462
10				6.55	5.90	4.461
11				3.28	5.91	4.462
12					5.90	4.461
13					5.91	4.462
14					5.90	4.461
15					5.91	4.462
16					2.95	4.461
17						4.462
18						4.461
19						4.462
20						4.461
21						2.231

the last three recovery classifications (27.5-year class, 31.5-year class, and 39-year class) listed in Table 62–1 are not included in Table 62–2. Depreciation for these three recovery periods is based on the straight-line method discussed in Unit 58.

MACRS depreciation calculations are based on the half-year convention. For practical purposes, this means that it is assumed that an asset was purchased halfway through the year and that only a half year of depreciation is allowable in the first year the asset is placed in service. Thus, the first-year depreciation rate is less than the second-year depreciation rate.

The calculation necessary to find first-year depreciation expense is a practical application of the portion formula presented in Unit 26. As illustrated in Example 1, the annual depreciation expense (portion) is found by multiplying the original cost (the base) by the rate of depreciation (the rate).

EXAMPLE 1   Calculating Depreciation Using MACRS

Rosales Fitness Center, Inc., purchased a new computer system for $2,300. If the modified accelerated cost recovery system is used, what is the first-year depreciation expense?

The recovery class is the 5-year class.

STEP 1   Determine the MACRS recovery classification. (Based on Table 62–1, a computer is included in the 5-year class.)

CALCULATOR TIP

Depress clear key
Enter 2,300   Depress ×
Enter 0.20   Depress =
Read the answer $460

The depreciation rate is 20 percent.

STEP 2   Determine the depreciation rate for the first year. (Based on Table 62–2, the depreciation rate for a 5-year asset is 20 percent.)

$2,300 × 0.20 = $460

STEP 3   To determine the first-year depreciation expense, multiply the original cost ($2,300) by the depreciation rate (20%).

Note that salvage value is ignored when the MACRS is used. Also, note that 20 percent has been converted to the decimal 0.20. To determine the depreciation expense amounts for the remaining years of asset life, multiply the original cost by the appropriate depreciation rate for each year from Table 62–2. Like other depreciation methods, book value is determined by subtracting total depreciation at the end of each year from the original cost. It is now possible to construct a depreciation schedule like the one illustrated in Figure 62–1.

HINT: You may want to review the material on portions in Unit 26 before completing this unit.

Use the Modified Accelerated Cost Recovery System to Complete the Following Problems

For all problems that do not work out evenly, carry your answer to two decimal places. To complete problems in this unit, you may want to use Tables 62–1 and 62–2.

1. T   F   The MACRS is sometimes referred to as the federal income tax method.

2. T   F   Most firms use the MACRS system for financial accounting purposes.

3. T   F   Salvage value is ignored when the MACRS is used.

FIGURE 62–1   Depreciation Schedule for a Computer That Cost $2,300 When MACRS Is Used (5-Year Class)

End of Year	Depreciation Rate (%)	Annual Depreciation	Total Depreciation	Book Value
1	20.00	$460.00	$ 460.00	$1,840.00
2	32.00	736.00	1,196.00	1,104.00
3	19.20	441.60	1,637.60	662.40
4	11.52	264.96	1,902.56	397.44
5	11.52	264.96	2,167.52	132.48
6*	5.76	132.48	2,300.00	.00

*Note that even though a computer is in the 5-year class, depreciation extends into the sixth year because the half-year convention is used with the modified accelerated cost recovery system.

4. What is the recovery classification for an automobile? _____

5. What is the recovery classification for residential rental property? _____

6. What is the recovery classification for office furniture? _____

7. What is the recovery classification for calculators? _____

8. What is the recovery classification for nonresidential real estate placed in service before May 13, 1993? _____

9. What is the first-year depreciation rate for an asset that is in the 3-year recovery class? _____

10. What is the second-year depreciation rate for an asset that is in the 5-year recovery class? _____

11. What is the fifth-year depreciation rate for an asset that is in the 7-year recovery class? _____

12. What is the first-year depreciation rate for an asset that is in the 15-year recovery class? _____

For problems 13 through 16, use MACRS to determine the first-year depreciation expense.

Type of Asset	Original Cost	First-Year Depreciation
Automobile	$11,000	**13.** _.20 × $11,000 = $2,200_
Typewriter	$ 650	**14.** _____
Telephone Equipment	$ 8,700	**15.** _____
Residential Real Estate	$88,200	**16.** _____

17. Mountain Plains Development, Inc., purchased a small office complex for $174,600 on February 10, 1995. If the firm uses the modified accelerated cost recovery system, what is the firm's first-year depreciation expense?

18. In problem 17, what is the book value for the office complex at the end of the first year?

19. Betty and Frank Martines purchased a three-bedroom rental house for $60,000 that they plan to lease for $650 a month. If they use the modified accelerated cost recovery system, what is their first-year depreciation expense?

20. In 1995, Southwest Produce purchased a new light truck that cost $12,500. If the firm uses the modified accelerated cost recovery system, what is the firm's first-year depreciation expense?

21. In problem 20, what is the book value for the truck at the end of the first year?

22. In problem 20, what is the firm's second-year depreciation expense?

23. In problem 20, what is the firm's third-year depreciation expense?

24. In problem 20, what is the firm's fourth-year depreciation expense?

25. In problem 20, what is the firm's fifth-year depreciation expense?

26. In problem 20, what is the firm's sixth-year depreciation expense?

INSTANT REPLAY

UNIT	IMPORTANT POINTS TO REMEMBER	EXAMPLES
UNIT 58 Calculating Straight-Line Depreciation	The following formula can be used to find the straight-line depreciation expense: Annual depreciation expense = $$\frac{\text{Original cost} - \text{Salvage value}}{\text{Useful life}}$$ The *book value* for an asset is the original cost minus the total depreciation to date.	Annual depreciation expense $$= \frac{\text{Original cost} - \text{Salvage value}}{\text{Useful life}}$$ $$= \frac{\$10{,}000 - \$1{,}000}{6}$$ $$= \frac{\$9{,}000}{6}$$ $= \$1{,}500$ $\$10{,}000$ original cost $- \ 1{,}500$ first-year depreciation $\ \$\ 8{,}500$ first-year book value
UNIT 59 Computing Units-of-Production Depreciation	The following two steps can be used to determine units-of-production depreciation: **STEP 1** Depreciation per unit $$= \frac{\text{Original cost} - \text{Salvage value}}{\substack{\text{Useful life}\\ \text{(in units produced)}}}$$ **STEP 2** Annual depreciation expense = Depreciation per unit × units produced	**STEP 1** Depreciation per unit $$= \frac{\text{Original cost} - \text{Salvage value}}{\text{Useful life (units)}}$$ $$= \frac{\$30{,}000 - 2{,}000}{100{,}000}$$ $$= \frac{\$28{,}000}{100{,}000}$$ $= \$0.28$ **STEP 2** Annual depreciation expense $= \$0.28 \times 25{,}000$ **units** $= \$7{,}000$
UNIT 60 Using the Sum-of-the-Years'-Digits Method of Depreciation	When using the sum-of-the-years'-digits method, the first step is to make a depreciation fraction. The denominator is the sum of the years' digits. The numerator in the depreciation fraction is the number of years remaining in the asset's useful life. Then the following formula can be used to determine depreciation: Annual depreciation expense = $$\left(\text{Original cost} - \text{Salvage value} \right) \times \text{Depreciation fraction}$$	Annual depreciation expense = $$\left(\text{Original cost} - \text{Salvage value} \right) \times \text{Depreciation fraction}$$ The asset has a useful life of 5 years. $= (\$15{,}000 - \$3{,}000) \times \frac{5}{15}$ $= \$12{,}000 \times \frac{5}{15}$ $= \$4{,}000$

continued from page 353

UNIT	IMPORTANT POINTS TO REMEMBER	EXAMPLES
UNIT 61 Using the Declining-Balance Method of Depreciation	To work a depreciation problem with the declining-balance method: (1) divide 100 percent by the number of years the asset will be used to obtain the straight-line rate; (2) multiply the straight-line rate by 2 or by 1.5 or by 1.25 to obtain the declining-balance rate; (3) multiply the original cost by the declining-balance rate; and (4) subtract the first-year depreciation amount from the original cost. For the remaining useful life of the asset, steps 3 and 4 are repeated *but* the preceding year's book value is substituted for the original cost.	(1) **100% ÷ 10 years = 10%** (2) **10% × 2 = 20%** (3) **$30,000 × 0.20 = $6,000** first-year depreciation (4) **$30,000 − $6,000 = $24,000** first-year book value
UNIT 62 Using the Modified Accelerated Cost Recovery System	To determine annual depreciation expense with MACRS, (1) determine the recovery class for the asset using Table 62–1, (2) determine the depreciation rate using Table 62–2, and (3) multiply the original cost by the depreciation rate.	Depreciate a copier that cost $1,200. (1) This asset is in the 5-year class. (2) The depreciation rate for the first year is 20 percent. (3) $ 1,200 original cost × 0.20 depreciation rate $240.00 first-year depreciation expense

CHAPTER 11	MASTERY QUIZ

DIRECTIONS *Round each answer to two decimal places or hundredths. (Each answer counts 5 points.)*

1. T F The salvage value of an asset is the total amount paid for the asset.

2. T F Both the sum-of-the-years'-digits and the declining-balance methods are accelerated methods of depreciation.

3. Quality Instruments purchased a photocopy machine for $1,250. The machine has a useful life of eight years and a salvage value of $100. If the straight-line method of depreciation is used, what is the first-year depreciation expense?

4. In problem 3, what is the first-year book value?

5. In problem 3, what is the second-year depreciation expense?

6. In problem 3, what is the second-year book value?

7. A new delivery truck cost $22,300. It has a salvage value of $2,500 and a useful life of 180,000 miles. If the units-of-production method is used, what is the depreciation amount for each unit produced?

8. In problem 7, what is the depreciation expense amount if the truck is driven 43,250 miles in one year?

9. If the sum-of-the-years'-digits method is used, what is the depreciation fraction for the first year for an asset with a useful life of seven years?

10. The Ace Printing Company purchased a new press that cost $124,000. The press has a salvage value of $8,500 and a useful life of six years. If the sum-of-the-years'-digits method is used, what is the first-year depreciation expense?

11. In problem 10, what is the first-year book value?

12. In problem 10, what is the second-year depreciation expense?

13. In problem 10, what is the second-year book value?

14. If the declining-balance method of depreciation is used, what is the depreciation rate for an asset with a useful life of 20 years?

15. Judy Guerra purchased a new Dodge minivan for business purposes. The van cost $17,400. It has a salvage value of $2,000 and an estimated life of five years. If the declining-balance method of depreciation is used, what is the first-year depreciation expense?

16. In problem 15, what is the first-year book value?

17. In problem 15, what is the second-year depreciation expense?

18. Boston Travel purchased a new delivery vehicle that cost $13,400. If the firm uses the modified accelerated cost recovery system, what is the firm's first-year depreciation expense? (Use Tables 62–1 and 62–2.)

19. In problem 18, what is the first-year book value?

20. In problem 18, what is the second-year depreciation expense?

CHAPTER 12

Taxes and Insurance

CHAPTER PREVIEW

We begin this chapter with taxes—a topic that is a practical concern for both individuals and business firms. We describe how to calculate sales taxes and property taxes in the first unit. Then in the next two units, we discuss income taxes paid to the federal and state governments. Then our focus shifts to insurance—a method used to protect against potential losses faced by business firms or individuals. Here we describe the topics of health insurance, life insurance, property insurance, and automobile insurance.

MATH TODAY

TAXES AND INSURANCE: WHO NEEDS THEM?

What do taxes and insurance have in common? Quite simply, taxes and insurance cost both businesses and individuals a lot of money. To make matters worse, too often business owners, managers, and individuals don't do a top-notch job of managing each of these areas. Before we begin Chapter 12, let's take a look at the numbers and see how they add up.

TAXATION

In one way or another, each of us helps pay for everything that government does—from regulating business to funding research into the causes and cures of cancer. And it takes a lot of money to run something as big as government. The federal government alone collects over $1,000 billion a year. As illustrated below, this vast sum of money came from a number of different taxes paid by both individuals and businesses.

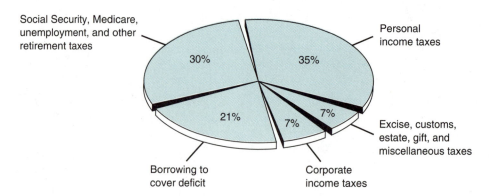

In addition to taxes paid to the federal government, we also pay taxes to 50 state governments, 3,042 county governments, 19,200 cities, 16,691 townships, 14,721 school districts, and 29,532 special taxing districts in this country.

INSURANCE

Americans like insurance. After all, it satisfies a basic need that everyone has—the need to minimize loss if a tragedy occurs. But satisfying this need is not cheap. Americans spend over $180 billion a year on insurance—coverage to protect homes, automobiles, health, and just about anything else that can be insured. Add to that amount over $220 billion paid by employers—over $2,700 per employee—for health care plans. That's a lot of money.

If you are like most people, you may gripe when an insurance bill arrives because the premium has gone up, and yet you write a check to pay the bill. But that may be the wrong approach. The experts suggest that you may save as much as 10 to 20 percent of what you pay for insurance by shopping for coverage. Just becoming familiar with different types of coverage can insure that you have proper protection and at the same time reduce your insurance premiums. According to the experts, the best advice when an overaggressive salesperson is really "pushing" a policy may be to JUST SAY NO.

Source: Based on information from *Your Federal Income Tax for Individuals,* Internal Revenue Service, Department of the Treasury, 1993, p. 351; "Labor Month in Review," *Monthly Labor Review,* February 1994, p. 2; Roger Thompson, "Keeping a Lid on Health Costs," *Nation's Business,* February 1994, pp. 48–49; "Homeowners' Insurance," *Consumer Reports,* October 1993, pp. 627–30: Sheryl Nance-Nash, "Insurance You Don't Need," *Money,* July 1993, pp. 78–79.

UNIT 63	SALES TAXES AND PROPERTY TAXES

LEARNING OBJECTIVES   *After completing this unit, you will be able to:*

1. find the dollar amount of sales tax.
2. determine the amount of property tax for a specific piece of real estate.

As pointed out in the opening case, each of us helps pay for everything that government does. We pay taxes to our local, state, and federal governments on the basis of what we earn, what we own, and even what we purchase. In this unit we cover two important taxes: sales taxes and property taxes.

SALES TAXES

Sales taxes are levied by both states and cities and are paid by the purchasers of consumer products. At the present time, 45 individual states charge sales taxes that range from 3 percent to 7 percent. Alaska, Delaware, Montana, New Hampshire, and Oregon have no statewide sales tax. In addition, most major cities also use sales taxes to finance their ongoing activities.

Retailers collect sales taxes as a specified percentage of the price of each taxed product. There are three commonly used methods to calculate sales tax amounts. First, most new, electronic cash registers can be programmed to automatically calculate sales tax amounts. After individual dollar amounts for each item are entered, the cash register prints a subtotal for all merchandise purchased. Then the cash register calculates the sales tax and prints the amount on the paper tape. Finally, the cash register calculates the total for the transaction.

Second, many retailers use tables to calculate sales tax amounts. Generally, sales tax tables are available from state taxing authorities or office supply stores. A portion of a sales tax table used in the state of Texas is illustrated in Table 63–1. This particular table includes both state and local sales taxes for a combined total tax of $8\frac{1}{4}$ percent. Using Table 63–1, we find the tax on a $3.75 purchase is $0.31 cents, as illustrated in Example 1.

EXAMPLE 1   Find the Sales Tax Amount and Total Cost for a $3.75 Purchase

---

The tax amount is $0.31   **STEP** [1]   In the sales tax table, locate the line for purchases that are at least $3.70 but less than $3.81.

$$\begin{array}{r} \$3.75 \\ + \ 0.31 \\ \hline \$4.06 \end{array}$$

STEP [2]   Add the tax amount found in step 1 ($0.31) to the subtotal for all merchandise purchased ($3.75).

CALCULATOR TIP

Depress clear key
Enter 0.31   Depress +
Enter 3.75   Depress =
Read the answer $4.06

Based on the information in Example 1, the retailer would charge the customer a total of $4.06.

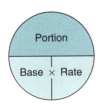

Finally, it is possible to use the portion formula ($P = B \times R$) presented in Unit 26 to determine sales tax. The sales tax amount (portion) can be found by multiplying the subtotal for all merchandise purchased (base) by the sales tax stated as a percent (rate).

EXAMPLE 2   Find the Sales Tax Amount and Total Cost for a $45 Purchase. Assume that the Sales Tax Is 7 Percent.

$45 × 0.07 = $3.15   **STEP** [1]   Multiply the subtotal for all merchandise purchased ($45) by the sales tax stated as a percent (7 percent).

$$\begin{array}{r} \$45.00 \\ + \ 3.15 \\ \hline \$48.15 \end{array}$$

STEP [2]   Add the tax amount found in step 1 ($3.15) to the subtotal for all merchandise purchased ($45).

CALCULATOR TIP

Depress clear key
Enter 45   Depress ×
Enter 0.07   Depress =
Read the answer $3.15
Enter 3.15   Depress +
Enter 45   Depress =
Read the answer $48.15

TABLE 63–1 A Portion of a Sales Tax Chart ($8\frac{1}{4}$ Percent)

Sales Price		Tax	Sales Price		Tax	Sales Price		Tax
0.07–	0.18	0.01	3.70–	3.81	0.31	7.34–	7.45	0.61
0.19–	0.30	0.02	3.82–	3.93	0.32	7.46–	7.57	0.62
0.31–	0.42	0.03	3.94–	4.06	0.33	7.58–	7.69	0.63
0.43–	0.54	0.04	4.07–	4.18	0.34	7.70–	7.81	0.64
0.55–	0.66	0.05	4.19–	4.30	0.35	7.82–	7.93	0.65
0.67–	0.78	0.06	4.31–	4.42	0.36	7.94–	8.06	0.66
0.79–	0.90	0.07	4.43–	4.54	0.37	8.07–	8.18	0.67
0.91–	1.03	0.08	4.55–	4.66	0.38	8.19–	8.30	0.68
1.04–	1.15	0.09	4.67–	4.78	0.39	8.31–	8.42	0.69
1.16–	1.27	0.10	4.79–	4.90	0.40	8.43–	8.54	0.70
1.28–	1.39	0.11	4.91–	5.03	0.41	8.55–	8.66	0.71
1.40–	1.51	0.12	5.04–	5.15	0.42	8.67–	8.78	0.72
1.52–	1.63	0.13	5.16–	5.27	0.43	8.79–	8.90	0.73
1.64–	1.75	0.14	5.28–	5.39	0.44	8.91–	9.03	0.74
1.76–	1.87	0.15	5.40–	5.51	0.45	9.04–	9.15	0.75
1.88–	1.99	0.16	5.52–	5.63	0.46	9.16–	9.27	0.76
2.00–	2.12	0.17	5.64–	5.75	0.47	9.28–	9.39	0.77
2.13–	2.24	0.18	5.76–	5.87	0.48	9.40–	9.51	0.78
2.25–	2.36	0.19	5.88–	5.99	0.49	9.52–	9.63	0.79
2.37–	2.48	0.20	6.00–	6.12	0.50	9.64–	9.75	0.80
2.49–	2.60	0.21	6.13–	6.24	0.51	9.76–	9.87	0.81
2.61–	2.72	0.22	6.25–	6.36	0.52	9.88–	9.99	0.82
2.73–	2.84	0.23	6.37–	6.48	0.53	10.00–	10.12	0.83
2.85–	2.96	0.24	6.49–	6.60	0.54	10.13–	10.24	0.84
2.97–	3.09	0.25	6.61–	6.72	0.55	10.25–	10.36	0.85
3.10–	3.21	0.26	6.73–	6.84	0.56	10.37–	10.48	0.86
3.22–	3.33	0.27	6.85–	6.96	0.57	10.49–	10.60	0.87
3.34–	3.45	0.28	6.97–	7.09	0.58	10.61–	10.72	0.88
3.46–	3.57	0.29	7.10–	7.21	0.59	10.73–	10.84	0.89
3.58–	3.69	0.30	7.22–	7.33	0.60	10.85–	10.96	0.90

Notice that the 7 percent sales tax has been changed to the decimal 0.07. Regardless of the method used to calculate sales tax amounts, retailers must forward the sales taxes they collect to the proper taxing authorities on the required reporting dates.

PROPERTY TAXES

Many cities use the property taxes that are levied on real estate (land, buildings, and houses) to finance the majority of their activities. Property taxes are usually computed as a portion of the assessed value of the real estate. The **assessed value** is determined by the local tax assessor as the fair market value of the property, a portion of its fair market value, or its replacement cost. The **fair market value** is the price at which a piece of property could be sold in the current real estate market. The **replacement cost** is the actual dollar amount needed to replace a building if it were totally destroyed using current construction costs.

Calculating property taxes is a practical application of the portion formula (P = B × R) presented in Unit 26. The property tax amount (portion) can be found by multiplying the assessed value (base) by the tax rate (rate). For example, suppose the city council has established a real estate tax rate of

$1.20 per $100 of assessed value. The tax on a home with an assessed value of $150,000 would be $1,800 as illustrated in Example 3.

EXAMPLE 3   Find the Property Tax Amount for Real Estate with an Assessed Value of $150,000 When the Tax Rate Is $1.20 per $100 of Assessed Value.

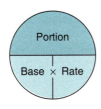

Portion

Base × Rate

$150,000 ÷ $100 = 1,500   **STEP** [1]   Since the tax rate is $1.20 *per $100* of assessed value, divide the assessed value ($150,000) by 100.

1,500 × $1.20 = $1,800   **STEP** [2]   Multiply the answer from step 1 (1,500) by the tax rate ($1.20).

CALCULATOR TIP

Depress clear key
Enter 150,000 Depress ÷
Enter 100   Depress =
Read the answer 1,500
Enter 1,500   Depress ×
Enter 1.20   Depress =
Read the answer $1,800

It is also quite common for property taxes to be calculated using tax rates based on $1,000 units of assessed value. Notice that in Example 4, the portion formula is again used to calculate the property tax amount.

EXAMPLE 4   Find the Property Tax Amount for Real Estate with an Assessed Value of $220,000 When The Tax Rate is $10.15 per $1,000 of Assessed Value.

$220,000 ÷ $1,000 = 220   **STEP** [1]   Since the tax rate is $10.15 *per $1,000* of assessed value, divide the assessed value ($220,000) by 1,000.

220 × $10.15 = $2,233   **STEP** [2]   Multiply the answer from step 2 (220) by the tax rate ($10.15).

It should be pointed out that in many localities it is common for property owners to pay city taxes, school taxes, county taxes, and in some cases taxes created by special taxing authorities like college districts or hospitals. Generally, property taxes are calculated using the same procedures regardless of which taxing authority collects the tax.

Complete the Following Problems

If necessary, round your answer to two decimal places. For problems 1–8, use the tax table illustrated in Table 63–1.

Subtotal of Purchases	Sales Tax Amount	Total Cost
$0.75	1. _____	2. _____
$3.90	3. _____	4. _____
$7.10	5. _____	6. _____
$9.00	7. _____	8. _____

For problems 9–20, assume that the sales tax rate is 7 percent.

Subtotal of Purchases	Sales Tax Amount	Total Cost
$ 9.50	**9.** $9.50 × .07 = $.665 = $.67	**10.** _____
$ 24.00	**11.** _____	**12.** _____
$ 36.80	**13.** _____	**14.** _____
$110.50	**15.** _____	**16.** _____
$149.95	**17.** _____	**18.** _____
$222.30	**19.** _____	**20.** _____

21. Barbara Ruiz purchased cosmetics valued at $72.40 at Sally's Beauty Supply. At the time of the purchase, the sales tax rate was 6.25 percent. What is the dollar amount of sales tax that Ms. Ruiz should pay the merchant?

$72.40 × .0625 = $4.525 = $4.53

22. On Monday, Jeff Mathis purchased two shirts. One shirt cost $24.50. The other shirt cost $19.95. He also purchased three ties that cost $11 each. Assuming that this purchase is subject to a 6 percent sales tax, what is the amount of the sales tax?

23. In problem 22, what is the total that Mr. Mathis should pay the retailer?

24. Assuming that Mr. Mathis pays for this purchase with a $100 bill, how much change should he receive?

25. Mary Turbyfil purchased the following items at Haltom Antique Mall.

    1 lace table cloth for $29.50
    6 stem goblets for $22 each
    1 cut glass pitcher for $110
    1 48-piece set of dishes for $155

Assuming that the merchant must charge a sales tax of 5 percent, what is the dollar amount of the sales tax?

26. In problem 25, what is the total amount that Mary must pay the merchant?

For problems 27–33, assume that the property tax rate is $1.15 per $100 of assessed value.

Assessed Value		Property Tax Amount
$ 50,000	**27.**	_____
$ 78,000	**28.**	_____
$ 91,000	**29.**	_____
$110,500	**30.**	_____
$145,200	**31.**	_____
$156,100	**32.**	_____
$240,300	**33.**	_____

34. Betsy and Marty Carralez own a home in Tucson. The home has an assessed value of $108,000. If the tax rate is $14.10 per $1,000 of assessed valuation, what is the dollar amount of property tax that this couple should pay this year?

35. The local school district has decided to raise the tax rate from $1.24 to $1.32 per $100 of assessed valuation. If a piece of commercial real estate has an assessed valuation of $330,000, based on the old tax rate, what is the amount of property tax that the owner must pay?

36. In problem 35, what is the amount of property tax the owner must pay if the new tax rate is used?

37. In problem 35, what is the total amount of increase for this property owner in the first year the new tax rate is used?

UNIT 64	PERSONAL INCOME TAXES

LEARNING OBJECTIVES

After completing this unit, you will be able to:

1. define the terms used by federal and state taxing authorities.
2. determine the federal income tax liability for an individual.

As pointed out in the opening case, it takes a lot of money to run something as big as government. Each year vast sums are spent for human services, salaries of government employees, operating expenses, and equipment and supplies that range from typewriter ribbons to aircraft. Most of the money comes from taxes.

FEDERAL INCOME TAXES

The individual (or personal) income tax is derived from the Sixteenth Amendment to the Constitution, which was enacted in 1913. It states that "The Congress shall have the power to lay and collect taxes on incomes, from whatever source derived, without apportioning among the several states and without regard to any census or enumeration." In 1914, the federal government collected an average of $0.28 per taxpayer. Today that average is more than $2,100 per person.

Every taxpayer must file an annual tax return by April 15 of each year for the previous calendar year. The process involves computing taxable income and then determining the amount of income tax owed. The following formula can be used to visualize the overall process.

Gross income *LESS* adjustments *LESS* deductions *Equals* taxable income

GROSS INCOME

Gross income is money received by an individual during a calendar year. For individuals, income generally includes wages, salaries, commissions, fees, dividends, and interest. Typically, taxpayers will receive a Form W-2 from employers that reports the amount of earnings, the amount of income tax withheld, and the amount of social security withheld during the tax year. Taxpayers may also receive a Form 1099-DIV and a Form 1099-INT that reports the amount of dividends and interest that were received during the tax year. In addition, there are other types of income that are taxable. For example, capital gains, rents, alimony, lottery winnings, and prizes are taxable in many cases.

EXAMPLE 1   Determine the Total Gross Income for Sherry and Ben Masters

During the past tax year, Sherry Masters earned $24,400. Ben earned $22,200. This couple also received interest that totaled $1,100 and dividends that totaled $580. The couple also won $2,500 in the state lottery. What is the gross income for the Masters?

$24,400	earnings for Sherry Masters
22,200	earnings for Ben Masters
1,100	interest income
580	dividend income
2,500	lottery winnings
$50,780	total gross income

CALCULATOR TIP

Depress clear key
Enter 24,400   Depress   +
Enter 22,200   Depress   +
Enter 1,100   Depress   +
Enter 580   Depress   +
Enter 2,500   Depress   =
Read the answer $50,780

Each of these amounts is itemized in the Income section on an Internal Revenue Form 1040 like the one illustrated in Table 64–1 on page 366.

LESS ADJUSTMENTS TO GROSS INCOME

Adjusted gross income (AGI) is gross income after certain reductions have been made. These reductions include contributions to an individual retirement account (IRA) or a Keogh retirement plan, self-employment tax, self-employment health insurance premiums, penalties for early withdrawal of savings, and alimony payments. All adjustments to gross income are itemized at the bottom of page one of Form 1040.

EXAMPLE 2 | Determine Adjusted Gross Income for Sherry and Ben Masters

---

During the past tax year, the Masters (the couple in Example 1) paid a $400 penalty on early withdrawal from a savings account. What is the adjusted gross income for the Masters?

$50,780	total gross income from Example 1
− 400	minus adjustments to income
$50,380	adjusted gross income

> **CALCULATOR TIP**
>
> Depress clear key
> Enter 50,780   Depress −
> Enter 400   Depress =
> Read the answer $50,380

Once the subtraction is completed, adjusted gross income is reported on the bottom line of the first page of Form 1040—see Table 64–1. This amount is also carried forward to the top line on the second page of Form 1040.

LESS DEDUCTIONS

A **deduction** is an amount subtracted from adjusted gross income to arrive at taxable income. Every taxpayer receives at least the standard deduction—a set amount on which no taxes are paid. As of 1993, single people received a standard deduction of $3,700 (married couples, filing jointly received $6,200). Many individuals qualify for more than the standard deduction if they itemize deductions. Typical itemized deductions include medical and dental expenses, taxes, gifts to charity, casualty and theft losses, moving expenses, allowable job-related expenses, and miscellaneous deductions. Taxpayers use IRS Schedule A to itemize deductions. The total from Schedule A is then reported at the top of the second page of Form 1040 as illustrated in Table 64–1.

In addition to deductions, all taxpayers are also entitled to exemptions. An **exemption** is a deduction from adjusted gross income for yourself, your spouse, and your children and qualified dependents. For 1993, taxable income was reduced by $2,350 for each exemption claimed. Under current tax law, Sherry and Ben Masters are entitled to two exemptions. With this information, it is possible to determine taxable income for the Masters.

EXAMPLE 3 | Determine Taxable Income for Sherry and Ben Masters

---

Because the itemized deductions for the Masters exceed the standard deduction amount, they choose to itemize the deductions listed below. This couple is also entitled to two exemptions. It is now possible to determine their taxable income.

> **CALCULATOR TIP**
>
> Depress clear key
> Enter 2,400   Depress +
> Enter 4,600   Depress +
> Enter 3,100   Depress =
> Read the answer $10,100
> Enter 50,380   Depress −
> Enter 10,100   Depress =
> Read the answer $40,280
> Enter 2,350   Depress ×
> Enter 2   Depress =
> Read the answer $4,700
> Enter 40,280   Depress −
> Enter 4,700   Depress =
> Read the answer $35,580

$ 2,400	taxes	**STEP**	1	Determine the total for itemized deductions.
4,600	interest			
3,100	contributions			
$10,100	total itemized deductions			

$50,380		**STEP**	2	Subtract total itemized deductions from step 1 ($10,100) from adjusted gross income ($50,380).
−10,100				
$40,280				

TABLE 64–1 Sample 1040 Tax Form for Sherry and Ben Masters.

Form **1040**

Department of the Treasury—Internal Revenue Service
U.S. Individual Income Tax Return (O) **1993**

IRS Use Only—Do not write or staple in this space

For the year Jan. 1–Dec. 31, 1993, or other tax year beginning _____ , 1993, ending _____ , 19 ___   OMB No. 1545–0074

Label
(See instructions on page 12.)

Use the IRS label. Otherwise, please print or type.

Your first name and Initial: *Sherry N.*   Last name: *Masters*

If a joint return, spouse's first name and initial: *Ben R.*   Last name: *Masters*

Home address (number and street). If you have a P.O. box, see page 12. *612 Harvest Valley*   Apt. no. *134*

City, town or post office, state, and ZIP code. If you have a foreign address, see page 12. *Houston, TX 77008*

Your social security number **429 51 3706**

Spouse's social security number **583 13 2946**

For Privacy Act and Paperwork Reduction Act Notice, see page 4.

Presidential Election Campaign
(See page 12.)

	Yes	No
Do you want $3 to go to this fund?	X	
If a joint return, does your spouse want $3 to go to this fund?	X	

Note: Checking "Yes" will not change your tax or reduce your refund.

Filing Status
(See page 12.)

Check only one box.

1 ☐ Single
2 ☒ Married filing joint return (even if only one had income)
3 ☐ Married filing separate return. Enter spouse's social security no. above and full name here. ▶
4 ☐ Head of household (with qualifying person). (See page 13.) If the qualifying person is a child but not your dependent, enter this child's name here. ▶
5 ☐ Qualifying widow(er) with dependent child (year spouse died ▶ 19 ___). (See page 13.)

Exemptions
(See page 13.)

If more than six dependents, see page 14.

6a ☒ **Yourself.** If your parent (or someone else) can claim you as a dependent on his or her tax return, do not check box 6a. But be sure to check the box on line 33b on page 2.

b ☒ **Spouse**

No. of boxes checked on 6a and 6b	*1*
No. of your children on 6c who: • lived with you • didn't live with you due to divorce or separation (see page 15)	*1*

c **Dependents:**

(1) Name (first, initial, and last name)	(2) Check if under age 1	(3) If age 1 or older, dependent's social security number	(4) Dependent's relationship to you	(5) No. of months lived in your home in 1993

Dependents on 6c and entered above

d If your child didn't live with you but is claimed as your dependent under a pre-1985 agreement, check here ▶ ☐

Add numbers entered on lines above: *2*

e Total number of exemptions claimed

Income

Attach Copy B of your Forms W-2, W2G, and 1099-R here.

If you did not get a W-2, see page 10.

If you are attaching a check or money order, put it on top of any Forms W-2, W2G, or 1099-R.

7	Wages, salaries, tips, etc. Attach Form(s) W-2	7	46,600	—
8a	**Taxable** interest income (see page 16). Attach Schedule B if over $400	8a	1,100	—
b	**Tax-exempt** interest (see page 17). DON't include on line 8a	8b		
9	Divident income. Attach Schedule B if over $400	9	580	—
10	Taxable refunds, credits, or offsets of state and local income taxes (see page 17)	10		
11	Alimony received	11		
12	Business income or (loss). Attach Schedule C or C-EZ	12		
13	Capital gain or (loss). Attach Schedule D	13		
14	Capital gain distributions not reported on line 13 (see page 17)	14		
15	Other gains or losses. Attach Form 4797	15		
16a	Total IRA distributions 16a ___ b Taxable amount (see page 18)	16b		
17a	Total pensions and annuities 17a ___ b Taxable amount (see page 18)	17b		
18	Rental real estate, royalties, partnerships, S corporations, trusts, etc. Attach Schedule E	18		
19	Farm income or (loss). Attach Schedule F	19		
20	Unemployment compensation (see page 19)	20		
21a	Social security benefits 21a ___ b Taxable amount (see page 19)	21b		
22	Other income. List type and amount—see page 20 *Lottery*	22	2,500	—
23	Add the amounts in the far right column for lines 7 through 22. This is your **total income** ▶	23	50,780	—

Adjustments to Income
(See page 20.)

24a	Your IRA deduction (see page 20)	24a		
b	Spouse's IRA deduction (see page 20)	24b		
25	One-half of self-employment tax (see page 21)	25		
26	Self-employed health insurance deduction (see page 22)	26		
27	Keogh retirement plan and self-employed SEP deduction	27		
28	Penalty on early withdrawal of savings	28	400	—
29	Alimony paid. Recipients's SSN ▶	29		
30	Add lines 24a through 29. These are your **total adjustments** ▶	30	400	—

Adjusted Gross Income

31	Subtract line 30 from line 23. This is your **adjusted gross income.** If this amount is less than $23,050 and a child lived with you, see page EIC-1 to find out if you can claim the "Earned Income Credit" on line 56 ▶	31	50,380	—

Cat. No. 11320B   Form **1040** (1993)

TABLE 64-1 Sample 1040 Tax Form—Concluded

Form 1040 (1993)				Page **2**

Tax Computation

(See page 23.)

32	Amount from line 31 (adjusted gross income)	32	50,380 —
33a	Check if: ☐ **You** were 65 or older, ☐ Blind; ☐ **Spouse** was 65 or older, ☐ Blind.		
	Add the number of boxes checked above and enter the total here . . . . ▶ 33a ☐		
b	If your parent (or someone else) can claim you as a dependent, check here. . . ▶ 33b ☐		
c	If you are married filing separately and your spouse itemizes deductions or you are a dual-status alien, see page 24 and check here . . . . ▶ 33c ☐		
34	Enter the **larger** of your: { **Itemized deductions** from Schedule A, line 26, **OR** **Standard deduction** shown below for yor filing status. **But if you checked any box on line 33a or b**, go to page 24 to find your standard deduction. If you checked **box 33c**, you standard deduction is zero. • Single—$3,700 • Head of household—$5,450 • Married filing jointly or Qualifying widow(er)—$6,200 • Married filing separately—$3,100 }	34	10,100 —
35	Subtract line 34 from line 32	35	40,280 —
36	If line 32 is $81,350 or less, multiply $2,350 by the total number of exemptions claimed on line 6e. If line 32 is over $81,350, see the worksheet on page 25 for the amount to enter	36	4,700 —

If you want the IRS to figure your tax, see page 24.

37	**Taxable Income.** Subtract line 36 from line 35. If line 36 is more than line 35, enter -0-. .	37	35,580 —
38	Tax. Check if from **a** ☒ Tax Table, **b** ☐ Tax Rate Schedules, **c** ☐ Schedule D Tax Work-sheet, or **d** ☐ Form 8615 (See page 25). Amount from Form(s) 8814 ▶ e _____	38	5,336 —
39	Additional taxes (see page 25). Check if from **a** ☐ Form 4970 **b** ☐ Form 4972	39	
40	Add lines 38 and 39 . . . . ▶	40	5,336 —

Credits

(See page 25.)

41	Credit for child and dependent care expenses. Attach Form 2441	41	
42	Credit for the elderly or the disabled. Attach Schedule R . .	42	
43	Foreign tax credit. Attach Form 1116	43	
44	Other credits (see page 26). Check if from **a** ☐ Form 3800 **b** ☐ Form 8396 **c** ☐ Form 8801 **d** ☐ Form (specify) _____	44	
45	Add lines 41 through 44	45	
46	Subtract line 45 from lie 40. If line 45 is more than line 40, enter -0- . . . . ▶	46	5,336 —

Other Taxes

47	Self-employment tax. Attach Schedule SE. Also, see line 25 .	47	
48	Alternative minimum tax. Attach Form 6251	48	
49	Recapture taxes (see page 26). Check if from **a** ☐ Form 4255 **b** ☐ Form 8611 **c** ☐ Form 8828	49	
50	Social security and Medicare tax on tip income not reported to employer. Attach form 4137 .	50	
51	Tax on qualified retirement plans, includint IRAs. If required, attach Form 5329	51	
52	Advance earned income credit payments from Form W-2 . . .	52	
53	Add lines 46 through 52. This is your **total tax** . . . . ▶	53	5,336 —

Payments

Attach Rorms W-2, W2G, and 1099-R on the front.

54	Federal income tax withheld. If any is from Form(s) 1099, check ▶ ☐	54	6,210 —
55	1993 estimated tax payments and amount applied from 1992 return	55	
56	Earned income credit. Attach Schedule EIC	56	
57	Amount paid with Form 4868 (extension request)	57	
58a	Excess social security, Medicare, and RRTA tax withheld (see page 28)	58a	
b	Deferral of additional 1993 taxes. Attach Form 8841 . . .	58b	
59	Other payments (see page 28). Check if from **a** ☐ Form 2439 **b** ☐ Form 4136	59	
60	Add lines 54 through 59. These are your **total payments** . . . . ▶	60	6,210 —

Refund or Amount You Owe

61	If line 60 is more than line 53, subtract line 53 from line 60. This is the amount you **OVERPAID** . ▶	61	874 —
62	Amount of line 61 you want **REFUNDED TO YOU** . . . . ▶	62	874 —
63	Amount of line 61 you want **APPLIED TO YOUR 1994 ESTIMATED TAX** ▶	63	
64	If line 53 is more than line 60, subtract line 60 from line 53. This is the **AMOUNT YOU OWE.** For details on how to pay, including what to write on your payment, see page 29 . . .	64	
65	Estimated tax penalty (see page 29). Also include on line 64	65	

Sign Here

Keep a copy of this return for your records.

Under penalties of perjury, I declare that I have examined this return and accompanying schedules and statements, and to the best of my knowledge and belief, they are true, correct, and complete. Declaration of preparer (other than taxpayer) is based on all information of which preparer has any knowledge.

Your signature	Date	Your occupation
Spouse's signature. If a joint return, BOTH must sign.	Date	Spouse's occupation

Paid Preparer's Use Only

Preparer's signature	Date	Check if self-employed	Preparer's social security no.
Firm's name (or yours if self-employed) and address		E.I. No.	
		ZIP code	

*U.S. Government Printing Office: 1993 — 345-188

TABLE 64–2 Partial Tax Table

1993 Tax Table

If line 6 (Form 1040EZ), line 22 (Form 1040A), or line 37 (Form 1040) is—		And you are—			
At least	But less than	Single	Married filing jointly *	Married filing separately	Head of a household
		Your tax is—			

32,000

At least	But less than	Single	Married filing jointly *	Married filing separately	Head of a household
32,000	32,050	6,094	4,804	6,569	5,119
32,050	32,100	6,108	4,811	6,583	5,133
32,100	32,150	6,122	4,819	6,597	5,147
32,150	32,200	6,136	4,826	6,611	5,161
32,200	32,250	6,150	4,834	6,625	5,175
32,250	32,300	6,164	4,841	6,639	5,189
32,300	32,350	6,178	4,849	6,653	5,203
32,350	32,400	6,192	4,856	6,667	5,217
32,400	32,450	6,206	4,864	6,681	5,231
32,450	32,500	6,220	4,871	6,695	5,245
32,500	32,550	6,234	4,879	6,709	5,259
32,550	32,600	6,248	4,886	6,723	5,273
32,600	32,650	6,262	4,894	6,737	5,287
32,650	32,700	6,276	4,901	6,751	5,301
32,700	32,750	6,290	4,909	6,765	5,315
32,750	32,800	6,304	4,916	6,779	5,329
32,800	32,850	6,318	4,924	6,793	5,343
32,850	32,900	6,332	4,931	6,807	5,357
32,900	32,950	6,346	4,939	6,821	5,371
32,950	33,000	6,360	4,946	6,835	5,385

33,000

At least	But less than	Single	Married filing jointly *	Married filing separately	Head of a household
33,000	33,050	6,374	4,954	6,849	5,399
33,050	33,100	6,388	4,961	6,863	5,413
33,100	33,150	6,402	4,969	6,877	5,427
33,150	33,200	6,416	4,976	6,891	5,441
33,200	33,250	6,430	4,984	6,905	5,455
33,250	33,300	6,444	4,991	6,919	5,469
33,300	33,350	6,458	4,999	6,933	5,483
33,350	33,400	6,472	5,006	6,947	5,497
33,400	33,450	6,486	5,014	6,961	5,511
33,450	33,500	6,500	5,021	6,975	5,525
33,500	33,550	6,514	5,029	6,989	5,539
33,550	33,600	6,528	5,036	7,003	5,553
33,600	33,650	6,542	5,044	7,017	5,567
33,650	33,700	6,556	5,051	7,031	5,581
33,700	33,750	6,570	5,059	7,045	5,595
33,750	33,800	6,584	5,066	7,059	5,609
33,800	33,850	6,598	5,074	7,073	5,623
33,850	33,900	6,612	5,081	7,087	5,637
33,900	33,950	6,626	5,089	7,101	5,651
33,950	34,000	6,640	5,096	7,115	5,665

34,000

At least	But less than	Single	Married filing jointly *	Married filing separately	Head of a household
34,000	34,050	6,654	5,104	7,129	5,679
34,050	34,100	6,668	5,111	7,143	5,693
34,100	34,150	6,682	5,119	7,157	5,707
34,150	34,200	6,696	5,126	7,171	5,721
34,200	34,250	6,710	5,134	7,185	5,735
34,250	34,300	6,724	5,141	7,199	5,749
34,300	34,350	6,738	5,149	7,213	5,763
34,350	34,400	6,752	5,156	7,227	5,777
34,400	34,450	6,766	5,164	7,241	5,791
34,450	34,500	6,780	5,171	7,255	5,805
34,500	34,550	6,794	5,179	7,269	5,819
34,550	34,600	6,808	5,186	7,283	5,833
34,600	34,650	6,822	5,194	7,297	5,847
34,650	34,700	6,836	5,201	7,311	5,861
34,700	34,750	6,850	5,209	7,325	5,875
34,750	34,800	6,864	5,216	7,339	5,889
34,800	34,850	6,878	5,224	7,353	5,903
34,850	34,900	6,892	5,231	7,367	5,917
34,900	34,950	6,906	5,239	7,381	5,931
34,950	35,000	6,920	5,246	7,395	5,945

35,000

At least	But less than	Single	Married filing jointly *	Married filing separately	Head of a household
35,000	35,050	6,934	5,254	7,409	5,959
35,050	35,100	6,948	5,261	7,423	5,973
35,100	35,150	6,962	5,269	7,437	5,987
35,150	35,200	6,976	5,276	7,451	6,001
35,200	35,250	6,990	5,284	7,465	6,015
35,250	35,300	7,004	5,291	7,479	6,029
35,300	35,350	7,018	5,299	7,493	6,043
35,350	35,400	7,032	5,306	7,507	6,057
35,400	35,450	7,046	5,314	7,521	6,071
35,450	35,500	7,060	5,321	7,535	6,085
35,500	35,550	7,074	5,329	7,549	6,099
35,550	35,600	7,088	5,336	7,563	6,113
35,600	35,650	7,102	5,344	7,577	6,127
35,650	35,700	7,116	5,351	7,591	6,141
35,700	35,750	7,130	5,359	7,605	6,155
35,750	35,800	7,144	5,366	7,619	6,169
35,800	35,850	7,158	5,374	7,633	6,183
35,850	35,900	7,172	5,381	7,647	6,197
35,900	35,950	7,186	5,389	7,661	6,211
35,950	36,000	7,200	5,396	7,675	6,225

36,000

At least	But less than	Single	Married filing jointly *	Married filing separately	Head of a household
36,000	36,050	7,214	5,404	7,689	6,239
36,050	36,100	7,228	5,411	7,703	6,253
36,100	36,150	7,242	5,419	7,717	6,267
36,150	36,200	7,256	5,426	7,731	6,281
36,200	36,250	7,270	5,434	7,745	6,295
36,250	36,300	7,284	5,441	7,759	6,309
36,300	36,350	7,298	5,449	7,773	6,323
36,350	36,400	7,312	5,456	7,787	6,337
36,400	36,450	7,326	5,464	7,801	6,351
36,450	36,500	7,340	5,471	7,815	6,365
36,500	36,550	7,354	5,479	7,829	6,379
36,550	36,600	7,368	5,486	7,843	6,393
36,600	36,650	7,382	5,494	7,857	6,407
36,650	36,700	7,396	5,501	7,871	6,421
36,700	36,750	7,410	5,509	7,885	6,435
36,750	36,800	7,424	5,516	7,899	6,449
36,800	36,850	7,438	5,524	7,913	6,463
36,850	36,900	7,452	5,531	7,927	6,477
36,900	36,950	7,466	5,542	7,941	6,491
36,950	37,000	7,480	5,556	7,955	6,505

37,000

At least	But less than	Single	Married filing jointly *	Married filing separately	Head of a household
37,000	37,050	7,494	5,570	7,969	6,519
37,050	37,100	7,508	5,584	7,983	6,533
37,100	37,150	7,522	5,598	7,997	6,547
37,150	37,200	7,536	5,612	8,011	6,561
37,200	37,250	7,550	5,626	8,025	6,575
37,250	37,300	7,564	5,640	8,039	6,589
37,300	37,350	7,578	5,654	8,053	6,603
37,350	37,400	7,592	5,668	8,067	6,617
37,400	37,450	7,606	5,682	8,081	6,631
37,450	37,500	7,620	5,696	8,095	6,645
37,500	37,550	7,634	5,710	8,109	6,659
37,550	37,600	7,648	5,724	8,123	6,673
37,600	37,650	7,662	5,738	8,137	6,687
37,650	37,700	7,676	5,752	8,151	6,701
37,700	37,750	7,690	5,766	8,165	6,715
37,750	37,800	7,704	5,780	8,179	6,729
37,800	37,850	7,718	5,794	8,193	6,743
37,850	37,900	7,732	5,808	8,207	6,757
37,900	37,950	7,746	5,822	8,221	6,771
37,950	38,000	7,760	5,836	8,235	6,785

38,000

At least	But less than	Single	Married filing jointly *	Married filing separately	Head of a household
38,000	38,050	7,774	5,850	8,249	6,799
38,050	38,100	7,788	5,864	8,263	6,813
38,100	38,150	7,802	5,878	8,277	6,827
38,150	38,200	7,816	5,892	8,291	6,841
38,200	38,250	7,830	5,906	8,305	6,855
38,250	38,300	7,844	5,920	8,319	6,869
38,300	38,350	7,858	5,934	8,333	6,883
38,350	38,400	7,872	5,948	8,347	6,897
38,400	38,450	7,886	5,962	8,361	6,911
38,450	38,500	7,900	5,976	8,375	6,925
38,500	38,550	7,914	5,990	8,389	6,939
38,550	38,600	7,928	6,004	8,403	6,953
38,600	38,650	7,942	6,018	8,417	6,967
38,650	38,700	7,956	6,032	8,431	6,981
38,700	38,750	7,970	6,046	8,445	6,995
38,750	38,800	7,984	6,060	8,459	7,009
38,800	38,850	7,998	6,074	8,473	7,023
38,850	38,900	8,012	6,088	8,487	7,037
38,900	38,950	8,026	6,102	8,501	7,051
38,950	39,000	8,040	6,116	8,515	7,065

39,000

At least	But less than	Single	Married filing jointly *	Married filing separately	Head of a household
39,000	39,050	8,054	6,130	8,529	7,079
39,050	39,100	8,068	6,144	8,543	7,093
39,100	39,150	8,082	6,158	8,557	7,107
39,150	39,200	8,096	6,172	8,571	7,121
39,200	39,250	8,110	6,186	8,585	7,135
39,250	39,300	8,124	6,200	8,599	7,149
39,300	39,350	8,138	6,214	8,613	7,163
39,350	39,400	8,152	6,228	8,627	7,177
39,400	39,450	8,166	6,242	8,641	7,191
39,450	39,500	8,180	6,256	8,655	7,205
39,500	39,550	8,194	6,270	8,669	7,219
39,550	39,600	8,208	6,284	8,683	7,233
39,600	39,650	8,222	6,298	8,697	7,247
39,650	39,700	8,236	6,312	8,711	7,261
39,700	39,750	8,250	6,326	8,725	7,275
39,750	39,800	8,264	6,340	8,739	7,289
39,800	39,850	8,278	6,354	8,753	7,303
39,850	39,900	8,292	6,368	8,767	7,317
39,900	39,950	8,306	6,382	8,781	7,331
39,950	40,000	8,320	6,396	8,795	7,345

40,000

At least	But less than	Single	Married filing jointly *	Married filing separately	Head of a household
40,000	40,050	8,334	6,410	8,809	7,359
40,050	40,100	8,348	6,424	8,823	7,373
40,100	40,150	8,362	6,438	8,837	7,387
40,150	40,200	8,376	6,452	8,851	7,401
40,200	40,250	8,390	6,466	8,865	7,415
40,250	40,300	8,404	6,480	8,879	7,429
40,300	40,350	8,418	6,494	8,893	7,443
40,350	40,400	8,432	6,508	8,907	7,457
40,400	40,450	8,446	6,522	8,921	7,471
40,450	40,500	8,460	6,536	8,935	7,485
40,500	40,550	8,474	6,550	8,949	7,499
40,550	40,600	8,488	6,564	8,963	7,513
40,600	40,650	8,502	6,578	8,977	7,527
40,650	40,700	8,516	6,592	8,991	7,541
40,700	40,750	8,530	6,606	9,005	7,555
40,750	40,800	8,544	6,620	9,019	7,569
40,800	40,850	8,558	6,634	9,033	7,583
40,850	40,900	8,572	6,648	9,047	7,597
40,900	40,950	8,586	6,662	9,061	7,611
40,950	41,000	8,600	6,676	9,075	7,625

* This column must also be used by a qualifying widow(er).

$2,350 × 2 = $4,700

STEP 3

Multiply the exemption amount ($2,350) by the number of exemptions that the Masters claim (2).

$40,280
– 4,700
$35,580

STEP 4

To determine taxable income, subtract the total for exemptions ($4,700) from the answer obtained in step 2 ($40,280).

TAX COMPUTATION

Generally, taxpayers who earn $100,000 or less use IRS tax tables to determine their tax amount.* A portion of the 1993 tax table is presented in Table 64–2. Based on the information in Table 64–2, and with taxable income of $35,580, the Masters will pay federal income tax totaling $5,336 as illustrated in Example 4.

EXAMPLE 4 Determine the Tax Amount for Sherry and Ben Masters

Sherry and Ben Masters are married and filing a joint tax return. If their taxable income is $35,580, what is the tax amount they should pay the IRS?

STEP 1

In the tax table (Table 64–2), locate the line for taxable income that is at least $35,500 but less than $35,600.

STEP 2

Read across the line until you reach the column for married filing jointly. The tax amount is $5,336.

It is now possible to compare the tax owed with estimated payments made during the tax year and with amounts withheld from employee's paychecks. If the amount of tax owed is less than the amount already paid, the taxpayer is entitled to a refund. If the amount of tax owed is greater than the amount already paid, the taxpayer must pay the difference.

STATE INCOME TAXES

All but seven states (Alaska, Florida, Nevada, South Dakota, Texas, Washington, and Wyoming) have a state income tax. In most states, the tax rate ranges from 1 to 10 percent and is based on the amount of adjusted gross income or the amount for taxable income as reported on your federal tax return. Individual states usually require that state income tax returns be filed when your federal income tax return is filed.

---

*Taxpayers with taxable income in excess of $100,000 must use a different set of Tax Rate Schedules furnished by the IRS.

Complete the Following Problems

If necessary, round your answer to two decimal places. Use Table 64–2 and the tax information presented in this chapter to answer all questions.

T F   **1.** Individual tax returns must be filed by April 15 of each year for the previous calendar year.

T F   **2.** Form 1099-INT is used to report wages, amount of income tax withheld, and Social Security withheld during the tax year.

T F   **3.** At the time of publication, the standard deduction for single taxpayers was $3,700.

T F   **4.** At the time of publication, a taxpayer was entitled to a $3,200 exemption.

T F   **5.** All fifty states now have a state income tax.

Taxable Income	Filing Status	Tax Amount
$40,040	single	**6.** _____
$33,420	married filing jointly	**7.** _____
$39,003	single	**8.** _____
$36,775	head of a household	**9.** _____
$37,089	married filing separately	**10.** _____
$32,004	single	**11.** _____
$38,290	single	**12.** _____
$40,152	married filing jointly	**13.** _____
$36,000	married filing jointly	**14.** _____
$39,655	head of household	**15.** _____

16. During 1995, Mark Ernest earned $26,500 at his job as a bookkeeper. During the same year, he received $1,250 in interest from certificates of deposit. He also earned $890 in dividends. What is the total gross income Mr. Ernest earned during the tax year?

$26,500 + $1,250 + $890 = $28,640

17. During the last tax year, The Lennoxes—Bob and Mary—had a combined salary of $76,400. They earned $2,100 in interest and had capital gains of $7,800. Finally, they had rental income of $2,650. During the same tax year, they were penalized $1,100 for an early withdrawal from a savings account. Also, Bob Lennox paid $4,800 in alimony to an ex-wife. What is the Lennoxes' adjusted gross income?

Personal Math Problem

Mike and Angie Gomez are trying to decide whether to take the standard deduction for a married couple filing jointly or itemize deductions on Schedule A. After examining their tax records, they determine that their itemized deductions are: taxes—$1,100; interest—$2,050; and contributions—$2,510.

18. What is the total of their itemized deductions?

19. In problem 18, should Mr. and Mrs. Gomez take the standard deduction or itemize deductions? Why? _____

_____

_____

_____

Adjusted Gross Income	Itemized Deduction Amount	Marital Status/ Number of Exemptions	Taxable Income	Tax Amount
$55,300	$ 9,800	Single/2	**20.** _____	**21.** _____
$63,400	$16,456	Married/3	**22.** _____	**23.** _____
$44,100	$ 4,500	Single/1	**24.** _____	**25.** _____
$46,556	$ 8,824	Married/2	**26.** _____	**27.** _____
$56,230	$12,260	Married/3	**28.** _____	**29.** _____

30. At the end of the last tax year, Darrel and LaTeisha Johnston determined that they had combined wages of $53,740. This couple also had itemized deductions totaling $9,100. If the couple is entitled to four exemptions, what is their taxable income?

31. In problem 30, what is the tax amount if the Johnstons file a joint return for a married couple?

Use the following information to (1) complete the blank IRS Form 1040 on pages 373–374 and (2) determine the total tax (line 38 on the second page of form 1040).

32. Peter Nettles and his wife Barbara both work for Heritage Proud Chicken, Inc. He works as the dispatcher in the shipping department and earns $28,600 a year. Barbara is an administrative assistant and earns $24,000 a year. Last year, the Nettleses received $1,080 interest and $756 in dividends. They have no additional income. The Nettleses have the following deductions: taxes—$2,750; Interest on their home—$3,600; gifts to charity—$4,000; and a tax-deductible casualty loss—$1,860. The Nettleses are entitled to two exemptions.

UNIT 65	CORPORATE INCOME TAXES

LEARNING OBJECTIVES   *After completing this unit, you will be able to:*

1. determine the federal income tax liability for a corporation.

As mentioned in the opening case, approximately 7 percent—almost $100 billion—of all the money received by the federal government is paid by corporations. Today, corporations pay federal income tax on their taxable income, which is what remains after deducting all legal business expenses from sales. Generally, the process involves computing gross sales, deducting allowable business expenses, and then determining the amount of corporate income tax owed. The following formula can be used to visualize the overall process.

Gross sales *LESS* allowable deductions *EQUALS* taxable income

GROSS SALES

Gross sales is the total amount of all goods and services sold during the tax year. From this amount are deducted the amounts for sales returns, sales allowances, and sales discounts. The remainder is called net sales. **Net sales** is the actual amount received by the firm for the goods and services it has sold, after adjustment for sales returns, allowances, and discounts.

Form **1040**	Department of the Treasury—Internal Revenue Service **U.S. Individual Income Tax Return** (O) 19		IRS Use Only—Do not write or staple in this space

For the year Jan. 1–Dec. 31, 1993, or other tax year beginning _____ , 1993, ending _____ , 19 ___ OMB No. 1545-0074

Label
(See instructions on page 12.)

Use the IRS label. Otherwise, please print or type.

L A B E L   H E R E

Your first name and Initial   Last name

If a joint return, spouse's first name and initial   Last name

Home address (number and street). If you have a P.O. box, see page 12.   Apt. no.

City, town or post office, state, and ZIP code. If you have a foreign address, see page 12.

Your social security number

Spouse's social security number

For Privacy Act and Paperwork Reduction Act Notice, see page 4.

Presidential Election Campaign (See page 12.)

Do you want $3 to go to this fund? . . . .
If a joint return, does your spouse want $3 to go to this fund? . . . . . .

Yes | No

Note: Checking "Yes" will not change your tax or reduce your refund.

Filing Status
(See page 12.)

Check only one box.

1 Single
2 Married filing joint return (even if only one had income)
3 Married filing separate return. Enter spouse's social security no. above and full name here. ▶
4 Head of household (with qualifying person). (See page 13.) If the qualifying person is a child but not your dependent, enter this child's name here. ▶
5 Qualifying widow(er) with dependent child (year spouse died ▶ 19). (See page 13.)

Exemptions
(See page 13.)

If more than six dependents, see page 14.

6a ☐ **Yourself.** If your parent (or someone else) can claim you as a dependent on his or her tax return, do not check box 6a. But be sure to check the box on line 33b on page 2.
b ☐ **Spouse** . . . .
c **Dependents:**

(1) Name (first, initial, and last name)	(2) Check if under age 1	(3) If age 1 or older, dependent's social security number	(4) Dependent's relationship to you	(5) No. of months lived in your home in 1993

No. of boxes checked on 6a and 6b

No. of your children on 6c who:
• lived with you
• didn't live with you due to divorce or separation (see page 15)
Dependents on 6c and entered above

d If your child didn't live with you but is claimed as your dependent under a pre-1985 agreement, check here ▶ ☐
e Total number of exemptions claimed . . . .

Add numbers entered on lines above

Income

Attach Copy B of your Forms W-2, W2G, and 1099-R here.

If you did not get a W-2, see page 10.

If you are attaching a check or money order, put it on top of any Forms W-2, W2G, or 1099-R.

7 Wages, salaries, tips, etc. Attach Form(s) W-2 . . . . . | 7
8a **Taxable** interest income (see page 16). Attach Schedule B if over $400 . . . . | 8a
b **Tax-exempt** interest (see page 17). DON't include on line 8a | 8b |
9 Dividend income. Attach Schedule B if over $400 . . . . | 9
10 Taxable refunds, credits, or offsets of state and local income taxes (see page 17) . . | 10
11 Alimony received . . . . | 11
12 Business income or (loss). Attach Schedule C or C-EZ . . | 12
13 Capital gain or (loss). Attach Schedule D . . . . | 13
14 Capital gain distributions not reported on line 13 (see page 17) . | 14
15 Other gains or losses. Attach Form 4797 . . . . | 15
16a Total IRA distributions | 16a | b Taxable amount (see page 18) | 16b
17a Total pensions and annuities | 17a | b Taxable amount (see page 18) | 17b
18 Rental real estate, royalties, partnerships, S corporations, trusts, etc. Attach Schedule E | 18
19 Farm income or (loss). Attach Schedule F . . . . | 19
20 Unemployment compensation (see page 19) . . . . | 20
21a Social security benefits | 21a | b Taxable amount (see page 19) | 21b
22 Other income. List type and amount—see page 20 _____ | 22
23 Add the amounts in the far right column for lines 7 through 22. This is your **total income** ▶ | 23

Adjustments to Income
(See page 20.)

24a Your IRA deduction (see page 20) . . . . | 24a |
b Spouse's IRA deduction (see page 20) . . . . | 24b |
25 One-half of self-employment tax (see page 21) . . | 25 |
26 Self-employed health insurance deduction (see page 22) | 26 |
27 Keogh retirement plan and self-employed SEP deduction | 27 |
28 Penalty on early withdrawal of savings . . . . | 28 |
29 Alimony paid. Recipients's SSN ▶ | 29 |
30 Add lines 24a through 29. These are your **total adjustments** . . . . . . ▶ | 30

Adjusted Gross Income

31 Subtract line 30 from line 23. This is your **adjusted gross income**. If this amount is less than $23,050 and a child lived with you, see page EIC-1 to find out if you can claim the "Earned Income Credit" on line 56 . . . . ▶ | 31

Cat. No. 11320B   Form **1040** (1993)

Form 1040 (19)   Page **2**

Tax Computation

(See page 23.)

32	Amount from line 31 (adjusted gross income)	32
33a	Check if: ☐ **You** were 65 or older, ☐ Blind; ☐ **Spouse** was 65 or older, ☐ Blind. Add the number of boxes checked above and enter the total here ▶ 33a ☐	
b	If your parent (or someone else) can claim you as a dependent, check here. ▶ 33b ☐	
c	If you are married filing separately and your spouse itemizes deductions or you are a dual-status alien, see page 24 and check here ▶ 33c ☐	
34	Enter the **larger** of your: **Itemized deductions** from Schedule A, line 26, **OR** **Standard deduction** shown below for yor filing status. **But if you checked any box on line 33a or b,** go to page 24 to find your standard deduction. If you checked **box 33c,** you standard deduction is zero. • Single—$3,700 • Head of household—$5,450 • Married filing jointly or Qualifying widow(er)—$6,200 • Married filing separately—$3,100	34
35	Subtract line 34 from line 32	35
36	If line 32 is $81,350 or less, multiply $2,350 by the total number of exemptions claimed on line 6e. If line 32 is over $81,350, see the worksheet on page 25 for the amount to enter	36
37	**Taxable Income.** Subtract line 36 from line 35. If line 36 is more than line 35, enter -0-	37
38	Tax. Check if from **a** ☐ Tax Table, **b** ☐ Tax Rate Schedules, **c** ☐ Schedule D Tax Work-sheet, or **d** ☐ Form 8615 (See page 25). Amount from Form(s) 8814 ▶ **e** _____	38
39	Additional taxes (see page 25). Check if from **a** ☐ Form 4970 **b** ☐ Form 4972	39
40	Add lines 38 and 39 ▶	40

If you want the IRS to figure your tax, see page 24.

Credits

(See page 25.)

41	Credit for child and dependent care expenses. Attach Form 2441	41	
42	Credit for the elderly or the disabled. Attach Schedule R	42	
43	Foreign tax credit. Attach Form 1116	43	
44	Other credits (see page 26). Check if from **a** ☐ Form 3800 **b** ☐ Form 8396 **c** ☐ Form 8801 **d** ☐ Form (specify) _____	44	
45	Add lines 41 through 44		45
46	Subtract line 45 from lie 40. If line 45 is more than line 40, enter -0- ▶		46

Other Taxes

47	Self-employment tax. Attach Schedule SE. Also, see line 25	47
48	Alternative minimum tax. Attach Form 6251	48
49	Recapture taxes (see page 26). Check if from **a** ☐ Form 4255 **b** ☐ Form 8611 **c** ☐ Form 8828	49
50	Social security and Medicare tax on tip income not reported to employer. Attach form 4137	50
51	Tax on qualified retirement plans, includint IRAs. If required, attach Form 5329	51
52	Advance earned income credit payments from Form W-2	52
53	Add lines 46 through 52. This is your **total tax** ▶	53

Payments

Attach Rorms W-2, W2G, and 1099-R on the front.

54	Federal income tax withheld. If any is from Form(s) 1099, check ▶ ☐	54	
55	1993 estimated tax payments and amount applied from 1992 return	55	
56	Earned income credit. Attach Schedule EIC	56	
57	Amount paid with Form 4868 (extension request)	57	
58a	Excess social security, Medicare, and RRTA tax withheld (see page 28)	58a	
b	Deferral of additional 1993 taxes. Attach Form 8841	58b	
59	Other payments (see page 28). Check if from **a** ☐ Form 2439 **b** ☐ Form 4136	59	
60	Add lines 54 through 59. These are your **total payments** ▶		60

Refund or Amount You Owe

61	If line 60 is more than line 53, subtract line 53 from line 60. This is the amount you **OVERPAID** ▶		61
62	Amount of line 61 you want **REFUNDED TO YOU** ▶		62
63	Amount of line 61 you want **APPLIED TO YOUR 1994 ESTIMATED TAX**	63	
64	If line 53 is more than line 60, subtract line 60 from line 53. This is the **AMOUNT YOU OWE.** For details on how to pay, including what to write on your payment, see page 29		64
65	Estimated tax penalty (see page 29). Also include on line 64	65	

Sign Here

Keep a copy of this return for your records.

Under penalties of perjury, I declare that I have examined this return and accompanying schedules and statements, and to the best of my knowledge and belief, they are true, correct, and complete. Declaration of preparer (other than taxpayer) is based on all information of which preparer has any knowledge.

Your signature	Date	Your occupation
Spouse's signature. If a joint return, BOTH must sign.	Date	Spouse's occupation

Paid Preparer's Use Only

Preparer's signature ▶	Date	Check if self-employed ☐	Preparer's social security no.
Firm's name (or yours if self-employed) and address ▶		E.I. No.	
		ZIP code	

*U.S. Government Printing Office: 1993 — 345-188

EXAMPLE 1 Find the Net Sales for Bankston Office Supply, Inc.

CALCULATOR TIP

Depress clear key
Enter 1,634,000 Depress –
Enter 25,400   Depress =
Read the answer
$1,608,600

During the past tax year, Bankston Office Supply, Inc., had gross sales that totaled $1,634,000. During the same period, the corporation had sales returns, sales allowances, and sales discounts totaling $25,400. What is the total net sales amount for Bankston Office Supply?

$1,634,000	gross sales
– 25,400	deductions for sales returns, allowances, and discounts
$1,608,600	net sales

For tax purposes, interest, fees, dividends, and rental income are also included in a firm's taxable income for a corporation.

ALLOWABLE DEDUCTIONS

For tax purposes, allowable deductions for a corporation are divided into two broad categories. The first category includes cost of goods that are available for sale to the firm's customers. The *cost of goods sold* is the total cost of merchandise sold during the tax year. The second category of allowable deductions includes the normal expenses of operating a business. Typical expenses in this category include: salaries and wages; repairs and maintenance; bad debts; rent; taxes; interest, depreciation, advertising, the cost of retirement plans, and the cost of employee benefit programs.

The dollar amounts for cost of goods sold and operating expenses are subtracted from net sales to determine taxable income.

EXAMPLE 2 Find the Taxable Income for Bankston Office Supply, Inc.

CALCULATOR TIP

Depress clear key
Enter 1,608,600 Depress –
Enter 868,700   Depress –
Enter 245,200   Depress =
Read the answer $494,700

For Bankston Office Supply, Inc. (the company in Example 1), cost of goods sold totaled $868,700. During the same tax year, Bankston operating expenses totaled $245,200. What is the taxable income for Bankston Office Supply?

$1,608,600	net sales (from Example 1)
– 868,700	cost of goods sold
– 245,200	operating expenses
$ 494,700	taxable income

TAX COMPUTATION

Beginning January 1, 1993, corporate taxable income is subject to the tax rate system illustrated in Table 65–1. Based on the tax rates presented in Table 65–1, the steps required to calculate the tax amount that Bankston Office Supply, Inc., must pay to the federal government are illustrated in Example 3.

TABLE 65–1   Corporate Tax Rates (Beginning on January 1, 1993)

Taxable Income	Tax Is
Not over $50,000	15%
Over $50,000 but not over $75,000	$ 7,500 + 25% of excess over $50,000
$75,000 but not over $100,000	$ 13,750 + 34% of excess over $75,000
$100,000 but not over $335,000	$ 22,250 + 39% of excess over $100,000
$335,000 but not over $10 million	$ 113,900 + 34% of excess over $335,000
$10 million but not over $15 million	$3,400,000 + 35% of excess over $10 million
$15 million but not over $18,333,333	$5,150,000 + 38% of excess over $15 million
Over $18,333,333	35%

EXAMPLE 3   Find the Tax Amount for Bankston Office Supply, Inc.

As determined in Example 2, Bankston Office Supply, Inc., has taxable income totaling $494,700.

The tax is $113,900 *plus* 34 percent of the excess over the first $335,000.	**STEP** 1	In the tax table (Table 65–1), locate the line for taxable income that is at least $335,000 but not over $10 million.
$494,700 −335,000 $159,700	**STEP** 2	Determine the dollar excess over $335,000.
$159,700 × 0.34 = $54,298	**STEP** 3	Multiply the excess amount from step 2 ($159,700) by the tax rate in step 1 (34 percent).
$113,900 + 54,298 $168,198	**STEP** 4	To determine the total tax amount, add the base tax amount from step 1 ($113,900) to the additional tax amount from step 3 ($54,298).

CALCULATOR TIP

Depress clear key
Enter 494,700 Depress −
Enter 335,000 Depress =
Read the answer $159,700
Enter 159,700 Depress ×
Enter 0.34   Depress =
Read the answer $54,298
Enter 113,900 Depress +
Enter 54,298   Depress =
Read the answer $168,198

It is now possible to compare the tax owed with the estimated tax payments made during the tax year. If the amount of tax owed is less than the amount already paid, the corporation is entitled to a refund. If the amount of tax owed is more than the amount already paid, the corporation must pay the difference.

Complete the Following Problems

If necessary, round your answer to two decimal places. Use Table 65–1 and the tax information presented in this chapter to answer all questions.

1. In your own words, define the formula used to determine taxable income for a corporation. _____

_____

2. T F   Gross sales are the actual amount received by the firm for the goods and services it has sold after adjustment for sales returns, allowances, and discounts.

3. T F   The cost of goods sold is the total cost for merchandise sold during the tax year.

4. T F   Under current tax law, the cost of employee benefit programs is an allowable deduction that can be used to reduce a corporation's tax amount.

5. During the last tax year, Martin Electronics Supply, Inc., sold electrical parts valued at $781,000. The firm had sales returns totaling $25,810. The firm also offered their customers sales discounts totaling $15,620. What is the total of the firm's sales returns and sales discounts?

6. In problem 5, what is the net sales amount for Martin Electronics Supply, Inc.?

Critical Thinking Problem

In 1995, Deep Down Foods, Inc., had gross sales of $2,340,000. The firm offered all customers a 2 percent cash discount if they paid for merchandise within 10 days. During 1995, approximately 65 percent of the firm's customers took advantage of the program, and it cost Deep Down $30,420 for cash discounts.

7. What is the total of the firm's net sales?

HINT: You may want to review the material on cash discounts in Unit 42 before answering this question.

8. What are the advantages and disadvantages of offering cash discounts?

Advantages _____

_____

Disadvantages _____

9. If you were president of Deep Down Foods, would you continue the cash discount program? Why? _____

_____

10. During the last tax year, Simpson Floor Covering had the following operating expenses: wages—$98,500; maintenance—$4,900; bad debts—$12,320; rent expense—$42,500; depreciation—$65,300; and advertising—$61,000. What are the total operating costs for Simpson Floor Covering?

11. In 1995, Hart Manufacturing, Inc., had net sales of $3,250,000. During the same year, the firm's cost of goods sold was $1,940,000. Operating expenses total $785,000. What is the firm's taxable income?

Use the following information to solve problems 12–15.

Sales	$325,000	
Less cost of goods sold	−160,000	
Subtotal	**12.**	_____
Less deductions		
Salaries	58,000	
Repairs	8,200	
Bad debt expense	2,400	
Advertising	15,600	
Depreciation	9,700	
Employee benefits	18,200	
Taxable income	**13.**	_____

Sales	$2,465,000	
Less cost of goods sold	−1,670,000	
Subtotal	**14.**	_____
Less deductions		
Salaries & wages	248,800	
Maintenance	15,420	
Bad debt expense	22,400	
Advertising	76,100	
Depreciation	62,560	
Employee benefits	84,350	
Taxable income	**15.**	_____

Taxable Income		Corporate Tax Amount
$ 125,000	16.	_____
$ 300,000	17.	_____
$1,200,000	18.	_____
$ 110,000	19.	_____
$ 562,000	20.	_____
$3,750,000	21.	_____
$ 42,000	22.	_____

23. Tejas Management Corporation—a local consulting firm—had net sales of $542,500 at the end of the current tax year. The firm also had operating expenses totaling $168,000. What is the amount of taxable income for Tejas Management?

24. In problem 23, how much corporate income tax does Tejas Management owe for this tax year?

Use the following financial information to answer questions 25–28.

Ford Chemical Supply
Financial Results 1995

Gross sales	$795,000	Salaries & wages	$77,200
Sales returns	11,000	Maintenance	6,900
Sales discounts	12,300	Advertising	76,100
Cost of goods sold	310,000	Depreciation	62,560

25. What is the gross sales amount for Ford Chemical Supply?

26. What is the net sales amount for Ford Chemical Supply?

27. What is the taxable income for Ford Chemical Supply?

28. What is the corporate tax that Ford Chemical owes for this tax year?

UNIT 66	HEALTH AND LIFE INSURANCE

LEARNING OBJECTIVES   *After completing this unit, you will be able to:*

1. determine health care premiums paid by the employer and the employee.
2. calculate the amount of reimbursement for health insurance claims.
3. calculate life insurance premiums.

Today, there are different types of insurance coverage available to cover just about every conceivable risk that an individual or business may encounter. In this unit, we examine two types of insurance coverage—health and life insurance. In the next unit, we look at property and automobile insurance.

HEALTH CARE INSURANCE

Health care insurance covers the cost of medical attention, including hospital care, physicians' and surgeons' fees, and related services. There are three ways in which individuals can obtain health insurance.

1. The traditional **fee-for-service** plans enable policyholders to choose their own physician and hospitals. Most of these plans require that policyholders first pay a specified deductible amount and then the

insurance company pays 70 to 80 percent of approved medical expenses.

2. A **health maintenance organization (HMO)** is a health insurance plan that directly employs or contracts with selected physicians, surgeons, and hospitals to provide policyholders with health care services in exchange for a fixed, prepaid monthly premium. With most HMOs, members pay no deductible. There may be a copayment ranging from $7 to $15 when HMO members visit the doctor's office or enter the hospital.

3. **Preferred Provider Organizations (PPOs)** consist of groups of doctors and hospitals that agree to provide health care at rates approved by the insurer. PPOs provide their members with essentially the same benefits as HMOs. However, unlike health maintenance organizations, PPOs allow members to choose from the list of approved doctors or hospitals when a medical need arises. Typically, the premium for PPOs is slightly higher than for HMOs.

THE COST OF HEALTH CARE

As a result of increasing costs, most employers will pay only a portion of health care costs, and employees must pay the remainder. For example, the union contract between Haskins Manufacturing and the firm's employees specifies that the company will pay 70 percent of health care premiums. The employee must pay the remaining 30 percent. In this situation, the portion formula ($P = B \times R$) can be used to find the amount of premium that the employer must pay. The employer's part (portion) is found by multiplying the total health care premium (base) by the percent paid by the employer (rate).

EXAMPLE 1   Determine the Employer's and Employee's Portion of Health Insurance Cost

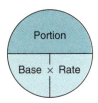

Under the current group health insurance policy, the total cost of health insurance for each worker employed by Haskins Manufacturing is $310 per month. If Haskins pays 70 percent of the cost, what is the dollar amount the employer and employee must pay?

$310 × 0.70 = $217   **STEP** [1]   Multiply the monthly premium ($310) by the percent paid by the employer (70 percent).

$310
−217
$ 93   **STEP** [2]   To determine the amount the employee must pay, subtract the employer's portion ($217) found in step 1 from the total monthly premium ($310).

CALCULATOR TIP

Depress clear key
Enter 310   Depress   ×
Enter 0.70   Depress   =
Read the answer $217
Enter 310   Depress   −
Enter 217   Depress   =
Read the answer $93

It is possible to extend employee coverage to include a spouse and dependent children. In almost every case, the additional premium for spouse and dependent children must be paid by the employee and not the employer.

REIMBURSEMENT FOR HEALTH CARE EXPENSES

With the exception of a small copayment that is charged with some HMO or PPO plans, there are no additional fees charged HMO or PPO members. With a fee-for-service health care plan, most policyholders must be concerned with their policy's deductible and the policy's coinsurance clause. The

deductible is a dollar amount that the insured must pay each year before benefits by the insurance company become payable. Typical annual deductible amounts range from $250 to $2,500. The **coinsurance clause** for a health insurance policy is a provision under which both the insured and the insurer share approved medical expenses. For most fee-for-service health plans, the coinsurance clause specifies that the insurer pays either 70 or 80 percent of approved medical expenses. Then, the policyholder must pay the remainder.

EXAMPLE 2 Determine the Amounts the Insurance Company and the Policyholder Must Pay in the Following Situation

---

Marty and Bob Winfrey have group health insurance provided by Marty's employer. There is a $500 deductible with this policy. Also, the coinsurance clause specifies that the insurer pay 80 percent of all approved medical expenses. If the Winfreys have approved medical expenses that total $2,900 during one calendar year, what is the dollar amount the insurer will pay? What is the dollar amount the Winfreys must pay?

$2,900 − 500 $2,400	**STEP** 1 Subtract the deductible amount ($500) from the total for approved medical expenses ($2,900).
$2,400 × 0.80 = $1,920	**STEP** 2 To determine the dollar amount the insurer will pay, multiply the answer from step 1 ($2,400) by the percent paid by the insurer (80 percent).
$2,400 −1,920 $ 480	**STEP** 3 To determine the amount the policyholder must pay, subtract the amount the insurer pays ($1,920) from the answer in step 1 ($2,400).

CALCULATOR TIP

Depress clear key
Enter 2,900   Depress −
Enter 500   Depress =
Read the answer $2,400
Enter $2,400   Depress ×
Enter 0.80   Depress =
Read the answer $1,920
Enter 2,400   Depress −
Enter 1,920   Depress =
Read the answer $480

It should be pointed out before leaving this example, the Winfreys pay $480 (the answer found in Step 3) *plus* the original $500 deductible amount. Thus, their out-of-pocket expense totals $980.

LIFE INSURANCE

Life insurance pays a stated amount of money on the death of the insured individual. Today, there are three main types of life insurance.

1. **Term insurance** provides protection to beneficiaries for a stated period of time. Because term insurance provides no other benefits, it is the least expensive form of life insurance.
2. **Whole life insurance** provides both protection to beneficiaries and savings. This type of policy has a cash value, which is an amount that is payable to the policyholder of a whole life insurance policy if the policy is canceled.
3. **Endowment life insurance** provides protection and guarantees the payment of a stated amount to the policyholder after a specified number of years. This type of policy has a higher cash value than the cash value of whole life policies.

TABLE 66–1 Annual Premiums for $1,000 of Term Life Insurance (Nontobacco Rates)

	$10,000 Policy			$50,000 Policy			$100,000 Policy	
Age	Male	Female	Age	Male	Female	Age	Male	Female
20	$ 8.52	$ 7.92	20	$ 2.62	$ 2.60	20	$ 1.39	$ 1.37
30	8.88	8.64	30	2.74	2.65	30	1.53	1.43
40	11.28	10.74	40	2.96	2.72	40	2.10	1.85
50	15.16	13.01	50	5.20	4.05	50	2.73	2.04
60	23.90	19.80	60	9.70	6.25	60	6.68	3.15
70	46.91	35.00	70	25.50	15.90	70	12.75	9.05

SPECIAL NOTE TO STUDENTS: The factors in Table 66–1 are based on $1,000 units of coverage for each insurance policy.

THE COST OF LIFE INSURANCE

When comparing the cost of the three different types of life insurance, term life insurance is the least expensive while endowment policies are the most expensive. In addition to type of policy, there are other factors that must be considered when determining the cost of insurance. The factor of age must be considered. Notice in Table 66–1, Annual Premiums for $1,000 of Term Life Insurance, that the older a person is, the higher the premium. On the average, older people are less likely to survive each year than younger people. Finally, females generally pay lower life insurance premiums than males of the same age because (on the average) they live longer.

Regardless of the type of life insurance, the same steps are used to calculate premiums. Using the information presented in Table 66–1, the yearly premium for a $50,000 term life insurance for a male who is 30 years old is $137. The steps required to determine the premium amount are illustrated in Example 3.

EXAMPLE 3 Determine the Annual Premium for a $50,000 Policy for a Male that Is 30 Years Old

CALCULATOR TIP

Depress clear key
Enter 50,000 Depress ÷
Enter 1,000 Depress =
Read the answer 50
Enter 50 Depress ×
Enter 2.74 Depress =
Read the answer $137

$50,000 ÷ $1,000 = 50 **STEP** 1 — Since Table 66–1 is based on $1,000 units, divide the policy amount ($50,000) by $1,000.

$50 \times \$2.74 = \$137$ **STEP** 2 — To determine the yearly premium, multiply the answer from step 1 (50) by the rate from Table 66–1 for a male that is 30 years old ($2.74).

Complete the Following Problems

If necessary, round your answer to two decimal places.

1. T F Fee-for-service health insurance plans allow policyholders to choose their own physician and hospital.

2. T F A type of health insurance that requires policyholders to pay a deductible yearly is called a health maintenance organization.

3. T F Premiums charged by preferred provider organizations are slightly lower than those charged by HMOs.

4. T F The typical deductible for fee-for-service health plans ranges between $250 to $2,500.

5. Last year, Mary Jenkins paid $289 a month for health coverage for herself and her husband. This year, the same coverage will cost $349 per month. What is the amount of premium increase?

6. As part of a new-employee benefits package, New Wave Software agreed to pay 72 percent of the health care premium for each employee during the next calendar year. If it costs $274 a month to insure each employee, what is the company's cost per employee?

7. If New Wave Software has 71 employees, what is the firm's monthly cost for health care coverage?

8. In problem 6, what is the employee's cost?

9. The union contract between U.S. Pipeworks and the firm's employees requires that the firm pay 92 percent of the cost for health insurance. If it costs $330 a month for coverage for each union worker, what is the company's cost per worker?

10. In problem 9, what is the worker's cost?

11. During 1995, Charlie Bell had medical expenses that total $4,190. If the deductible for his health insurance plan is $750, what is the amount that should be submitted for reimbursement?

For problems 12–19, assume that the deductible in each problem has already been satisfied.

Approved Medical Expenses	Coinsurance Clause	Dollar Amount Paid by Insurer	Policyholder Portion
$ 6,700	80%	**12.** _____	**13.** _____
$ 2,500	70%	**14.** _____	**15.** _____
$11,000	75%	**16.** _____	**17.** _____
$ 5,250	80%	**18.** _____	**19.** _____

20. The year of 1995 was not a good year for the Browns. Because both Yowanda and Marcus have been in the hospital, their approved medical expenses total $33,980. Fortunately, the Browns have health care that provides that the insurer will pay 80 percent once a $2,500 deductible has been paid by the Browns. What is the amount that the insurance company will pay in this case?

21. In problem 20, what is the amount the Browns will pay?

22. T F Whole life insurance provides the least expensive form of life insurance.

23. T F Of the three types of life insurance described in this chapter, endowment life insurance provides the largest cash value.

24. T F Generally, females pay higher life insurance premiums than males.

Use Table 66–1, to answer problems 25–38.

Life Insurance Amount	Age	Male/Female	Table Factor per $1,000
$100,000	20	Female	**25.** _____
$ 50,000	40	Male	**26.** _____
$ 10,000	50	Male	**27.** _____
$ 50,000	60	Female	**28.** _____
$100,000	30	Male	**29.** _____

Life Insurance Amount	Age	Male/Female	Annual Premium Amount
$100,000	30	Male	**30.** _____
$ 50,000	40	Female	**31.** _____
$ 10,000	70	Male	**32.** _____
$ 50,000	50	Male	**33.** _____
$100,000	30	Female	**34.** _____
$100,000	40	Female	**35.** _____

Personal Math Problem

Pete and Cindy Shellenberger have been married for three years. Peter is 40 years old and Cindy is 30 years old. They have one two-year-old daughter and are expecting another child. Finally, they are a two-income family with each making about $25,000 a year. They were told by a professional financial planner that they should each purchase a $100,000 term life insurance policy.

36. What is the annual premium for the policy for Pete Shellenberger?

37. What is the annual premium for the policy for Cindy Shellenberger?

38. If you were either Pete or Cindy Shellenberger, would you purchase one policy to insure Pete, one policy to insure Cindy, or two policies that would insure each marital partner? Why? _____

_____

_____

_____

_____

UNIT 67	FIRE AND AUTOMOBILE INSURANCE

LEARNING OBJECTIVES

After completing this unit, you will be able to:

1. calculate premiums for property and automobile coverage.
2. calculate the required amount of coverage and actual reimbursement when the insurance company requires coinsurance.

Becoming familiar with different types of property and automobile insurance can insure that you have proper protection and at the same time reduce your insurance premiums.

FIRE INSURANCE

Fire insurance provides protection against partial or complete loss of a building and/or its contents when that loss is caused by fire or lightning. Premiums depend on the construction of the building, its use and contents, whether risk-reduction devices (such as smoke alarms and sprinkler systems) are installed in the building, and where the building is located. Typical rates for fire insurance are illustrated in Table 67–1. Using the information in this table, the total annual premium (building and contents) for an $80,000 brick building is $448, as illustrated in Example 1.

TABLE 67–1 Fire Insurance Rates for $100 of Coverage—Buildings and Contents (Annual Rates)

| | Brick | | Brick Veneer | |
	Building	Contents	Building	Contents
City	$0.24	$0.32	$0.29	$0.38
Rural	0.36	0.46	0.44	0.57

| | Stucco | | Frame | |
	Building	Contents	Building	Contents
City	$0.35	$0.48	$0.51	$0.67
Rural	0.53	0.72	0.77	0.98

EXAMPLE 1 Calculate the Premium for an $80,000 Brick Building Located Within the City Limits

$80,000 ÷ $100 = 800   **STEP** 1   Since Table 67–1 is based on $100 units, divide the value of the building ($80,000) by $100.

800 × $0.24 = $192   **STEP** 2   To determine the premium for the building, multiply the answer from step 1 (800) by the rate from Table 67–1 for a brick building located within the city limits ($0.24).

800 × $0.32 = $256   **STEP** 3   To determine the premium for contents, multiply the answer from step 1 (800) by the rate from Table 67–1 for a brick building located within the city limits ($0.32).

$192
+256
$448   **STEP** 4   To determine the total premiums, add the premiums for building ($192) and contents ($256).

COINSURANCE CLAUSE

To reduce their insurance premiums, individuals and businesses sometimes insure property for less than its actual cash value. Their theory is that fire rarely destroys a building completely—thus they need not buy full coverage. However, if the building is partially destroyed, they expect their insurance to cover all the damage. This places an unfair burden on the insurance company, which receives less than the full premium but must cover the full loss. To avoid this problem, insurance companies include a coinsurance clause in most fire insurance policies.

A **coinsurance clause** is a part of a fire insurance policy that requires the policyholder to purchase coverage at least equal to a specified percentage of the replacement cost of the property to obtain full reimbursement for losses. In most cases, the required percentage is 80 percent of the replacement cost. The portion formula ($P = B \times R$) can be used to determine the required amount of coverage. The replacement cost is the base. The coinsurance requirement stated as a percent is the rate.

EXAMPLE 2   Determine the Required Amount of Coverage for an Insurance Policy with an 80 Percent Coinsurance Clause

---

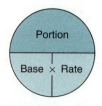

A building has a replacement cost of $200,000. The insurance policy has a coinsurance clause that requires the property be insured for 80 percent of its replacement cost.

$200,000 × 0.80 = $160,000   **STEP** ☐1   To determine the required amount of coverage, multiply the replacement cost for the building ($200,000) by the percent required for coinsurance (80 percent).

CALCULATOR TIP

Depress clear key
Enter 200,000   Depress   ×
Enter .80   Depress   =
Read the answer $160,000

When a building is insured for less than 80 percent of the replacement cost, the following coinsurance formula can be used to determine the amount the insurance company will pay for a loss.

$$\frac{\text{Actual coverage}}{\text{Required coverage}} \times \begin{array}{c}\text{Amount}\\\text{of}\\\text{damage}\end{array} = \begin{array}{c}\text{Insurance}\\\text{company}\\\text{payment}\end{array}$$

If the building in Example 2 is insured for $120,000 and is partially destroyed and damage amounts to $70,000, the insurance company will pay only $52,500. The steps used to calculate the insurance company payment are illustrated in Example 3.

EXAMPLE 3   Find the Amount the Insurance Company Will Pay When a Building Is Insured for Less Than 80 Percent

---

STEP ☐1   State the coinsurance formula.

$$\frac{\text{Actual coverage}}{\text{Required coverage}} \times \begin{array}{c}\text{Amount}\\\text{of}\\\text{damage}\end{array} = \begin{array}{c}\text{Insurance company}\\\text{payment}\end{array}$$

$120,000 ÷ 160,000 = 0.75   **STEP** ☐2   Divide the amount of actual coverage (120,000) by the required amount of coverage ($160,000) determined in Example 2.

CALCULATOR TIP

Depress clear key
Enter 120,000   Depress   ÷
Enter 160,000   Depress   =
Read the answer 0.75
Enter 70,000   Depress   ×
Enter 0.75   Depress   =
Read the answer $52,500

70,000 × 0.75 = $52,500   **STEP** ☐3   Multiply the amount of damage ($70,000) by the answer from step 2 (0.75).

Had the building in Example 2 been insured for at least 80 percent of the replacement cost, the insurance company would have paid the entire dollar amount of the loss ($70,000).

AUTOMOBILE COVERAGE

The main coverages for automobile insurance can be grouped into two categories—liability coverage and property damage coverage.

Automobile liability insurance is insurance that covers financial losses resulting from injuries or damages caused by the insured vehicle. Most automobile policies have a split-liability limit that contains three numbers. For example, the liability limits stated on a typical policy are 50/100/50. The first two numbers indicate the maximum amounts the company will pay for bodily injury. Bodily injury liability pays medical bills and other costs in the event that an injury or death results from an automobile accident in which the policyholder is at fault. In the above example, the insurance company will pay up to $50,000 to each person injured and up to $100,000 to all those injured in a single accident. Property damage liability coverage pays for the repair of damage that the insured vehicle does to the property of another person. In the above example, the third number (50) indicates that the insurance company will pay up to $50,000 for property damage.

Automobile property damage insurance is insurance that covers damage to the insured vehicle. **Collision insurance** pays for the repair of damage to the insured vehicle as a result of an accident. Most collision policies include a deductible amount—anywhere from $50 up—that the policyholder must pay. The insurance company then pays either the remaining cost of the repairs or the actual cash value of the vehicle, whichever is less.

Comprehensive insurance covers damage to the insured vehicle caused by fire, theft, hail, dust storm, vandalism, and almost anything else except normal wear and tear. Like collision coverage, comprehensive insurance also has a deductible. In addition to collision and comprehensive insurance, it is also possible to purchase coverage for uninsured motorists, towing, and rental car reimbursement.

THE COST OF AUTOMOBILE INSURANCE

The premium you pay for automobile insurance is influenced by several factors. The main factors are automobile type, where the car is primarily driven, individual's driving record, and age of the driver. Typical rates for automobile liability are illustrated in Table 67–2. Using the information in this table, the semiannual premium for 50/100/50 liability coverage for Mike Boston is $566.32 as illustrated in Example 4.

TABLE 67–2 Liability Premiums for Bodily Injury and Property Damage—Auto Class 6B (Semiannual Rates)

Class	Bodily Injury					Property Damage			
	20/40	25/50	50/100	100/200	100/300	15	25	50	100
1A	101.33	111.46	129.70	146.93	156.05	62.48	63.73	66.85	69.35
1B	116.55	128.21	149.18	169.00	179.49	71.93	73.37	76.97	79.84
1C	101.33	111.46	129.70	146.93	156.05	62.48	63.73	66.85	69.35
2A–1	291.90	321.09	373.63	423.26	449.53	180.08	183.68	192.69	199.89
2A–2	177.45	195.20	227.14	257.30	273.27	109.20	111.38	116.84	121.21

TABLE 67–3   Premiums for Collision and Comprehensive Insurance Coverage —Auto Class 6B (Semiannual Rates)

Class	Collision			Comprehensive		
	$100 Ded.	$250 Ded.	$500 Ded.	$100 Ded.	$250 Ded.	$500 Ded.
1A	$120	$ 98	$ 76	$60	$48	$38
1B	139	111	83	72	57	43
1C	120	98	76	60	48	38
2A–1	226	180	145	86	69	52
2A–2	210	168	126	79	63	47

EXAMPLE 4   Determine the Liability Premium for Mike Boston

Mike Boston is married and under the age of 21. Therefore, the insurance company includes Mr. Boston in the 2A-1 class. If he chooses 50/100/50 liability coverage, what is his premium?

CALCULATOR TIP

Depress clear key
Enter 373.63   Depress   +
Enter 192.69   Depress   =
Read the answer $566.32

The semiannual premium is $373.63.

STEP [1]   In Table 67–2 locate the line for the 2A-1 class and move across that row to the column for 50/100 bodily injury coverage.

The semiannual premium is $192.69.

STEP [2]   In Table 67–2, locate the line for the 2A-1 class and move across that row to the column for 50 property damage.

$373.63
+192.69
$566.32

STEP [3]   Add the answers from step 1 ($373.63) and step 2 ($192.69).

Table 67–3 can be used to find the premium for collision and comprehensive coverage. Using the information in this table, Mike Boston will pay $249 for collision and comprehensive coverage with a $250 deductible.

EXAMPLE 5   Determine the Premium for Collision and Comprehensive Coverage for Mike Boston

The semiannual premium is $180.

STEP [1]   In Table 67–3, locate the line for the 2A-1 class and move across that row to the column for collision coverage with $250 deductible.

The semiannual premium is $69.

STEP [2]   In Table 67–3, locate the line for the 2A-1 class and move across that row to the column for comprehensive coverage with $250 deductible.

$180
+ 69
$249

STEP [3]   Add the answers from step 1 ($180) and step 2 ($69).

It is now possible to determine the total semiannual premiums for all coverage purchased by Mike Boston.

EXAMPLE 6 Determine Total Automobile Insurance Premiums

$373.63	bodily injury
192.69	property damage
180.00	collision
69.00	comprehensive
$815.32	total semiannual insurance premium

Complete the Following Problems

If necessary, round your answer to two decimal places.

T F **1.** Premiums for a building located in the city or a rural setting are the same.

T F **2.** The type of materials used to construct a building does not influence the premiums paid for fire insurance.

T F **3.** The typical coinsurance clause requires that property be insured for 80 percent of its value.

Use Table 67–1 for problems 4–21.

Replacement Value	Type of Construction	Location	Premium for Building	Premium for Contents
$ 70,000	Brick	City	4. _$168_ $70,000 ÷ $100 = 700 700 × $.24 = $168	5. _700 × $.32 = $224_
$120,000	Stucco	Rural	6. _____	7. _____
$ 45,000	Frame	Rural	8. _____	9. _____
$160,000	Brick Veneer	City	10. _____	11. _____
$ 67,500	Frame	Rural	12. _____	13. _____
$112,400	Brick	City	14. _____	15. _____

Replace-ment Value	Type of Construction	Location	Premium for Building	Premium for Contents
$ 82,300	Brick Veneer	City	16. _____	17. _____

18. The premium for the building portion of a fire policy is $320. The premium for the contents portion is $405. What is the total premium for this policy?

19. You are an employee in a small insurance office. The agent asks you to talk with two customers—Nancy and Lucas Antonio—who are interested in purchasing fire insurance on a brick veneer building located in a major city. The replacement value for the building is $248,700. What is the premium for coverage of the building?

20. In problem 19, what is the premium for coverage of the building's contents?

21. In problem 19, what is the total insurance premium for the building and contents?

22. If the coinsurance clause requires that policyholders insure their property for 70 percent of the replacement value, what is the required amount of coverage for a building that has a replacement value of $152,000?

23. The All-Star Insurance Company includes a coinsurance clause in their fire insurance policies. If the coinsurance clause requires that policyholders insure their property for 80 percent of the replacement value, what is the required amount of coverage for a building that has a replacement value of $210,000?

24. A frame building has a replacement value of $100,000, but the policyholder has decided to insure the building for $70,000. Assuming there is a coinsurance clause that requires that the property be insured for 80 percent of its replacement value, how much will the insurance company pay if the building is partially destroyed and damage amounts to $40,000?

25. A brick veneer building has a replacement value of $250,000, but the policyholders have decided to insure the building for $150,000. Assuming there is a coinsurance clause that requires that the property be insured for 80 percent of its replacement value, how much will the insurance company pay if the building is partially destroyed and damage amounts to $110,000?

26. A stucco building has a replacement value of $90,000, but the policyholders have decided to insure the building for $50,400. Assuming there is a coinsurance clause that requires that the property be insured for 70 percent of its replacement value, how much will the insurance company pay if the building is partially destroyed and damage amounts to $32,000?

27. T F Automobile liability insurance is insurance that covers damage to the insured vehicle.

28. T F Comprehensive insurance pays for the repair of damage to the insured vehicle as a result of an accident.

Use Table 67–2 and Table 67–3 to answer questions 30–50.

Liability Coverage	Driver Class	Premium for Bodily Injury	Premium for Property Damage
20/40/15	1A	29. _____	30. _____
25/50/50	1C	31. _____	32. _____
50/100/50	2A-2	33. _____	34. _____
20/40/50	2A-1	35. _____	36. _____
50/100/100	1B	37. _____	38. _____

Driver Class	Deductible	Premium for Collision	Premium for Comprehensive
1A	$250	39. _____	40. _____
2A-1	$500	41. _____	42. _____
1B	$100	43. _____	44. _____
2A-2	$500	45. _____	46. _____

Personal Math Problem

Jane Shu is trying to decide which deductible to choose for her collision and comprehensive insurance. According to her insurance agent, she is in the 1B class.

47. What is the premium for collision and comprehensive coverage if she chooses a deductible of $100?

48. What is the premium for collision and comprehensive coverage if she chooses a deductible of $500?

49. If you were Ms. Shu, which deductible would you choose? Why? _____

_____

_____

_____

50. You are an assistant in the Drive-Right Insurance Agency. Mathew Brokin, the agent, asks you to determine the policy premium for Emerson Peterson. According to the agent, Mr. Peterson is classified as 1C and would like 50/100/50 liability coverage. Mr. Peterson would also like collision and comprehensive coverage with a $500 deductible. What is Mr. Peterson's total premium?

INSTANT REPLAY

UNIT	IMPORTANT POINTS TO REMEMBER	EXAMPLES
UNIT 63 Sales Taxes and Property Taxes	Sales tax amounts can be automatically calculated by cash register, by using sales tax tables, or by using the portion (P = B × R) formula.	$35 purchase and a 5 percent sales tax **$35 × 0.05 = $1.75** **$35 + $1.75 = $36.75**
	In most cases, property taxes are calculated on a portion of the assessed value of the property. Calculating property taxes is a practical application of the portion (P = B × R) formula.	$75,000 building when the tax rate is $1.22 per $100 **$75,000 ÷ $100 = 750** **750 × $1.22 = $915**
UNIT 64 Personal Income Taxes	Basically, gross income *less* adjustments *less* deductions equals taxable income. Taxpayers use IRS tax tables to determine their tax amount.	**$50,100** gross income **1,100** less adjustments **8,200** less itemized deductions **2,350** less 1 exemption **$38,450** taxable income Based on Table 64–2 for a single taxpayer, the tax is $7,900.

continued from page 394

UNIT	IMPORTANT POINTS TO REMEMBER	EXAMPLES
UNIT 65 Corporate Income Tax	Corporations pay federal income tax on their taxable income. Generally, the process involves computing net sales, deducting cost of goods sold and allowable business expenses. Then it is possible to determine the amount of corporate income tax owed.	$3,200,000 net sales 1,900,000 less cost of goods sold 750,000 less operating expenses $ 550,000 taxable income Based on Table 65–1, the corporate tax is $187,000.
UNIT 66 Health and Life Insurance	The portion formula (P = B × R) can be used to determine the amount of health care premium the employer and employee pay.	Monthly health premium $180; employer pays 80% **$180 × 0.80 = $144** (employer) **$180 – $144 = $ 36** (employee)
	When a claim is made, a deductible must be subtracted from allowable medical expenses. Then the portion formula (P = B × R) can be used to determine the amount the insurer and policyholder will pay.	$500 deductible; approved medical expenses $3,200; and a coinsurance clause of 70% **$3,200 – $500 = $2,700** **$2,700 × 0.70 = $1,890**
	The type of policy, age, and whether the policyholder is male or female affect the premiums paid for life insurance. Actual premiums are calculated using tables where rates per $1,000 of coverage are listed.	Policyholder is a female age 40; Coverage amount is $50,000. **$50,000 ÷ $1,000 = 50** (Table 66–1) **50 × $2.72 = $136**
UNIT 67 Fire and Automobile Insurance	Premiums for fire insurance depend on the construction of the building, whether risk-reduction devices are installed, and where the building is located. Actual premiums for building and contents are calculated using tables where rates per $100 of coverage are listed.	Building is brick with a replacement value of $130,000, located in a city. **$130,000 ÷ $100 = 1,300** (Table 67–1) **1,300 × $0.24 = $312** (building) **1,300 × 0.32 = $416** (contents) **$312 + $416 = $728** total premium

continued from page 395		
UNIT	**IMPORTANT POINTS TO REMEMBER**	**EXAMPLES**

UNIT 67
Fire and Automobile Insurance (concluded)

The following coinsurance formula can be used to determine the insurance company payment.

$$\frac{\text{Actual coverage}}{\text{Required coverage}} \times \begin{array}{c}\text{Amount}\\\text{of}\\\text{damage}\end{array} = \begin{array}{c}\text{Insurance}\\\text{company}\\\text{payment}\end{array}$$

Premiums for automobile insurance are influenced by the factors of automobile type, where the car is driven, individual's driving record, and age of the driver. Actual premiums are calculated by using tables for liability (bodily injury and property), collision, and comprehensive coverage.

Replacement value, $120,000; coinsurance clause 80%; actual coverage $80,000; amount of damages $40,000

$120,000 × 0.80 = $96,000
Required

$\frac{\$80,000}{\$96,000}$ × **$40,000 =** Insurance company payment

0.83 × 40,000 = $33,200

Driver is in the 1A class; liability coverage is 50/100/50; collision and comprehensive coverage has a $500 deductible.

(Table 67–2)
$129.70 bodily injury
$ 66.85 property damage

(Table 67–3)
$76 collision
$38 comprehensive

$129.70
　66.85
　76.00
　38.00
$310.55 **total semiannual premium**

CHAPTER 12	MASTERY QUIZ

DIRECTIONS: *Round each answer to two decimal places or hundredths. (Each answer counts 5 points.)*

1. On Monday, Bobbie Nations purchased three blouses at $19.98 each; one belt for $11.00; and one skirt for $39.50. What is the sales tax amount if the sales tax rate is 6 percent?

2. In problem 1, what is the total amount of the purchase including the sales tax?

For problems 3–5, assume the property tax rate is $1.18 per $100 of assessed value.

Assessed Value	Property Tax Amount
$ 62,000	**3.** _____
$153,400	**4.** _____
$114,200	**5.** _____

6. During the past tax year, Susan Deitz earned $18,900. She also received dividends that totaled $320 and interest that totaled $570. What is the gross income for Ms. Deitz?

For problems 7–9, use Table 64–2.

Taxable Income	Filing Status	Tax Amount
$35,210	Single	**7.** _____
$37,455	Married filing jointly	**8.** _____
$36,142	Single	**9.** _____

10. During 1995, Jackson Clothing, Inc., had gross sales of $658,900. The corporation also had sales returns that total $17,200 and sales allowances that total $8,150. What is the firm's net sales?

For problems 11–12, use Table 65–1.

11. What is the corporate income tax for a corporation that has taxable income of $189,000?

12. What is the corporate income tax for a corporation that has taxable income of $366,000?

13. According to the current union contract, the Swedelund Plastics Corporation must pay 90 percent of the health care cost for workers. If each worker's health care coverage cost $324 a month, what is the Swedelund's portion?

14. A health care policy requires that the insurance company pay 80 percent of all approved medical expenditures. If a policyholder has approved medical expenditures of $6,420, what is the amount the insurance company will pay?

For problems 15–16, use Table 66–1.

Life Insurance Amount	Age	Male/Female	Annual Premium Amount
$100,000	30	Female	**15.** _____
$ 50,000	60	Male	**16.** _____

17. Peter and Cindy Tolbert want to purchase coverage for a frame building that has a replacement value of $72,000. The building is located in the city. What is the total premium for the building and contents? (Use Table 67–1.)

18. A brick building has a replacement value of $400,000, but the policyholder has decided to insure the building for $280,000. Assuming there is a coinsurance clause that requires that the property be insured for 80 percent of its replacement value, how much will the insurance company pay if the building is partially destroyed and damage amounts to $120,000?

19. Susan Clemson wants to purchase 100/200/100 liability coverage. According to her insurance agent, her classification is 2A-2. What is the premium for bodily injury and property damage? (Use Table 67–2.)

20. Robert Grey is classified as 1C and wants to purchase collision and comprehensive insurance coverage. If Mr. Grey chooses a $250 deductible, what is the total premium for collision and comprehensive coverage? (Use Table 67–3.)

SPECIAL NOTE TO STUDENTS: To help you build a foundation for success in math, we have included a cumulative review at the end of every three or four chapters in the text. If you do not score at least 80 on this cumulative review, you may want to review the type of problems that you missed.

DIRECTIONS: *If necessary, round each answer to two decimal places or hundredths. (Each answer counts 5 points.)*

1. The owners of Mesquite Metal Works borrowed $12,000 at $9\frac{1}{4}$ percent for 80 days from Mesquite State Bank. What is the dollar amount of the interest? (Use a banker's year.)

2. What is the maturity date for a 120-day loan originating on June 12?

3. Find the number of days between July 2 and November 10.

National Grocer's Supply agreed to accept a $15,600 note from one of its customers. The 90-day note was dated May 8. As part of the financial arrangement, the customer agreed to pay 8 percent interest until maturity. (Use a banker's year.)

4. What is the amount of interest that the customer must pay to National Grocer's Supply?

5. What is the maturity value for this note?

6. If National Grocer's Supply discounts this note on June 20, how many days are there in the discount period?

7. If the bank charges National Grocer's Supply 11 percent to discount the note, what is the dollar amount of the bank discount?

8. Find the compound principal value for a $22,000 investment that pays 6 percent interest compounded semiannually for 4 years. (Use Table 54–1.)

9. Find the present value amount that can be invested at 5 percent compounded annually in order to obtain $40,000 in 8 years. (Use Table 55–1.)

10. What is the monthly payment amount for principal and interest for an $82,500 home mortgage if the interest rate is 9 percent and the repayment period is 20 years? (Use Table 56–1)

11. Find the total repayment amount for a 30-year home mortgage if the monthly payment is $910.

12. Universal Printing purchased a new computerized press that cost $15,400. The press has a useful life of five years and a salvage value of $1,000. If the straight-line method of depreciation is used, what is the first-year depreciation expense?

13. The Flower Source Company purchased a new refrigerated storage unit that cost $4,000. It is estimated that the unit has a useful life of eight years with a $500 salvage value. If the sum-of-the-years' digits method is used, what is the first-year depreciation expense?

14. Martin Gonzales purchased a new Ford truck for business purposes. The truck cost $15,200. It has an estimated life of five years with no salvage value. If the declining-balance method is used, what is the first-year depreciation expense?

15. In problem 14, what is the first-year depreciation expense if Mr. Gonzales uses the modified accelerated cost recovery system? (Use Table 62–1 and 62–2.)

16. On Tuesday, Bill Peterson purchased two shirts at $29.95 and three pairs of pants at $34.95. What is the sales tax amount if the sales tax rate is 7 percent?

17. The assessed value for a piece of real estate is $144,000. If the property tax rate is $1.24 per $100 of assessed value, what is the property tax amount?

18. Maxine and John McCormick had taxable income of $35,600. If this married couple files a joint return, what is the tax amount they should pay? (Use Table 64–2.)

19. What is the annual premium for a $50,000 term life insurance policy when the insured is male and 40 years old? (Use Table 66–1.)

20. A brick building has a replacement value of $300,000, but the policyholder has decided to insure the building for $180,000. Assuming there is a coinsurance clause that requires that the property be insured for 80 percent of its replacement value, how much will the insurance company pay if the building is partially destroyed and damage amounts to $100,000?

Evaluation of
Financial Statements

CHAPTER PREVIEW

We begin this chapter with an overview of two important financial statements: the income statement and the balance sheet. By analyzing each of these statements, managers and lenders can obtain answers to a variety of questions about a firm's ability to do business and stay in business, its ability to repay its debts, and its profitability. Investors can also analyze the balance sheet and income statement to determine whether a company is a good investment. Then we examine several important financial ratios that measure a company's financial health. These ratios can be compared with the company's past ratios, with those of competitors, and with industry averages.

MATH TODAY

ANYTIME, ANYWHERE, THE MAGIC OF AT&T

American Telephone & Telegraph (AT&T) was the first company to receive *two* Malcolm Baldrige National Quality Awards. Presented by the U.S. Department of Commerce, this award is given to U.S. companies with superior products, excellent service, and exceptional employee and customer relations. The firm, always known for its world-class research and development laboratories, is now in a position to use the most advanced technology to market state-of-the-art telecommunications products and services.

In 1992, the last year for which complete financial results were available, the firm had sales revenues of over $64.9 billion and net income in excess of $3.8 billion. Although both sales revenues and earnings hit record highs, the annual report for AT&T provides more financial insights that both managers and stockholders can use to evaluate the company. For example, AT&T had a net profit margin of 5.9%, a debt to asset ratio of 46.1%, and the firm's earnings per share were $2.86. Of particular interest to stockholders, the price of a share of stock increased from $39 to $51 a share—a 30 percent increase in value in just 12 months. In addition, the company paid dividends totaling $1.32 a share.

Financial analysts are quick to point out the connection between quality, innovation, and financial performance. And because of AT&T's commitment to quality and innovation, the financial picture for the firm will be strong for years to come.

Source: For more information, see *AT&T 1992 Annual Report;* Burt Ziegler, "American Telephone & Multimedia?" *Business Week,* September 6, 1993, pp 78–79; Thomas Jaffee, "Ma Bellwether?" *Forbes,* March 1, 1993, p. 144, *Moody's Handbook of Common Stocks Fall, 1993,* Moody's Investors Service, 99 Church Street, New York, NY 10007; Gary Hoover, *Hoover's Handbook of American Business 1993,* The Reference Press, 1992, p. 116.

UNIT 68	ANALYZING THE INCOME STATEMENT

LEARNING OBJECTIVES   *After completing this unit, you will be able to:*

1. read and interpret an income statement.
2. calculate the percent of increase or decrease for the same amounts taken from two different accounting periods.
3. calculate the percent of net sales for each item listed on a firm's income statement.

The true story behind the information contained in an annual report is not the upbeat letter from the chairman of the board or in the glossy pictures—it's in a corporation's financial statements. Two financial statements—the income statement and balance sheet—are considered essential tools for managing a business. In this unit, we discuss the income statement. In Unit 69, we examine the balance sheet.

An **income statement** is a summary of a firm's revenues and expenses during a specified accounting period. This statement may be prepared monthly, quarterly, semiannually, or annually. Table 68–1 shows the income statement for Southwest Window Tint and Alarm, Inc. Note that it consists of four sections.

REVENUES

The revenues section of the income statement begins with gross sales. **Gross sales** are the total amount of all goods and services sold during the accounting period. From this amount are deducted the amounts for sales returns, sales allowances, and sales discounts. The difference is called net sales. **Net sales** are the actual amount received by the firm for the goods and

TABLE 68–1 Southwest Window Tint and Alarm, Inc.

Southwest Window Tint and Alarm, Inc.
Income Statement
(For the year ended December 31, 1994)

Revenues			
Gross sales		$510,000	
Less sales returns and allowances	$ 8,000		
Less sales discounts	3,100	11,100	
Net sales			$498,900
Cost of goods sold			
Beginning inventory, Jan. 1, 1994		$ 60,000	
Purchases	$360,000		
Less purchase discounts	15,000		
Net purchases		345,000	
Cost of goods available for sale		$405,000	
Less ending inventory, Dec. 31, 1994		70,000	
Cost of goods sold			$335,000
Gross profit on sales			$163,900
Operating expenses			
Salaries		$65,000	
Advertising		21,000	
Sales promotion		11,000	
Rent		4,500	
Depreciation		9,000	
Utilities		1,200	
Insurance		1,100	
Miscellaneous		1,000	
Total Operating Expenses			113,800
Net income before taxes			$ 50,100
Less federal income taxes			7,525
Net income after taxes			42,575

services it has sold, after adjustment for returns, allowances, and discounts. For Southwest, net sales are $498,900.

COSTS OF GOODS SOLD

The standard method of determining the costs of goods sold may be summarized as follows.

$$\text{Costs of goods sold} = \text{Beginning inventory} + \text{Purchases} - \text{Ending inventory}$$

According to Table 68–1, Southwest began its accounting period with merchandise inventory that cost $60,000. During the period, the firm purchased merchandise worth $345,000. Thus, during the year, Southwest had goods available for sale valued at $60,000 + $345,000 = $405,000. At the end of the accounting period, Southwest had an ending inventory of $70,000. The cost of goods sold by Southwest was therefore $405,000 *less* $70,000, or $335,000.

A firm's **gross profit** is its net sales *less* the cost of goods sold. For Southwest, gross profit on sales was $163,900.

OPERATING EXPENSES

A firm's **operating expenses** are those costs that do not result directly from the purchase or manufacture of the products it sells. Typical expenses in this section of the income statement include salaries, advertising and other promotional expenses, all costs of operating retail stores, administrative offices, and factories. As illustrated in Table 68–1, total operating expenses for Southwest are $113,800.

NET INCOME

Net income is the profit earned (or the loss suffered) by a firm during an accounting period. As illustrated in Table 68–1, Southwest's net income before taxes totals $50,100. Based on current federal rates, Southwest must pay corporate income tax of $7,525. This leaves Southwest with a net income after taxes of $42,575.

PERCENT OF INCREASE OR DECREASE

Often, managers and owners of a business compare individual amounts for two or more accounting periods. This type of horizontal analysis compares each item in one year with the same item for the previous year. To calculate the percent of increase or decrease, follow these steps:

1. Find the dollar amount of increase or decrease by subtracting the smaller number from the larger number.
2. Divide the dollar amount of increase or decrease by the total amount for the earlier period.
3. Convert the decimal answer to a percent.

EXAMPLE 1　If 1994 Net Sales Were $498,900 and 1993 Net Sales Were $476,200, What Is the Percent of Increase?

$498,900
−476,200
$ 22,700

STEP 1　Find the amount of increase or decrease by subtracting the smaller number from the larger number.

$22,700 ÷ $476,200 = 0.048

STEP 2　Divide the dollar amount of increase or decrease from step 1 by the individual amount for the earlier period.

0.048 = 4.8%

STEP 3　Convert the decimal answer to a percent.

CALCULATOR TIP

Depress clear key
Enter 498,900　Depress　−
Enter 476,200　Depress　=
Read the answer $22,700
Enter 22,700　　Depress　÷
Enter 476,200　Depress　=
Read the answer 0.048

This calculation is a practical application of the rate formula $(R = P \div B)$ presented in Unit 27. The amount of increase or decrease is the *portion.* The amount for the earlier period is the *base.*

PERCENT OF NET SALES

The percent of net sales for each account included in the income statement can be found by dividing the dollar amount for each account by the total dollar amount for net sales. To calculate the percent of net sales, follow these steps:

1. Divide the individual amount by the total amount for net sales.
2. Convert the decimal answer to a percent.

EXAMPLE 2 | Determine the Percent of Net Sales for the Salary Expense When Net Sales Are $498,900 and Salaries Are $65,000

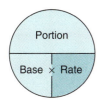

$65,000 ÷ $498,900 = 0.13	**STEP** 1 Divide the individual amount by the total amount for net sales.
0.13 = 13%	**STEP** 2 Convert the decimal answer to a percent.

CALCULATOR TIP

Depress clear key
Enter 65,000   Depress ÷
Enter 498,900   Depress =
Read the answer 0.13

This calculation is a practical application of the rate formula (R = P ÷ B) presented in Unit 27. The individual amount is the *portion*. The net sales amount is the *base*.

Complete the Following Problems

For all problems that do not work out evenly, carry your answers to two decimal places.

1. In your own words, describe the purpose of an income statement. _____

_____

_____

2. T   F   For practical purposes, gross sales and net sales mean the same thing.

3. T   F   Cost of goods sold equals beginning inventory plus purchases minus ending inventory.

4. T   F   The federal government does not tax the net income earned by a corporation.

5. During 1994, Office Products of Western Kansas had gross sales of $864,250. Sales returns were $21,400. Sales allowances were $12,210. Sales discounts were $34,600. What were the firm's net sales?

$864,250 – $21,400 – $12,210 – $34,600 = $796,040

6. On January 1, Stanley Pharmacy and Drugs had a beginning inventory valued at $24,500. During the year, the drug store purchased goods valued at $184,600. At the end of the year, the firm's ending inventory was valued at $32,180. What was the firm's cost of goods sold?

7. During the month of October, the Trails Inn Motel had the following expenses: (a) salaries—$2,456.78; (b) advertising—$785.40; (c) utilities—$1,900.24; (d) insurance—$950.00. What is the total of the motel's operating expenses?

8. During the month of February, Armadillo Country Store had net sales of $56,450. The firm's cost of goods sold was $21,456. The firm also had operating expenses of $7,911. What is the firm's net income?

HINT: Before completing the following problems, you may want to review the material in Unit 27, finding the rate.

9. 1994 net sales for K & L Carpets were $224,789. 1993 net sales were $195,670. What is the amount of change?

10. In problem 9, what is the percent of increase or decrease?

11. During 1994, Allied Distributing had insurance expense amounting to $8,677. During 1993, the firm had insurance expense totaling $9,700. What is the amount of change?

12. In problem 11, what is the percent of increase or decrease?

13. Income Statements for Cypress Springs Boat Sales for 1994 and 1993 are provided. Determine the amount of change and percent of increase or decrease for each account listed.

Cypress Springs Boat Sales Income Statement
(For the years ended December 31, 1994 and 1993)

	1994	1993	Amount of Change	% of Change
Net sales	$360,000	$290,000	_____	_____
Cost of goods sold	190,000	$150,000	_____	_____

	1994	1993	Amount of Change	% of Change
Gross profit on sales	$170,000	$140,000	_____	_____
Operating expenses	68,000	59,000	_____	_____
Net income before taxes	$102,000	$81,000	_____	_____

14. Net sales for Cade's Building Materials were $562,400. If the firm's costs of goods sold total $188,700, what is the percent of net sales for cost of goods sold?

15. During March, Box Plumbing, Inc., had net sales of $64,782. During this same period, this business also had rent expense of $1,250. What is the percent of net sales for rent expense?

16. An income statement for Heritage Furniture and Appliance is provided. Determine the percent of net sales for each account listed.

Heritage Furniture and Appliance
Income Statement
(For the year ended December 31, 1994)

	Dollar Amount	% of Net Sales
Net sales	$784,000	*100%*
Cost of goods sold	313,600	_____
Gross profit on sales	470,400	_____
Less operating expenses		
Salaries expense	186,000	_____
Advertising	52,000	_____
Sales promotion	13,000	_____
Rent	3,900	_____
Depreciation	82,000	_____
Utilities	2,100	_____
Insurance	3,200	_____
Miscellaneous	1,000	_____
Net income before taxes	$127,200	_____

Critical Thinking Problem

The account balances for Mount Pleasant Optical for the month of May 1994 are as follows:

Net sales	$29,450
Cost of goods sold	11,200
Operating expenses	8,500

17. From the above information, prepare an income statement for this business.

18. If net income for 1993 was $11,200, given the information above, what is the amount of increase or decrease for net income?

19. Given the above information, what is the percent of increase or decrease for net income?

20. In your own words, describe what you would do to improve the financial performance of Mt. Pleasant Optical if you were the firm's owner. _____

_____

_____

UNIT 69	ANALYZING THE BALANCE SHEET

LEARNING OBJECTIVES: *After completing this unit, you will be able to:*

1. read and interpret a balance sheet.
2. calculate the percent of increase or decrease for amounts taken from two different accounting periods.
3. calculate the percent of total assets for each item listed in the asset section of a balance sheet.
4. calculate the percent of total liabilities and owners' equity for each item listed in the equity section of the balance sheet.

TABLE 69–1 Southwest Window Tint and Alarm, Inc.

Southwest Window Tint and Alarm, Inc.
Balance Sheet
(December 31, 1994)

Assets

Current assets

Cash	$ 64,000	
Marketable securities	20,000	
Accounts receivable	35,000	
Notes receivable	28,000	
Inventory	70,000	
Prepaid expenses	2,000	
Total current assets		$219,000

Fixed assets

Delivery equipment	$125,000		
Less accumulated depreciation	20,000	$105,000	
Store equipment	75,000		
Less accumulated depreciation	15,000	60,000	
Total fixed assets			165,000
Total assets			$384,000

Liabilities and owners' equity

Current liabilities

Accounts payable	62,000	
Notes payable	42,000	
Salaries payable	5,900	
Taxes payable	2,100	
Total current liabilities		$112,000

Long-term liabilities

Mortgage payable on store equipment	30,000	
Total long-term liabilities		30,000
Total liabilities		$142,000

Owners' equity

Common stock, 20,000 shares at $10 par value	$200,000	
Retained earnings	42,000	
Total owners' equity		$242,000
Total liabilities and owners' equity		$384,000

A **balance sheet** (or **statement of financial position**) is a summary of a firm's assets, liabilities, and owners' equity accounts at a particular time. A balance sheet is prepared at least once a year. Most firms also have balance sheets prepared semiannually, quarterly, or monthly.

Table 69–1 shows the balance sheet for Southwest Window Tint and Alarm, Inc. Note that assets are reported at the top of the statement, followed by liabilities and owners' equity. Let us work through the accounts in Table 69–1, from top to bottom.

ASSETS

Assets are the items of value that a firm owns—cash, inventories, land, equipment, buildings, and the like. Generally, assets are divided into two categories. **Current assets** are assets that will be used or converted to cash in one year or less. Typical current assets include cash, securities, accounts receivable, notes receivable, merchandise inventory, and prepaid expenses. For Southwest, current assets total $219,000.

Fixed assets are assets that will be held or used for a period longer than one year. They generally include land, buildings, and equipment. Note that the values of the fixed assets listed in Table 69–1 are decreased by their accumulated depreciation. Depreciation—a topic covered in Chapter 11—is the process of apportioning the cost of a fixed asset over the period during which it will be used. The amount allotted to each year is an expense for that year, and the value of the asset must be reduced by that expense amount. For example, the cost of Southwest's delivery equipment ($125,000) has been reduced by $20,000. Its value at this time is thus $125,000 less $20,000, or $105,000. For Southwest, fixed assets total $165,000. Now it is possible to total both current and fixed assets. As illustrated in Table 69–1, total assets are $384,000.

LIABILITIES AND OWNERS' EQUITY

The liability and owners' equity accounts complete the balance sheet. **Liabilities** are the firm's debts and obligations—what it owes to others. Generally, these accounts are separated into two groups—current and long-term—on the balance sheet. A firm's **current liabilities** are debts that will be repaid in one year or less. Typical current liabilities include accounts payable, notes payable, salaries payable, and taxes payable. For Southwest, current liabilities total $112,000.

Long-term liabilities are debts that need not be repaid for at least one year. Southwest lists only a $30,000 mortgage payable in this group. Bonds and other long-term loans would be included here as well, if they existed. As illustrated in Table 69–1, current and long-term liabilities total $142,000.

Owners' equity is the difference between a firm's assets and its liabilities—what would be left over for the firm's owners if assets were used to pay off liabilities. To finance the cost of business start-up, Southwest sold common stock. Its value is shown as its par value ($10) times the number of shares originally sold (20,000). In addition, $42,000 of Southwest's earnings have been reinvested in the business since it was founded. Thus, owners' equity totals $242,000.

As the two grand totals show, Southwest's assets and the sum of its liabilities and owners' equity are balanced or equal at $384,000.

PERCENT OF INCREASE OR DECREASE

Often, managers and owners use horizontal analysis to analyze each balance sheet item in one year with the same item for the previous year. To calculate the percent of increase or decrease, follow these steps:

1. Find the dollar amount of increase or decrease by subtracting the smaller number from the larger number.
2. Divide the dollar amount of increase or decrease by the total amount for the earlier period.
3. Convert the decimal answer to a percent.

EXAMPLE 1	If the Cash Balance in 1994 Was $64,000 and the Cash Balance in 1993 Was $52,000, What Is the Percent of Increase?

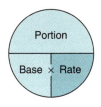

$64,000 −52,000 $12,000	**•STEP** 1	Find the amount of increase or decrease by subtracting the smaller number from the larger number.
$12,000 ÷ $52,000 = 0.23	**STEP** 2	Divide the dollar amount of increase or decrease by the individual amount for the earlier period.
0.23 = 23%	**STEP** 3	Convert the decimal answer to a percent.

CALCULATOR TIP

Depress clear key
Enter 64,000 Depress −
Enter 52,000 Depress =
Read the answer $12,000
Enter 12,000 Depress ÷
Enter 52,000 Depress =
Read the answer 0.23

This calculation is a practical application of the rate formula ($R = P ÷ B$) presented in Unit 27. The amount of increase or decrease is the *portion.* The amount for the earlier period is the *base.*

PERCENT OF TOTAL ASSETS OR PERCENT OF TOTAL LIABILITIES AND OWNERS' EQUITY

The percent of total assets for each account included in the balance sheet can be found by dividing the dollar amount for the account by the dollar amount for total assets. To calculate the percent of total assets, follow these steps:

1. Divide the individual amount by the amount for total assets.
2. Convert the decimal answer to a percent.

EXAMPLE 2	Determine the Percent of Total Assets for Inventory When Total Assets Are $384,000 and Inventory Is $70,000

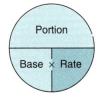

$70,000 ÷ $384,000 = 0.18	**STEP** 1	Divide the individual amount by the amount for total assets.
0.18 = 18%	**STEP** 2	Convert the decimal answer to a percent.

CALCULATOR TIP

Depress clear key
Enter 70,000 Depress ÷
Enter 384,000 Depress =
Read the answer 0.18

It is also possible to find the percent for each liability and owners' equity account by dividing the dollar amount for each individual account by the total amount for liabilities and owners' equity.

EXAMPLE 3　Determine the Percent of Total Liabilities and Owners' Equity for Account Payable When Total Liabilities and Owners' Equity Are $384,000 and Accounts Payable Is $62,000

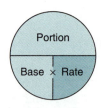

| Portion |
| Base × Rate |

$62,000 ÷ $384,000 = 0.16　**STEP** 1　Divide the individual amount by the amount for total liabilities and owners' equity.

0.16 = 16%　**STEP** 2　Convert the decimal answer to a percent.

CALCULATOR TIP

Depress clear key
Enter 62,000　　Depress ÷
Enter 384,000　Depress =
Read the answer 0.16

These calculations are a practical application of the rate formula (R = P ÷ B) presented in Unit 27. The individual amount is the *portion*. The amount for total assets or total liabilities and owners' equity is the *base*.

Complete the Following Problems

For all problems that do not work out evenly, carry your answers to two decimal places.

 1. In your own words, describe the purpose of the balance sheet. _____

_____

_____

_____

2. T　F　The terms *balance sheet* and *statement of financial position* mean the same thing.

3. T　F　Current assets are assets that will be held or used for a period longer than one year.

4. T　F　The total for assets should equal the total of liabilities and owners' equity.

5. At the end of 1994, Daniels Medical Laboratory had the following assets: (a) cash—$45,000, (b) marketable securities—$25,000, (c) accounts receivable—$32,500; and (d) prepaid expenses—$2,000. What is the firm's total current assets?

$45,000 + $25,000 + $32,500 + $2,000 = $104,500

6. McArthur Dry Goods owns a building. The original cost for the building was $125,000. Accumulated depreciation totals $48,000. Given this information, what amount should be reported on the firm's balance sheet for the building?

7. Tucson Electric Supply has current assets valued at $67,500. The firm's fixed assets total $89,400. What is the total for all assets?

8. Lockwood Nursery and Landscape has current liabilities that total $29,570. The firm also has a long-term mortgage with a balance of $55,600. What is the total of the firm's liabilities?

9. Back in 1985, the managers of Hess Furniture, Inc., sold 10,000 shares of stock to finance the business. The stock was sold with a par value of $25 a share. In addition, $120,000 of the firm's earnings have been reinvested in the business since it was started. What is the amount of the owners' equity?

10. Shelton's Western Wear has assets valued at $310,000. The firm also has liabilities that total $74,000. What is the amount of owners' equity?

11. The account balances for Mount Pleasant Optical for the month of May are as follows:

Current assets	$14,500
Fixed assets	45,200
Current liabilities	8,100
Long-term liabilities	10,000
Owners' equity	41,600

In the space below, use the above information to prepare a balance sheet for this business.

HINT: Before completing the following problems, you may want to review the material in Unit 27, finding the rate.

12. 1994 current assets were $56,798. 1993 current assets were $48,650. What is the amount of change?

13. In problem 12, what is the percent of increase or decrease?

14. At the end of 1994, C & S Electric had accounts payable totaling $25,670. At the end of 1993, the firm's accounts payable were $33,400. What is the amount of change?

15. In problem 14, what is the percent of increase or decrease?

16. Balance sheets for Omaha Industries for 1994 and 1993 are provided. Determine the amount of change and percent of increase or decrease for each account listed.

Omaha Industries, Inc.
Balance Sheet
(December 31, 1994 and 1993)

	1994	1993	Amount of Change	% of Change
Assets				
Current assets	$120,000	98,000	_____	_____
Fixed assets	180,000	170,000	_____	_____
Total assets	300,000	268,000	_____	_____
Liabilities and Owners' Equity				
Current liabilities	$85,000	72,000	_____	_____
Long-term liabilities	52,000	40,000	_____	_____
Owners' equity	163,000	156,000	_____	_____
Total liabilities and owners' equity	300,000	268,000	_____	_____

17. Total assets for Hidden Valley Apartments are $756,200. If the firm's cash account totals $34,800, what is the percent of total assets for the cash account?

18. At the end of December, Baxter's Auto Supply had total liabilities and owners' equity valued at $228,300. If the firm's accounts payable total $49,750, what is the percent of total liabilities and owners' equity for accounts payable?

19. A balance sheet for French Antiques of Georgia is provided. Determine the percent of total assets (or of total liabilities and owners' equity) for each account listed.

French Antiques of Georgia
Balance Sheet
(December 31, 1994)

	Dollar Amount	% of Total Assets
Current assets		
Cash	$ 70,000	
Accounts receivable	110,000	
Inventory	240,000	
Total current assets	$420,000	
Fixed assets		
Store equipment	54,000	
Delivery equipment	22,000	
Total fixed assets	$ 76,000	
Total assets	$496,000	100%

	Dollar Amount	% of Total Liabilities and Owners' Equity
Liabilities and Owners' equity		
Current liabilities		
Accounts payable	$105,000	
Total current liabilities	$105,000	
Owners' equity		
Common stock	$280,000	
Retained earnings	111,000	
Total owners' equity	$391,000	
Total liabilities and owners' equity	$496,000	100%

Critical Thinking Problem

Last year, Mathews Construction Company lost $34,600. At the end of the year, the firm had a cash balance of $38,000. It also had current liabilities totaling $42,000 that must be paid within the next 30 days.

20. What effect would the $34,600 loss have on the firm's balance sheet? _____

_____

_____

_____

21. If you were Ben Mathews—the owner of Mathews Construction Company—

what would you do to work out of this situation? _____

_____

_____

_____

_____

UNIT 70   RATIOS—TESTS OF PROFITABILITY

LEARNING OBJECTIVES: *After completing this unit, you will be able to:*

1. calculate a net profit margin.
2. determine return on equity.
3. compute earnings per share.

As we have seen, a firm's income statement summarizes a firm's operations during one accounting period. Its balance sheet provides a "financial picture" of the firm at a particular time. Both can provide answers to a variety of questions about the firm's profitability, its value as an investment, and its ability to repay its debts. Even more information can be obtained by calculating the financial ratios discussed in this unit.

A **financial ratio** is a number that shows the relationship between two elements of a firm's financial statements. In this unit, we discuss three ratios that measure a firm's profitability. The information required to form these ratios is found in the income statement (Table 68–1) and the balance sheet (Table 69–1) for Southwest Window Tint and Alarm, Inc.

NET PROFIT MARGIN

Net profit margin is a financial ratio that is calculated by dividing net income after taxes by net sales. To calculate this ratio, you may want to use the following formula:

$$\text{Net profit margin} = \frac{\text{Net income after taxes}}{\text{Net sales}}$$

EXAMPLE 1   Calculate the Net Profit Margin for Southwest Window Tint and Alarm

Net income after taxes = $42,575	**STEP** 1	Determine the net income after taxes for Southwest Window Tint and Alarm from Table 68–1.
Net sales = $498,900	**STEP** 2	Determine the net sales amount for Southwest Window Tint and Alarm from Table 68–1.
$42,575 ÷ $498,900 = 0.085	**STEP** 3	Divide the net income after taxes amount by the net sales amount.
0.085 = 8.5%	**STEP** 4	Convert the decimal answer to a percent.

CALCULATOR TIP

Depress clear key
Enter 42,575   Depress ÷
Enter 498,900   Depress =
Read the answer 0.085

The net profit margin indicates how effectively the firm is transforming sales into profits. Today, the average net profit margin for all business firms is between 4 and 5 percent. With a net profit margin of 8.5 percent, Southwest is above average. A low net profit margin (less than 4 or 5 percent) can be increased by reducing expenses or by increasing sales.

RETURN ON EQUITY

Return on equity, sometimes called return on investment, is a financial ratio calculated by dividing net income after taxes by owners' equity. The following formula can be used.

$$\text{Return on equity} = \frac{\text{Net income after taxes}}{\text{Owners' equity}}$$

EXAMPLE 2 Calculate the Return on Equity for Southwest Window Tint and Alarm, Inc.

CALCULATOR TIP
Depress clear key
Enter 42,575   Depress ÷
Enter 242,000 Depress =
Read the answer 0.176

Net income after taxes = $42,575   **STEP** 1 Determine the net income after taxes for Southwest Window Tint and Alarm from Table 68–1.

Owners' equity = $242,000   **STEP** 2 Determine the total for owners' equity for Southwest Window Tint and Alarm from Table 69–1.

$42,575 ÷ $242,000 = 0.176   **STEP** 3 Divide the net income after taxes amount by the total for owners' equity.

0.176 = 17.6%   **STEP** 4 Convert the decimal answer to a percent.

Return on equity indicates how much income is generated by each dollar of equity. Southwest is providing income of about 18 cents per dollar invested in the business. The average return on equity is between 12 and 15 percent for all firms. The only practical way to improve return on equity is to increase net income after taxes. This means reducing expenses, increasing sales, or both.

EARNINGS PER SHARE

Earnings per share is calculated by dividing net income after taxes by the number of shares of common stock outstanding. The following formula can be used.

$$\text{Earnings per share} = \frac{\text{Net income after taxes}}{\text{Number of shares of common stock}}$$

EXAMPLE 3 Calculate the Earnings per Share for Southwest Window Tint and Alarm, Inc.

CALCULATOR TIP
Depress clear key
Enter 42,575   Depress ÷
Enter 20,000   Depress =
Read the answer $2.13

Net income after taxes = $42,575   **STEP** 1 Determine the net income after taxes for Southwest Window Tint and Alarm from Table 68–1.

Number of shares of common stock = 20,000   **STEP** 2 Determine the number of shares of common stock for Southwest Window Tint and Alarm from Table 69–1.

$42,575 ÷ 20,000 = $2.13 **STEP** [3] Divide the net income after taxes amount by the number of shares of common stock.

Earnings per share is obviously a measure of the amount earned (after taxes) per share of common stock owned by investors. There is no meaningful average for this measure, mainly because the number of outstanding shares of a firm's stock is subject to change. As a general rule, however, an increase in earnings per share is a good sign for any corporation.

Complete the Following Problems

For all problems that do not work out evenly, carry your answers to three decimal places.

1. What is the formula for calculating net profit margin?

2. What is the average net profit margin for all firms?

3. If you were the financial manager of a firm, what steps could you take to improve a low net profit margin? _____

_____

_____

4. Megatron Food Imports is a local corporation that sells vitamin supplements throughout the United States. Last year, Megatron had net income after taxes of $135,640 on net sales of $2,086,768. What is the firm's net profit margin?

5. Groom's Automotive Repair had net sales of $745,690. After expenses and taxes were deducted, the firm earned $31,319. What is the firm's net profit margin?

6. What is the formula for determining return on equity?

7. What is the average return on equity for all firms?

8. If you were the manager of a firm, what steps could you take to improve a low return on equity? _____

_____

_____

9. Elite Artworks, Inc., is an art gallery located in San Francisco. During the last 12 months, the firm increased net income after taxes by 22 percent to a record $356,000. If owners' equity totals $2,086,768, what is the firm's return on equity?

10. Metroplex Gold and Silver Exchange was incorporated in 1988. Originally, four people purchased stock valued at $450,000. Since incorporation, $160,000 of the firm's earnings have been reinvested in the business. If net income after taxes is $63,440, what is the firm's return on equity?

11. What is the formula for computing earnings per share?

12. According to the text, there is no meaningful average for earnings per share. Why? _____

_____

_____

_____

13. If you were the manager of a firm, what steps could you take to increase earnings per share? _____

_____

_____

_____

14. Sweet Cream Dairy Products, Inc., earned $78,500 last year. If this corporation has issued 165,000 shares of stock, what is the firm's earnings per share?

15. During one 12-month accounting period, the Home Depot, Inc., reported income after taxes of $362,863,000. During this period, there was an average of 444,989,000 shares of common stock owned by stockholders. What was the firm's earning per share?

TABLE 70–1   Income Statement for Wal-Mart Stores, Inc.

Consolidated Statements of Income

Wal-Mart Stores, Inc. and Subsidiaries
(Amounts in thousands except per share data)   ***Fiscal year ended January 31,***

	1993
Revenues:	
Net sales	$55,483,771
Rentals from licensed departments	36,035
Other income-net	464,758
	55,984,564
Cost and Expenses:	
Cost of sales	44,174,685
Operating, selling, and general and administrative expenses	8,320,842
Interest Costs:	
Debt	142,649
Capital leases	180,049
	52,818,225
Income Before Income Taxes	3,166,339
Provision For Federal and State Income Taxes:	
Current	1,136,918
Deferred	34,627
	1,171,545
Net Income	$ 1,994,794

Source: *1993 Annual Report,* Wal-Mart Stores, Inc., Bentonville, AR 72760, p. 10.

Critical Thinking Problem

Because of its motivated employees, well-managed physical distribution system, and shrewd buying practices, Wal-Mart is now the largest retailer in the United States. It is also one of the most profitable firms in the country. The 1993 income statement for Wal-Mart Stores, Inc.—the latest year for which complete financial results are available—is presented in Table 70–1. Use the information in Table 70–1 to answer the questions below.

16. What is the net profit margin for Wal-Mart?

17. What is the return on equity for Wal-Mart? (Note: owners' equity for this accounting period was $8,759,180,000.)

18. What are the earnings per share amount for Wal-Mart? (Note: Stockholders had purchased 2,299,638,000 shares of stock when this income statement was prepared.)

Personal Math Problem

19. Based on your calculations, would you want to invest in Wall Stores, Inc.?

_____

_____

_____

_____

_____

_____

_____

UNIT 71	RATIOS—TESTS OF FINANCIAL HEALTH

LEARNING OBJECTIVES: *After completing this unit, you will be able to:*

1. calculate the current ratio.
2. compute the acid-test ratio.
3. determine inventory turnover.
4. find the debt to assets ratio.

Three financial ratios—the current ratio, the acid-test ratio, and the debt to assets ratio—permit managers (and lenders and investors) to evaluate the ability of a firm to pay its liabilities.

CURRENT RATIO

A firm's **current ratio** is computed by dividing current assets by current liabilities. To calculate this ratio, you may want to use the following formula:

$$\text{Current ratio} = \frac{\text{Current assets}}{\text{Current liabilities}}$$

EXAMPLE 1 Calculate the Current Ratio for Southwest Window Tint and Alarm

CALCULATOR TIP

Depress clear key
Enter 219,000 Depress ÷
Enter 112,000 Depress =
Read the answer 1.96

Current assets = $219,000   **STEP** [1] Determine the current asset amount for Southwest Window Tint and Alarm from Table 69–1.

Current liabilities = $112,000   **STEP** [2] Determine the current liability amount for Southwest Window Tint and Alarm from Table 69–1.

$219,000 ÷ $112,000 = 1.96   **STEP** [3] Divide the current asset amount by the current liability amount.

This ratio means that Southwest has $1.96 of current assets for every $1 of current liabilities. The average current ratio for all industries is 2.0, but it varies greatly from industry to industry. A low current ratio can be improved by repaying current liabilities, by converting current liabilities to long-term liabilities, or by increasing the firm's cash balance by reducing dividend payments to stockholders.

ACID-TEST RATIO

The **acid-test ratio** is calculated by subtracting inventories from the current asset amount and dividing the total by current liabilities. It is similar to the current ratio, except that the value of the firm's inventories does not enter into the calculation. Inventories are "removed" from current assets because they are not converted into cash as easily as other current assets. The following formula can be used.

$$\text{Acid-test ratio} = \frac{\text{Current assets} - \text{inventory}}{\text{Current liabilities}}$$

EXAMPLE 2   Calculate the Acid-Test Ratio for Southwest Window Tint and Alarm

Current assets = $219,000   **STEP** ☐1   Determine the current
Inventory = $70,000   asset amount and the
                                                 inventory amount for
                                                 Southwest Window Tint
                                                 and Alarm from Table
                                                 69–1.

$219,000 − $70,000 = $149,000   **STEP** ☐2   Subtract the inventory
                                                 amount from the current
                                                 asset amount.

CALCULATOR TIP

Depress clear key
Enter 219,000   Depress   −
Enter 70,000   Depress   =
Read the answer $149,000
Enter 149,000   Depress   ÷
Enter 112,000   Depress   =
Read the answer 1.33

Current liabilities = $112,000   **STEP** ☐3   Determine the current
                                                 liability amount for
                                                 Southwest Window Tint
                                                 and Alarm from Table
                                                 69–1.

$149,000 ÷ $112,000 = 1.33   **STEP** ☐4   Divide the total from step
                                                 2 by the current liability
                                                 amount.

The average acid-test ratio for all businesses is 1.0. Southwest is above average with an acid-test ratio of 1.33. To increase a low ratio (less than 1.0), a firm would have to repay current liabilities, obtain additional cash from investors, or convert current liabilities to long-term debt.

INVENTORY TURNOVER

A firm's inventory turnover is the number of times the firm sells and replaces its merchandise inventory in one year. **Inventory turnover** is approximated by dividing the cost of goods sold in one year by the average dollar value of inventory.

The average dollar value of inventory can be found by adding the beginning and ending inventory values (as reported on a firm's income statement) and dividing the total by 2. Once average inventory is known, you can use the following formula to calculate this ratio.

$$\text{Inventory turnover} = \frac{\text{Cost of goods sold}}{\text{Average inventory}}$$

EXAMPLE 3 Determine the Inventory Turnover for Southwest Window Tint and Alarm

Beginning inventory = $60,000 Ending inventory = $70,000	**STEP** 1 Determine the amounts for beginning inventory and ending inventory for Southwest Window Tint and Alarm from Table 68–1.

$ 60,000
+ 70,000
$130,000

STEP 2   Add the beginning and ending inventory amounts.

130,000 ÷ 2 = $65,000

STEP 3   To find the average inventory, divide the answer from step 2 ($130,000) by 2.

Cost of goods sold = $335,000

STEP 4   Determine the cost of goods sold for Southwest Window Tint and Alarm from Table 68–1.

$335,000 ÷ $65,000 = 5.15
times a year

STEP 5   Divide the cost of goods sold ($335,000) by the average inventory ($65,000).

CALCULATOR TIP

Depress clear key
Enter 60,000   Depress   +
Enter 70,000   Depress   =
Read the answer $130,000
Enter 130,000   Depress   ÷
Enter 2   Depress   =
Read the answer $65,000
Enter 335,000   Depress   ÷
Enter 65,000   Depress   =
Read the answer 5.15

Southwest Window Tint and Alarm sells and replaces its merchandise inventory 5.15 times each year, or about once every 70 days.

The higher a firm's inventory turnover, the more effectively it is using the money invested in inventory. The average inventory turnover for all firms is about 9 times per year, but turnover rates vary widely from industry to industry. For example, supermarkets may have turnover rates of 20 or higher, whereas turnover rates for furniture stores are generally well below the national average. The quickest way to improve inventory turnover is to order merchandise in smaller quantities at more frequent intervals.

DEBT TO ASSETS RATIO

The **debt to assets ratio** is calculated by dividing total liabilities by total assets. The following formula can be used.

$$\text{Debt to assets ratio} = \frac{\text{Total liabilities}}{\text{Total assets}}$$

EXAMPLE 4 Calculate the Debt to Assets Ratio for Southwest Window Tint and Alarm

Total liabilities = $142,000	**STEP** 1 Determine the total liability amount for Southwest Window Tint and Alarm from Table 69–1.

Total assets = $384,000   STEP ☐2   Determine the total asset amount for Southwest Window Tint and Alarm from Table 69–1.

$142,000 ÷ $384,000 = 0.37   STEP ☐3   Divide the total liability amount by the total asset amount.

0.37 = 37%   STEP ☐4   Convert the decimal answer to a percent.

Southwest's debt to assets ratio of 37 percent means that slightly more than one-third of its assets are financed by creditors. For all businesses, the average debt to assets ratio is 33 percent.

The lower this ratio is, the more assets the firm has to back up its borrowing. A high debt to assets ratio can be reduced by restricting both short-term and long-term borrowing, by securing additional financing from stockholders, or by reducing dividend payments to stockholders.

Complete the Following Problems

For all problems that do not work out evenly, carry your answers to two decimal places.

1. What is the formula for calculating the current ratio?

2. What is the average current ratio for all firms?

3. If you were the manager of a firm, what steps could you take to improve a low current ratio?

_____

_____

_____

4. Kinder Sales, Inc., is an importer/exporter of electronic components. At the end of last year, Kinder Sales had current assets valued at $184,500. The firm also had current liabilities that totaled $110,230. What is the firm's current ratio?

5. Martha Rollins, the owner of Applewood Wholesale Fashions, has just been turned down for a short-term loan by her banker. According to the banker, her current ratio was "too low." When she got back to the office, she looked at her balance sheet and determined that the firm's current assets were $174,500 and her current liabilities were $148,200. Based on this information, what is the current ratio for Applewood Wholesale Fashions?

6. What is the formula for determining acid-test ratio?

7. What is the average acid-test ratio for all firms?

8. If you were the manager of a firm, what steps could you take to improve a low acid-test ratio?

_____

_____

_____

9. At the end of the year, Exquisite Custom Furniture had total current assets valued at $456,000. The firm's inventory was valued at $240,000. If the firm's current liabilities totaled $238,000, what is the firm's acid-test ratio?

10. The accountant for Fisher-Wholesale Nursery and Landscaping just delivered the end-of-the-year financial statements. The following information was contained in the firm's balance sheet.

Cash	$32,400
Marketable securities	10,000
Accounts receivable	6,700
Inventory	45,600
Prepaid expenses	3,500
Accounts payable	49,200 _____

Debra Gleason, the firm's accountant, is concerned that the firm's acid-test ratio is too low. What is the firm's acid-test ratio?

11. What is the formula for determining inventory turnover?

12. What is the average for inventory turnover for all firms?

13. If you were the manager of a firm, what steps could you take to improve a low inventory turnover? _____

_____

_____

14. At the beginning of the year, Matlock's Nursery and Garden Supply had inventory valued at $32,000. At the end of the year, the firm had inventory valued at $51,000. If Matlock's cost of goods sold was $456,500, what is the firm's inventory turnover?

15. The following information was included in the income statement for Marston Floral Supply.

Beginning inventory	$ 12,000
Purchases	256,000
Ending inventory	26,700
Cost of goods sold	241,300

What is the inventory turnover for Marston Floral Supply?

16. What is the formula for computing the debt to assets ratio?

17. What is the average debt to asset ratio for all firms?

 18. If you were the manager of a firm, what steps could you take to improve a high debt to assets ratio?

_____

_____

19. Lenox Manufacturing, Inc., has liabilities that total $510,000. If total assets are $2,125,000, what is the firm's debt to assets ratio?

20. The following information was contained in the most recent balance sheet for Town View Apartments.

	Assets	**Liabilities**
Current assets	$ 24,520	
Fixed assets	867,800	
Current liabilities	35,610	
Long-term liabilities	210,000	
Owners' equity	646,710	

Based on the above information, what is the debt to asset ratio for this apartment complex?

Critical Thinking Problem

Cleveland-based American Greetings Corporation is the world's largest publicly owned manufacturer and distributor of greeting cards. The 1993 statement of financial condition for American Greetings—the latest year for which complete financial results are available—is presented in Table 71–1. Use the information contained in Table 71–1 to answer the questions below.

21. What is the current ratio for American Greetings?

22. What is the acid-test ratio for American Greetings?

23. What is the debt to asset ratio for American Greetings?

Personal Math Problem

24. Based on your calculations, would you want to invest in American Greetings Corporation? _____

_____

_____

_____

_____

_____

_____

TABLE 71–1 Statement of Financial Position for the American Greetings Corporation

CONSOLIDATED STATEMENT OF FINANCIAL POSITION
February 28, 1993
(thousands of dollars)

ASSETS	1993
Current Assets	
Cash and equivalents	$ 235,186
Trade accounts receivable, less allowances for sales returns of $72,054 and for doubtful accounts of $13,816	276,932
Inventories:	
Raw material	44,469
Work in process	30,171
Finished products	204,010
	278,650
Less LIFO reserve	84,887
	193,763
Display material and factory supplies	34,360
Total inventories	228,123
Deferred and refundable income taxes	66,339
Prepaid expense	105,277
Total current assets	911,857
Other Assets	248,991
PROPERTY, PLANT AND EQUIPMENT	
Land	6,182
Buildings	258,511
Equipment and fixtures	443,548
	708,241
Less accumulated depreciation and amortization	320,689
Property, plant and equipment—net	387,552
Total Assets	$1,548,400

LIABILITIES AND SHAREHOLDERS' EQUITY	1993
Current Liabilities	
Debt due within one year	$ 113,986
Accounts payable	113,684
Payrolls and payroll taxes	54,099
Retirement plans	17,409
Dividends payable	7,837
Income taxes	23,191
Total current liabilities	330,206
Long-Term Debt	169,381
Deferred Income Taxes	96,278
Shareholders' Equity	
Common shares—par value $1:	
Class A—34,357,286 shares issued less 33,236 Treasury shares in 1993	34,324
Class B—3,032,261 shares issued less 905,371 Treasury shares in 1993	2,127
Capital in excess of par value	259,093
Treasury stock	(28,152)
Cumulative translation adjustment	(11,580)
Retained earnings	696,723
Total shareholders' equity	952,535
	$1,548,400

Source: *1993 Annual Report,* American Greetings Corporation, 10500 American Road, Cleveland, Ohio, 44144.

INSTANT REPLAY

UNIT	IMPORTANT POINTS TO REMEMBER	EXAMPLES
UNIT 68 Analyzing the Income Statement	Revenues less cost of goods sold less operating expenses equals profit or loss. It is possible to calculate percent of increase or decrease for two different accounting periods. It is also possible to calculate percent of net sales for any account listed in an income statement.	'94 sales = **$100,000** '93 sales = <u>**85,000**</u> Amt. chg.= **$ 15,000** **$15,000 ÷ $85,000 = .176 or 17.6 percent.** **$24,500 (advertising) ÷ $192,400 (net sales) = .127 or 12.7 percent.**
UNIT 69 Analyzing the Balance Sheet	Assets equals liabilities plus owners' equity. It is possible to calculate percent of increase or decrease for two different accounting periods. It is also possible to calculate percent of total assets (or percent of total liabilities and owners' equity) for any account listed in a balance sheet.	'94 assets = **$450,000** '93 assets = <u>**360,000**</u> Amt. chg. = **90,000** **$90,000 ÷ $360,000 = .25 or 25 percent.** **$26,000 (Cash) ÷ $450,000 (Total Assets) = .058 or 5.8 percent.**
UNIT 70 Ratios—Tests of Profitability	Net profit margin = net income after taxes divided by net sales. Return on equity = net income after taxes divided by owners' equity. Earnings per share = net income after taxes divided by the number of shares of common stock outstanding.	**$19,000 ÷ $380,000 = .05 or 5 percent.** **$19,000 ÷ $135,714 = .14 or 14 percent.** **$19,000 ÷ 30,000 = 0.63 cents a share.**
UNIT 71 Ratios—Tests of Financial Health	Current ratio = current assets divided by current liabilities. Acid-test ratio = current assets minus inventory divided by current liabilities. Inventory turnover = cost of goods sold divided by average inventory. Debt to asset ratio = total liabilities divided by total assets.	**$76,000 ÷ $30,000 = 2.5 to 1.** **$76,000 – 40,000 ÷ 30,000 = 1.2 to 1.** **$270,000 ÷ $30,000 = 9 times a year.** **$120,000 ÷ $400,000 = .30 or 30 percent.**

NOTES

CHAPTER 13	MASTERY QUIZ

DIRECTIONS: *Round each answer to two decimal places or hundredths. (Each answer counts 5 points.)*

1. T F   Net sales *plus* operating expenses equals net income or loss.

2. T F   Fixed assets are assets that will be held or used for a period longer than one year.

3. T F   When profits are reinvested in a firm, the total for owners' equity will decrease.

4. T F   The average net profit margin for all firms is between 10 and 15 percent.

5. T F   Before calculating a current ratio, the dollar amount of inventory must be subtracted from the total current asset amount.

6. On January 1, Orser Fabrics had a beginning inventory of $24,560. During the year, the firm purchased merchandise valued at $227,000. On December 31, the firm's ending inventory was valued at $29,900. What was the firm's cost of goods sold?

7. During the month of October, All-Star Sports Equipment had net sales of $126,750. The firm's cost of goods sold totaled $47,890. Also, the firm had expenses that totaled $37,800. What is the firm's net income before taxes?

8. In 1994, Lawson Lighting had net sales of $910,000. In 1993, the firm's net sales were $1,134,500. What is the amount of change?

9. In problem 8, what is the percent increase or decrease?

10. During February, Northstar Publishing had net sales of $124,500. During this same period, Northstar had salary expense of $32,100. What is the percent of net sales for salary expense?

11. Bryan's Menswear has assets valued at $240,000. The firm also has liabilities that total $90,200. What is the amount of owner's equity?

12. In 1994, total assets for Handy Dry-Cleaners were valued at $75,400. In 1993, total assets were $65,200. What is the amount of change?

13. In problem 12, what is the percent of increase or decrease?

14. Last year, total assets for Allen & Stone Women's Fashions were $248,000. During the same accounting period, accounts receivable totaled $21,900. What is the percent of total assets for accounts receivable?

Use the following accounting information for Fowler Fence and Patio Company to answer the questions below.

Cash	$ 41,200
Inventory	48,700
Current assets	120,500
Total assets	235,000
Current liabilities	64,700
Total liabilities	82,400
Net income after taxes	42,100
Net sales	526,250
Owners' equity	296,479

15. What is the net profit margin for the Fowler Company?

16. What is the return on equity for the Fowler Company?

17. What is the earnings per share for the Fowler Company? (Note: the company has issued 25,000 shares of stock.)

18. What is the current ratio for the Fowler Company?

19. What is the acid-test ratio for the Fowler Company?

20. What is the debt to asset ratio for the Fowler Company?

CHAPTER 14

Annuities, Stocks, Bonds, and Mutual Funds

CHAPTER PREVIEW

In this chapter, we discuss the methods that are used by successful investors to evaluate potential investments. We begin by discussing how annuities can help you obtain the money needed to fund an investment program. In the remainder of the chapter, we discuss stocks, bonds, and mutual funds—investment alternatives that serious investors often use to reach their investment goals.

MATH TODAY

ESTABLISHING AN INVESTMENT PROGRAM FOR THE JACKSONS

In the spring of 1991, Julie and Bob Jackson bought their family a Ford Mustang convertible. The couple enjoyed the car and liked to take rides with the top down. Their two sons—aged 19 and 17—borrowed the car whenever they could. And within two months of purchasing the convertible, the Jacksons had driven it almost 6,000 miles.

At the time they purchased the Mustang convertible, the Jacksons had about $21,000 invested in three different annuities. The interest they earned on their annuities had been falling since early in 1990 as the economy improved and interest rates for similar investments declined. As a result, they were considering a shift in investments—perhaps to stocks, bonds, or mutual funds. Their enjoyment of their convertible, along with the general hoopla over the return of ragtops, led them to consider Ford Motor Company as a possible investment.

The Jacksons decided that Ford must be rolling in profits from the sale of automobiles like the Mustang convertible and the "rounded-look" cars. Bob researched the firm's common stock and found that its price had been as low as $25 per share in 1990, but had risen to about $49 by the end of the year. During the first part of 1991, the stock's price had dropped to $39 a share. Using some of the money from their annuities, the Jacksons could buy 500 shares. And they did exactly that.

Bob and Julie probably would not have spent almost $20,000 in any other way without first shopping carefully and learning all they could about their purchase. But like many other amateur investors, they seemed to feel it was reasonable to jump into the stock market almost blindly. At this point, the Jacksons had made two common investing errors. First, they had arbitrarily decided to invest without considering what they wanted to accomplish through their investing. Second, they had not considered any investment alternatives other than the Ford stock.

But what about Bob and Julie Jackson and their Ford stock? In spite of their lack of investing knowledge, the Jacksons seemed to have done well. The market price of their stock began to increase. By the middle part of 1994, it had reached $52 a share. After much deliberation and hair pulling, they sold their stock and realized a profit of about $6,000.

On the surface, it would seem that the Jacksons were just lucky. But their investment in Ford was just the beginning. After this first stock investment, they continued to invest in stocks, bonds, and mutual funds whenever they could afford it. As a result, their portfolio in 1995 consisted of 7 different investments valued at more than $48,000. According to Julie Jackson, the secret of their success was that they realized the necessity of evaluating *any* investment option, and they didn't try to make a big killing overnight.

Source: Based on information from *Moody's Handbook of Common Stocks,* Spring 1994 (New York: Moody's Investors Services, Inc, 1994); and *The Wall Street Journal,* June 21, 1994, p. C4.

UNIT 72　DETERMINING THE FUTURE VALUE OF AN ANNUITY

LEARNING OBJECTIVES　*After completing this unit, you will be able to:*

1. find the dollar value of an ordinary annuity at the end of a specified number of years.
2. find the dollar value of an annuity due at the end of a specified number of years.
3. find the dollar value of an annuity when deposits are made more than once a year.

Before you begin an investment program, you must have some money to invest. And while this fact may seem obvious, unfortunately the money needed for an investment program doesn't magically appear. That's why the first unit in this chapter on investments covers the topic of annuities. With an **annuity**, a series of equal payments are deposited in an interest-bearing account for a specified number of years. The money accumulated in an annuity can then be used to fund an investment program, make a major purchase, or for anything else that you value more than the money in the annuity.

ORDINARY ANNUITIES

There are two types of annuities. We begin with the steps required to determine the value of an ordinary annuity. With an **ordinary annuity,** all deposits (payments) are made at the end of the period. The simple interest formula ($I = P \times R \times T$) presented in Unit 49 can be used to calculate the interest amount and dollar value for an ordinary annuity. For example, let's assume that you deposit $1,000 at the end of each year for three years in an ordinary annuity that pays 6 percent interest. At the end of three years, your deposits *plus* interest total $3,183.60.

EXAMPLE 1 Find the Value of a Three-Year Ordinary Annuity that Pays 6 Percent Interest When Annual Deposits Are $1,000

CALCULATOR TIP

Depress clear key
Enter 1,000   Depress ×
Enter 0.06   Depress =
Read the answer $60
Enter 1,000   Depress +
Enter 60   Depress +
Enter 1,000   Depress =
Read the answer $2,060
Enter 2,060   Depress ×
Enter 0.06   Depress =
Read the answer $123.60
Enter 2,060   Depress +
Enter 123.60   Depress +
Enter 1,000   Depress =
Read the answer $3,183.60

Dollar value at the end of year 1 = $1,000

STEP 1 No interest is paid on the $1,000 deposit for year 1 because this is an ordinary annuity and the deposit is made at the end of the first year.

$1,000 × 0.06 = $60

STEP 2 To determine the interest paid at the end of year 2, multiply the value of the annuity at the end of year 1 ($1,000) by the interest rate (6 percent).

$1,000
+   60
+1,000
$2,060

STEP 3 To determine the value of the annuity at the end of year 2, add the value at the end of year 1 ($1,000) *plus* the interest calculated in step 2 ($60) *plus* the deposit for year 2 ($1,000).

$2,060 × 0.06 = $123.60

STEP 4 To determine the interest paid at the end of year 3, multiply the value of the annuity at the end of year 2 ($2,060) by the interest rate (6 percent).

$2,060.00
+   123.60
+1,000.00
$3,183.60

STEP 5 To determine the value of the annuity at the end of year 3, add the value of the annuity at the end of year 2 ($2,060) *plus* the interest calculated in step 4 ($123.60) *plus* the deposit made at the end of year 3 ($1,000).

TABLE 72–1 Ordinary Annuity Table for $1

Period	1%	2%	3%	4%	5%	6%	7%	8%	9%	10%	11%
1	1.000	1.000	1.000	1.000	1.000	1.000	1.000	1.000	1.000	1.000	1.000
2	2.010	2.020	2.030	2.040	2.050	2.060	2.070	2.080	2.090	2.100	2.110
3	3.030	3.060	3.091	3.122	3.153	3.184	3.215	3.246	3.278	3.310	3.342
4	4.060	4.122	4.184	4.246	4.310	4.375	4.440	4.506	4.573	4.641	4.710
5	5.101	5.204	5.309	5.416	5.526	5.637	5.751	5.867	5.985	6.105	6.228
6	6.152	6.308	6.468	6.633	6.802	6.975	7.153	7.336	7.523	7.716	7.913
7	7.214	7.434	7.662	7.898	8.142	8.394	8.654	8.923	9.200	9.487	9.783
8	8.286	8.583	8.892	9.214	9.549	9.897	10.260	10.637	11.028	11.436	11.859
9	9.369	9.755	10.159	10.583	11.027	11.491	11.978	12.488	13.021	13.579	14.164
10	10.462	10.950	11.464	12.006	12.578	13.181	13.816	14.487	15.193	15.937	16.722
11	11.567	12.169	12.808	13.486	14.207	14.972	15.784	16.645	17.560	18.531	19.561
12	12.683	13.412	14.192	15.026	15.917	16.870	17.888	18.977	20.141	21.384	22.713
13	13.809	14.680	15.618	16.627	17.713	18.882	20.141	21.495	22.953	24.523	26.212
14	14.947	15.974	17.086	18.292	19.599	21.015	22.550	24.215	26.019	27.975	30.095
15	16.097	17.293	18.599	20.024	21.579	23.276	25.129	27.152	29.361	31.772	34.405
16	17.258	18.639	20.157	21.825	23.657	25.673	27.888	30.324	33.003	35.950	39.190
17	18.430	20.012	21.762	23.698	25.840	20.213	30.840	33.750	36.974	40.545	44.501
18	19.615	21.412	23.414	25.645	28.132	30.906	33.999	37.450	41.301	45.599	50.396
19	20.811	22.841	25.117	27.671	30.539	33.760	37.379	41.446	46.018	51.159	56.939
20	22.019	24.297	26.870	29.778	33.066	36.786	40.995	45.762	51.160	57.275	64.203
25	28.243	32.030	36.459	41.646	47.727	54.865	63.249	73.106	84.701	98.347	114.410
30	34.785	40.588	47.575	56.085	66.439	79.058	94.461	113.280	136.310	164.490	199.020
40	48.886	60.402	75.401	95.026	120.800	154.760	199.640	259.060	337.890	442.590	581.830
50	64.463	84.579	112.800	152.670	209.350	290.340	406.530	573.770	815.080	1,163.900	1,668.800

In Example 1, the owner of the annuity made total deposits that equal $3,000 ($1,000 each year for 3 years). To find the interest paid on this annuity over the three-year period, subtract the total investment from the value of the annuity at the end of year 3.

EXAMPLE 2   Find the Interest Amount When the Total Investment Is $3,000 and the Value of the Annuity at the End of Year 3 Is $3,183.60

$3,183.60	value of the annuity at the end of year 3
−3,000.00	total investment (3 deposits × $1,000)
$ 183.60	interest earned over the three-year period

CALCULATOR TIP

Depress clear key
Enter 3,183.60 Depress −
Enter 3,000   Depress =
Read the answer $183.60

The procedure in Example 1 works well when the annuity is for two or three years. When annuities must be calculated for many periods, you may find that it is easier to use an annuity table like the one shown in Table 72–1. The table shows the annuity value for $1 for different interest rates and time periods.

To use an ordinary annuity table, read down the left side of the table to locate the correct number of periods. Then move across that row to find the interest rate. Finally, multiply this table factor by the annual deposit (payment) amount to find the value of the annuity.

EXAMPLE 3 | Find the Value of a 10-Year Ordinary Annuity that Pays 7 Percent Interest When Annual Deposits Are $500—Use Table 72–1

The table factor is 13.816

STEP 1 | In the ordinary annuity table, locate the number of periods (10) and move across that row to the column for the interest rate (7 percent).

$500 × 13.816 = $6,908

STEP 2 | Multiply the annual deposit amount ($500) by the table factor (13.816).

CALCULATOR TIP

Depress clear key
Enter 500   Depress ×
Enter 13.816   Depress =
Read the answer $6,908

ANNUITY DUE

With an ordinary annuity, all deposits (payments) are made at the end of each investment period. Thus, there is no interest paid during the first investment period for an ordinary annuity. With an **annuity due,** all deposits (payments) are made at the beginning of each investment period *and* interest is paid for the first investment period. For example, let's assume that you deposit $2,000 in an annuity due that pays 5 percent interest for three years. At the end of three years, your deposits *plus* interest equals $6,620.25.

EXAMPLE 4 | Find the Value of a Three-Year Annuity Due that Pays 5 Percent Interest When Annual Deposits Are $2,000

$2,000 × 0.05 = $100

STEP 1 | To determine the interest paid at the end of year 1, multiply the first year's deposit ($2,000) by the interest rate (5 percent).

$2,000
+ 100
$2,100

STEP 2 | To determine the value of the annuity due at the end of year 1, add the first year deposit ($2,000) plus the interest calculated in step 1 ($100).

$2,100 + $2,000
× 0.05 = $205

STEP 3 | To determine the interest paid at the end of year 2, add the value of the annuity at the end of year 1 ($2,100) *plus* the deposit for year 2 ($2,000) and multiply by the interest rate (5 percent).

$2,100
+2,000
+ 205
$4,305

STEP 4 | To determine the value of the annuity at the end of year 2, add the value at the end of year 1 ($2,100) *plus* the deposit for year 2 ($2,000) *plus* the interest calculated in step 3 ($205).

$4,305 + $2,000
× 0.05 = $315.25

STEP 5 | To determine the interest paid at the end of year 3, add the value of the annuity at the end of year 2 ($4,305) *plus* the deposit for year 3 ($2,000) and multiply by the interest rate (5 percent).

CALCULATOR TIP

Depress clear key
Enter 2,000   Depress ×
Enter 0.05   Depress =
Read the answer $100
Enter 2,000   Depress +
Enter 100   Depress =
Read the answer $2,100
Enter 2,100   Depress +
Enter 2,000   Depress ×
Enter 0.05   Depress =
Read the answer $205
Enter 2,100   Depress +
Enter 2,000   Depress +
Enter 205   Depress =
Read the answer $4,305
Enter 4,305   Depress +
Enter 2,000   Depress ×
Enter 0.05   Depress =
Read the answer $315.25
Enter 4,305   Depress +
Enter 2,000   Depress +
Enter 315.25   Depress =
Read the answer $6,620.25

$4,305.00
+2,000.00
+   315.25
$6,620.25

STEP [6] To determine the value of the annuity at the end of year 3, add the value of the annuity at the end of year 2 ($4,305) *plus* the deposit for year 3 ($2,000) *plus* the interest calculated in step 5 ($315.25).

It is also possible to use Table 72–1 and the steps in Example 5 to calculate the value of an annuity due.

EXAMPLE 5   Find the Value of a 7-Year Annuity Due that Pays 5 Percent Interest When Annual Deposits Are $5,000—Use Table 72–1

CALCULATOR TIP

Depress clear key
Enter 5,000   Depress ×
Enter 9.549   Depress =
Read the answer $47,745
Enter 47,745   Depress –
Enter 5,000   Depress =
Read the answer $42,745

7 + 1 = 8 investment periods

STEP [1] Determine the number of investment periods (7) and add one additional investment period. This adjustment is made because this example is for an annuity due.

The table factor is 9.549

STEP [2] In Table 72–1, locate the number of periods from step 1 (8) and move across that row to the column for the interest rate (5 percent).

$5,000 × 9.549 = $47,745

STEP [3] Multiply the annual deposit amount ($5,000) by the table factor (9.549).

$47,745
– 5,000
$42,745

STEP [4] Subtract one annual deposit ($5,000) from the answer in step 3 ($47,745). This adjustment is made because this example is for an annuity due.

WORKING ANNUITY PROBLEMS WHEN COMPOUNDING OCCURS MORE THAN ONCE A YEAR

Deposits to an annuity can be made more than once a year. In fact, annuities where semiannual, quarterly, or monthly deposits are made are quite common. Two additional steps are necessary when compounding occurs more than once a year. First, *multiply* the *time* factor by the number of times compounding occurs. That is, multiply by 2 if compounding occurs twice a year; multiply by 4 if compounding occurs four times a year; and multiply by 12 if compounding occurs twelve times a year. Second, *divide* the *interest rate* by the number of times compounding occurs. Now it is possible to use the answers found in steps 1 and 2 to find an interest factor from an annuity table like Table 72–1.

EXAMPLE 6   Find the Value of a 6-Year Ordinary Annuity that Pays 8 Percent Interest When $3,000 Deposits Are Made Every Six Months (Semiannually)—Use Table 72–1

CALCULATOR TIP

Depress clear key
Enter 3,000   Depress ×
Enter 15.026   Depress =
Read the answer $45,078

6 × 2 = 12 periods   STEP |1| Multiply the time factor (6) by the number of times compounding occurs (2).

8% ÷ 2 = 4%   STEP |2| Divide the interest rate (8 percent) by the number of times compounding occurs (2).

The table factor is 15.026   STEP |3| In Table 72–1, locate the number of periods (12) and move across that row to the column for the interest rate (4 percent).

$3,000 × 15.026 = $45,078   STEP |4| Multiply the deposit amount ($3,000) by the table factor (15.026).

HINT: You may want to review the material on simple interest in Unit 49 before completing this unit. Since the procedures for using compound interest and present value tables are the same as the procedures for using annuity tables, you may also want to review the material in Chapter 10.

Complete the Following Problems

If necessary, round your answer to two decimal places.

  1. In your own words, define an ordinary annuity.

_____

_____

  2. In your own words, define an annuity due.

_____

_____

Use the Five-Step Procedure on Page 437 and the Information Below to Complete Problems 3–5

In 1995, Joe and Mary Martinez began to deposit $3,000 a year in an ordinary annuity that pays 4 percent interest.

3. What is the dollar value of the ordinary annuity at the end of the first year?

4. How much interest is paid on the ordinary annuity in the second year?

5. What is the dollar value of the ordinary annuity at the end of the second year?

Use the Five-Step Procedure on Page 437 and the Information Below to Complete Problems 6–10

According to the Millses, it is past time to begin an investment program. They have talked about it for what seems like an eternity, but they never seem to have enough money left over to buy stocks, bonds, or mutual funds. Now, after putting it off for over five years, they have taken that first step. They have begun depositing $2,000 a year in an ordinary annuity that pays 6 percent interest.

6. What is the dollar value of the ordinary annuity at the end of the first year?

7. How much interest is paid on the ordinary annuity in the second year?

8. What is the dollar value of the ordinary annuity at the end of the second year?

9. How much interest is paid on the ordinary annuity in the third year?

10. What is the dollar value of the ordinary annuity at the end of the third year?

11. Five years ago, T & M Manufacturing Company began to save money that would eventually be used for expansion of facilities. For each of the five years, they deposited $17,500 in an annuity. The value of an annuity (deposits *plus* interest) at the end of Year 5 is $100,564. What is the amount of interest that T & M Manufacturing has earned during the five-year period?

Use Table 72–1 to Answer Problems 12–18. Each Problem in This Group Is an Ordinary Annuity

Annual Deposit Amount	Interest Rate	Number of Years	Dollar Value at the End of the Last Year
$ 1,000	8%	8	**12.** *$1,000 × 10.637 = $10,637*
$ 500	5%	10	**13.** _____
$ 2,200	6%	6	**14.** _____

Annual Deposit Amount	Interest Rate	Number of Years	Dollar Value at the End of the Last Year
$ 3,000	10%	4	**15.** _____
$ 2,400	4%	12	**16.** _____
$10,000	7%	7	**17.** _____
$ 850	9%	20	**18.** _____

Use the Six-Step Procedure on Pages 439–440 and the Information Below to Complete Problems 19–25.

In 1995, Jan and Peter Beckworth began to save for a new car. Their goal was to have $20,000 that could be used to pay cash for their next car. To accomplish their goal, they decide to deposit $5,200 each year for three years in an annuity due that pays 7 percent interest.

19. How much interest is paid on the annuity due at the end of the first year?

$5,200 × .07 = $364

20. What is the dollar value of the annuity due at the end of the first year?

21. How much interest is paid on the annuity due in the second year?

22. What is the dollar value of the annuity due at the end of the second year?

23. How much interest is paid on the annuity due in the third year?

24. What is the dollar value of the annuity due at the end of the third year?

25. Did Mr. and Mrs. Beckworth accomplish their goal of obtaining $20,000 to finance a new-car purchase?

Use Table 72–1 to Complete Problems 26–32. Each Problem in This Group Is an Annuity Due.

Annual Deposit Amount	Interest Rate	Number of Years		Dollar Value at the End of the Last Year
$ 2,000	7%	5	**26.**	_____
$ 500	5%	10	**27.**	_____
$ 2,200	6%	6	**28.**	_____
$ 3,000	10%	15	**29.**	_____
$ 2,400	4%	12	**30.**	_____
$10,000	8%	4	**31.**	_____
$12,500	9%	5	**32.**	_____

33. What is the dollar value of an *ordinary annuity* that earns 4 percent interest when semiannual deposits of $750 are made for 8 years?

34. Jackson and Marie Brown want to purchase a new home. To save enough money for the down payment, they decide to make quarterly deposits of $1,200 every three months for the next three years in an *ordinary annuity* that pays 8 percent interest. At the end of three years, how much money will the Browns have accumulated for a down payment for their new home?

35. The owners of Plano Tool & Die Company want to purchase new, state-of-the-art computer-aided manufacturing (CAD) equipment in four years, and they estimate that it will cost $250,000. If they make semiannual deposits of $25,000 every six months for the next four years in an *ordinary annuity* that pays 6 percent interest, how much money will they have accumulated at the end of four years?

36. If the CAD equipment in problem 35 costs $250,000, how much additional financing will be needed to complete this equipment purchase?

| UNIT 73 | COMMON AND PREFERRED STOCKS |

LEARNING OBJECTIVES *After completing this unit, you will be able to:*

1. calculate dividends, current yields, and price-to-earnings ratios.
2. read stock quotations.
3. determine the costs involved when buying and selling stocks.

In the last unit, we described how annuities could be used to obtain the money needed to start an investment program. In this unit and the next two, we discuss calculations that good investors use to evaluate investments in stocks, bonds, and mutual funds. Let's begin with some basics about stocks.

COMMON AND PREFERRED STOCKS

There are two types of stock—common and preferred stocks. A share of **common stock** represents the most basic form of corporate ownership. Generally, corporations sell common stock to finance their business start-up costs and help pay for their ongoing business activities. Owners of common stock may vote on corporate matters, share in dividend income, and sell their stock to other investors. When compared to common stockholders, owners of **preferred stock** have a priority claim on any dividends paid by the corporation. If the board of directors approves dividend payments, holders of preferred stock must receive their dividends before holders of common stock are paid any dividends. And when compared with common stockholders, preferred stockholders have first claim (after creditors) on corporate assets if the firm is dissolved *or* declares bankruptcy.

Stock investments, as illustrated by the opening case for this chapter, can be very good investments that increase in value. However, you should consider at least two factors before investing in stocks. First, a stockholder who decides to sell his or her stock must sell it to another investor. In many cases, a stockholder sells a stock because she or he thinks its price is going to decrease in value. The purchaser, on the other hand, buys that stock because he or she thinks the stock's price is going to increase. This creates a situation in which either the seller or the buyer is going to lose money.

Second, a corporation is under no legal obligation to pay dividends to stockholders. Dividends are paid out of earnings, but if a corporation that usually pays dividends should have a bad year, its board of directors can vote to omit dividend payments in order to help pay necessary business expenses. Corporations can also retain earnings so that additional financing is available for expansion, research and development, or other business activities.

CALCULATING DIVIDEND AMOUNTS

One of the first factors to consider when evaluating a stock investment is dividend income. A **dividend** is a distribution of money, stock, or other property that a corporation pays to stockholders. In most cases, dividends are paid in cash and paid on a per-share basis every three months or quarterly.

EXAMPLE 1   Determine the Total Amount of Quarterly Dividends Paid to a Shareholder Who Owns 210 Shares of Quaker Oats

---

CALCULATOR TIP

Depress clear key
Enter 0.43   Depress   ×
Enter 210   Depress   =
Read the answer $90.30

Mike Cantu owns 210 shares of common stock of the Quaker Oats Company. If the company pays a quarterly dividend of $0.43, what is his total dividend amount for this quarter?

$ 0.43	dollar amount of dividend per share
× 210	number of shares of common stock
$90.30	total dividend amount for this quarter

The dividend paid on a share of preferred stock is known before the stock is purchased. It is stated on the stock certificate either as an actual dollar amount or a specific percent of the par value of the stock. The **par value** of a stock is an assigned (and often arbitrary) dollar value printed on the stock certificate. When dividends are stated as a percent of the par value, the portion formula (P = B × R) presented in Unit 26 can be used to determine the dividend amount. The dividend amount (portion) is found by multiplying the par value (base) by the dividend stated as a percent (rate).

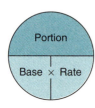

Portion

Base × Rate

EXAMPLE 2   Calculate the Annual Dividend for One Share of Pitney Bowes Preferred Stock

---

CALCULATOR TIP

Depress clear key
Enter 50   Depress   ×
Enter 0.04   Depress   =
Read the answer $2

Pitney Bowes, a U.S. manufacturer of office and business equipment, issued preferred stock with a par value of $50 that pays 4 percent annually. What is the annual dividend amount for one share of preferred stock?

P = B × R   **STEP** ☐1   State the portion formula.

$50 × 0.04 = $2   **STEP** ☐2   To determine the amount of dividend for each share of preferred stock, multiply the par value ($50) by the dividend stated as a percent (4 percent).

From the stockholder's standpoint, the higher the dividend paid for a share of common or preferred stock, the better.

CALCULATING THE CURRENT YIELD

One of the most common calculations that investors use to monitor the value of their investments is the current yield. The **current yield** is the annual dividend amount for one share of stock divided by the current market value of one share of stock. For example, let's assume that you purchase stock in Ford Motor Company. Let's also assume that Ford pays an annual dividend of $1.80 and is currently selling for $59 a share. The current yield is 3.1 percent as illustrated in Example 3.

EXAMPLE 3   Calculate the Current Yield for a Share of Ford Motor Company Stock

CALCULATOR TIP

Depress clear key
Enter 1.80   Depress ÷
Enter 59   Depress =
Read the answer 0.0305 = 3.1%

$1.80 ÷ $59 = 0.0305 = 3.1%   Divide the annual dividend for one share ($1.80) by the market value of one share ($59).

Like dividends, the higher the current yield for a share of stock, the better.

CALCULATING THE PRICE-EARNINGS (PE) RATIO

The price-to-earnings ratio is a key factor that serious investors use to evaluate stock investments. The **price-earnings ratio (PE ratio)** is the price of a share of stock divided by the corporation's earnings per share of stock for the last 12 months. For example, assume that a share of Avon Products has a current market value of $56 a share. Also, assume that the corporation earns $3.46 per share.* The PE ratio is 16.2 as illustrated below.

EXAMPLE 4   Calculate the PE Ratio for Avon Products

CALCULATOR TIP

Depress clear key
Enter 56   Depress ÷
Enter 3.46   Depress =
Read the answer 16.2

$56 ÷ $3.46 = 16.2   Divide the price of a share of stock ($56) by the amount of earnings per share ($3.46).

For most corporations, price-earnings ratios range between 5 and 20. A low price-earnings ratio indicates that a stock may be a good investment and a high price-earnings ratio indicates that it may be a poor investment.

READING STOCK QUOTATIONS

Transactions involving stocks listed on the New York Stock Exchange, American Stock Exchange, and other exchanges are reported in tables like the top section of Figure 73–1. Stocks are listed alphabetically. Your first task is to move down the table to find the stock you're interested in. Then, to read the transaction report, or stock quotation, you read across the table. Parts of a dollar are traditionally quoted as fractions rather than as cents. Thus $\frac{1}{8}$ means $0.125, or 12.5 cents, and $\frac{3}{4}$ means $0.75, or 75 cents. The first row in the table in Figure 73–1 gives detailed information about Heinz. (The numbers above each column in Figure 73–1 refer to the explanatory notes below the illustration in the figure.)

*Earnings per share were discussed in Unit 70 of Chapter 13.

FIGURE 73–1   How to Read Stock Quotations Contained in the Newspaper

52 Weeks		Stock	Sym	Div	Yld %	PE	Vol 100s	Hi	Lo	Close	Net Chg
Hi	Lo										
39⅞	30¾	Heinz	HNZ	1.32	4.0	14	5235	33⅝	33	33⅜	– ¼
30	22¾	HeleneCur	HC	.24	.8	19	250	29⅜	28¾	29¼	+ ⅝
27¼	23⅝	HellerFnl pfA		2.03	8.3	...	43	24⅜	24⅜	24⅜	– ¼
37¼	24¾	HelmPayne	HP	.50f	1.9	27	1521	27³⁄₁₆	26¼	26¼	+ ⅛
121½	73½	Hercules	HPC	2.24	2.0	22	797	112	110¼	110⅝	–2⅜

1. Highest price paid for one share during the past 52 weeks.
2. Lowest price paid for one share during the past 52 weeks.
3. Abbreviated name of the corporation; *pf* denotes a preferred stock.
4. Symbol used to identify the corporation on the exchange.
5. Total dividends paid per share during the last 12 months.
6. Percent of yield based on the current dividend and current price of the stock.
7. Price-earnings ratio: the price of a share of the stock divided by the corporation's earnings per share of stock during the last 12 months.
8. Number of shares traded during the day, expressed in hundreds of shares.
9. Highest price paid for one share during the day.
10. Lowest price paid for one share during the day.
11. Price paid in the last transaction for the day.
12. Difference between the price paid for the last share today and the price paid for the last share on the previous day.

Source: *The Wall Street Journal,* June 21, 1994, p. C4.

If a corporation has more than one stock issue, the common stock is always listed first. Preferred stock, as indicated by the letters *pf* in the stock column, is listed below the firm's common stock issue.

DETERMINING THE COST OF BUYING AND SELLING STOCK

Once an investor and her or his account executive have decided on a particular stock to buy or sell, the investor gives the account executive an order for that transaction. Commission charges are based on the number of shares and the value of stock bought and sold. On the trading floor, stocks are traded in round lots. A **round lot** is a unit of 100 shares of a particular stock. An **odd lot** is fewer than 100 shares of a particular stock. Brokerage firms generally charge higher fees for trading in odd lots, primarily because several odd lots must be combined into round lots before they can actually be traded through a stock exchange. For most brokerage firms, commission charges range between 2 percent and 3 percent for buying *or* selling stock.

When buying stock, three steps are needed to determine the total cost of the transaction. First, the number or shares that the investor wants to purchase must be multiplied by the price for one share of stock. Second, the dollar amount of commission must be determined. The portion formula ($P = B \times R$) presented in Unit 26 can be used to determine the dollar amount of commission. Third, the commission must be added to the cost of the stock.

EXAMPLE 5   Find the Total Cost for a Stock Purchase

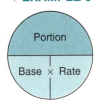

Brenda Watkins purchased 100 shares of Coca-Cola stock at the current market price of $40 a share. If the brokerage firm charges 2 percent of the purchase price, what is the total cost for this stock transaction?

$100 \times \$40 = \$4,000$	**STEP** 1	Multiply the number of shares purchased by the price for one share of stock.
$\$4,000 \times 0.02 = \$80$	**STEP** 2	Multiply the cost of the stock from step 1 ($4,000) by the percent of commission (2 percent).
$\begin{array}{r}\$4,000 \\ +\quad 80 \\ \hline \$4,080\end{array}$	**STEP** 3	Add the cost of the stock from step 1 ($4,000) and the amount of commission from step 2 ($80).

CALCULATOR TIP

Depress clear key
Enter 100   Depress ×
Enter 40   Depress =
Read the answer $4,000
Enter 4,000   Depress ×
Enter 0.02   Depress =
Read the answer $80
Enter 4,000   Depress +
Enter 80   Depress =
Read the answer $4,080

To determine the proceeds or the amount the investor receives when selling stock, the commission amount is subtracted from the total selling price.

EXAMPLE 6   Find the Proceeds When an Investor Sells Stock

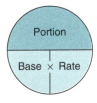

At the end of six months, Brenda Watkins decides to sell her 100 shares of Coca-Cola stock. At that time, the market price was $49\frac{1}{2}$. If the brokerage firm charges 2 percent of the selling price, what is the dollar amount of the proceeds that Ms. Watkins will receive?

$100 \times \$49.50 = \$4,950$	**STEP** 1	Multiply the number of shares sold by the price for one share of stock.
$\$4,950 \times 0.02 = \$99$	**STEP** 2	Multiply the value of the stock that was sold in step 1 ($4,950) by the percent of commission (2 percent).
$\begin{array}{r}\$4,950 \\ -\quad 99 \\ \hline \$4,851\end{array}$	**STEP** 3	Subtract the commission amount from step 2 ($99) from the value of the stock that was sold in step 1 ($4,950).

CALCULATOR TIP

Depress the clear key
Enter 100   Depress ×
Enter 49.50   Depress =
Read the answer $4,950
Enter 4,950   Depress ×
Enter 0.02   Depress =
Read the answer $99
Enter 4,950   Depress −
Enter 99   Depress =
Read the answer $4,851

To determine the amount of profit or loss for a stock transaction, one additional step is necessary. The difference between the total cost of purchasing stock and the proceeds received when selling stock must be determined.

EXAMPLE 7   Find the Total Amount of Profit or Loss for Brenda Watkins' Stock Transaction

CALCULATOR TIP

Depress clear key
Enter 4,851   Depress −
Enter 4,080   Depress =
Read the answer $771

$\begin{array}{r}\$4,851 \\ -4,080 \\ \hline \$\ \ 771\end{array}$ To determine the amount of profit or loss, find the difference between the total cost of purchasing stock and the proceeds when selling stock.

450

Because Ms. Watkins sold her Coca-Cola stock for more than she paid for it, she made a total profit of $771, as illustrated in Example 7. If she had sold her stock for less than she paid for it, she would have experienced a loss.

Complete the Following Problems

Unless otherwise indicated, round your answer to two decimal places.

T F **1.** Common stock is the most basic form of corporate ownership.

T F **2.** When compared to preferred stockholders, common stockholders have a priority claim to dividends.

T F **3.** A corporation has a legal obligation to pay dividends to stockholders.

T F **4.** Dividends are usually paid in cash.

T F **5.** A higher PE ratio is better than a lower PE ratio.

T F **6.** For stock transactions, commission is charged only when stock is purchased.

For problems 7–11, determine the total dividend amount for each problem.

Stock	Dollar Amount of Dividend per Share	Number of Shares	Total Dividend Amount
Home Depot	$0.16	100	**7.** $0.16 × 100 = $16
Jenny Craig	0.60	75	**8.**
Disney	0.30	250	**9.**
Mobil	3.40	142	**10.**
Reader's Digest	1.40	35	**11.**

12. Metroplex Electric issued preferred stock with a par value of $50. If the dividend is 6 percent of the par value, what is the annual dividend amount for one share of preferred stock?

$50 × .06 = $3

13. General South Telephone issued preferred stock with a par value of $100. If the dividend is 8 percent, what is the annual dividend amount for one share of preferred stock?

14. Northstar Manufacturing, Inc., used a preferred stock issue to finance a corporate expansion. This preferred stock issue paid $7\frac{1}{2}$ percent on the stock's par value of $80. What is the dividend amount for one share of preferred stock?

15. In your own words, define current yield.

_____

_____

For problems 16–22, determine the current yield for each problem. Use three decimal places for this set of problems.

Stock	Dollar Amount of Dividend Per Share	Market Value for One Share	Current Yield
Quaker Oats	$2.12	$75	**16.** _____
Chemical Bank	1.52	39	**17.** _____
American Express	1.00	27	**18.** _____
H.R. Block	1.12	43	**19.** _____
Maytag	0.50	18	**20.** _____
Pep Boys	0.17	31	**21.** _____
Polaroid	0.60	33	**22.** _____

23. In your own words, define price-earnings ratio.

_____

_____

_____

24. The price for one share of stock for Ralston Purina is $37. If the company earns $3.00 a share, what is the PE ratio? (Carry your answer to three decimal places.)

25. The price for one share of stock for Quaker State is $15. If the company earns $0.55 a share, what is the PE ratio? (Use 3 decimal places.)

26. The price for one share of stock for Air Products is $42\frac{1}{4}$. If the company earns $1.28 a share, what is the PE ratio? (Use 3 decimal places.)

Use the following information to answer questions 27–34.

52 Weeks Hi	Lo	Stock	Sym	Div	Yld %	PE	Vol 100s	Hi	Lo	Close	Net Chg
35¾	27¼	KeyCorp	KEY	1.28f	3.9	11	3930	33⅛	33	33⅛	+ ⅛
28⅞	26½	KeyCorp pf		2.50	9.4	...	7	26¾	26½	26⅝	...
15⅛	7¾	KeystnCon	KES		...	cc	104	15	14⅞	14⅞	– ⅛
29½	20¾	KeystnInt	KII	.74	3.4	18	219	21¾	21⅜	21⅝	– ¼
58¼	44⅝	KimbClark	KMB	1.76	3.1	17	1141	56⅛	55⅜	55⅞	+ ⅛
39¼	31½	KimcoRlty	KIM	2.00	5.5	21	25	36⅝	36½	36½	...
n 24⅞	21¼	KimcoRlty dep pf		1.94	8.4	...	27	23¼	23	23	– ¼
3⅛	1⅞	KimminsEnvr	KVN		...	10	94	2	1⅞	2	+ ⅛
44⅛	32½	KingWorld	KWP		...	15	1896	41¾	41	41¾	– ⅜
10	8⅞	KleinBenAus	KBA	.72a	7.8	...	135	9⅜	9¼	9¼	– ⅛
25	14¾	Kmart	KM	.96	6.1	dd	12019	15¾	15¼	15⅝	+ ⅛

27. What is the closing price for a share of KimbClark?

28. What is the dividend for a share of KimbClark?

29. What is the 52-week high for KimcoRlty?

30. What is the dividend for Key Corp common stock?

31. What is the 52-week low price for KeystnInt?

32. What is the closing price for a share of preferred stock for Key Corp?

33. What is the symbol for Kmart?

34. What is the PE ratio for King World?

For problems 35–40, determine the commission amount for each transaction.

Cost of Stock	Commission Stated as a Percent	Dollar Amount of Commission
$ 5,000	2%	**35.** _____
$ 7,400	3%	**36.** _____
$10,300	2%	**37.** _____
$ 2,540	$1\frac{1}{2}$%	**38.** _____
$15,400	$2\frac{1}{4}$%	**39.** _____
$ 3,300	2%	**40.** _____

41. Motorola common stock is selling for $46 a share. If an investor purchases 100 shares and the brokerage firm charges 2 percent commission, what is the total cost for this stock transaction?

42. J.C. Penney stock is selling for $50 a share. If an investor purchases 300 shares and the brokerage firm charges $1\frac{3}{4}$ percent commission, what is the total cost for this stock transaction?

43. Ross Michael purchased 70 shares of LA Gear at $6\frac{1}{2}$ a share. Because this is an odd-lot transaction, the brokerage firm charges $2\frac{1}{2}$ percent commission. What is the total cost for this stock transaction?

44. Kitty Barrow has decided to sell her 130 shares of Baltimore Gas & Electric stock. If the stock is selling for $21 and the brokerage firm charges 2 percent commission, what is the dollar amount of the proceeds that Ms. Barrow will receive?

45. Back in 1992, Alvin and Patty Pyle paid a total of $3,060 for 200 shares of Centex stock. Over the next three years, the stock went as high as $48 a share and as low as $26 a share. Then in 1995, the Pyles decided to sell their stock when the price reached $38 a share. If the brokerage firm charges 2 percent commission, what is the total amount of profit or loss for this stock transaction?

Personal Math Problem

In January 1992, Walter Evens purchased 200 shares of Burlington Industries. At that time, the stock was selling for $22 a share plus he paid 2 percent commission to the brokerage firm for helping him buy the stock. Three years later, he decided to sell his Burlington stock and sold the stock for $18 a share. When he sold his stock, he paid 3 percent commission.

46. What is the total cost when Mr. Evens purchased his stock?

47. What is the dollar amount of the proceeds that Mr. Evens received when he sold his stock?

48. Did Mr. Evens have a profit or loss for this transaction?

49. What is the total amount of profit or loss for this stock transaction?

50. If you had been Mr. Evens, would you have sold the Burlington Industries stock at this time? Why or why not?

_____

_____

_____

UNIT 74   BONDS

LEARNING OBJECTIVES   *After completing this unit, you will be able to:*

1. calculate bond interest amounts, current yields, and bond prices.
2. read bond newspaper quotations.
3. determine the costs involved when buying and selling bonds.

A bond investment, like any other investment, must be carefully evaluated before you decide to invest. The material in this unit will help you evaluate investments in bonds.

DIFFERENT TYPES OF BONDS

A **corporate bond** is a corporation's written pledge to repay a specified amount of money with interest. Today, the three most popular corporate bond investments are:

1. Debentures—corporate bonds that are backed only by the reputation of the issuing corporation.
2. Mortgage bonds—corporate bonds that are secured by various assets of the corporation that issued the bonds.
3. Convertible bonds—corporate bonds that can be exchanged at the bondholder's option for a specified number of shares of the corporation's stock.

In addition to corporations, the U.S. government and state and local governments issue bonds to obtain financing. The main reason why investors choose bonds issued by the U.S. government is that most investors consider them risk-free. Because of the decreased risk of default, they offer lower interest rates than corporate bonds. Many bonds issued by state and local governments are tax-free, which enables owners to earn income that is exempt from federal income taxes.

CALCULATING INTEREST

Between the time of purchase and the maturity date, bondholders receive interest payments—usually every six months—at the stated interest rate. To calculate the dollar amount of interest, multiply the face value of the bond by the interest rate. The usual face value of a bond is $1,000, although the face value of some bonds may be as high as $50,000. The interest formula ($I = P \times R \times T$) presented in Unit 49 can be used to determine the interest amount as illustrated in Example 1.

EXAMPLE 1 Find the Dollar Amount of Interest for a $1,000 Bond that Pays 6.25 Percent Annual Interest Issued by Zenith Corporation

$1,000 \times 0.0625 = \$62.50$ Multiply the face value ($1,000) by the interest rate (6.25 percent). Be sure to convert the percent to a decimal (0.0625).

Since interest for bonds is usually paid every six months or semiannually, one additional step is needed to determine the actual amount a bondholder would receive at the end of each six-month period.

EXAMPLE 2 Determine the Amount of Interest that a Bondholder Would Receive for Each Six-Month Period

$\$62.50 \div 2 = \$31.25$ Divide the interest amount found in Example 1 ($62.50) by the number of six-month periods in one year (2).

FIGURE 74–1 How to Read Bond Quotations Contained in the Newspaper

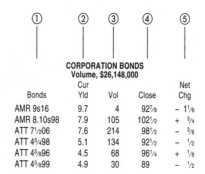

Bonds	Cur Yld	Vol	Close	Net Chg
AMR 9s16	9.7	4	92⅞	– 1⅛
AMR 8.10s98	7.9	105	102½	+ ¾
ATT 7½06	7.6	214	98½	– ⅜
ATT 4¾98	5.1	134	92½	– ½
ATT 4⅜96	4.5	68	96¼	+ ⅛
ATT 4⅜99	4.9	30	89	– ½

CORPORATION BONDS
Volume, $26,148,000

1. The abbreviated name of the corporation issuing bond; the bond's interest rate; and the year of maturity.
2. Percent of yield based on the annual interest and the current price of the bond. The letters *cv* indicate that the bond is convertible into a specified number of shares of common stock.
3. Number of individual bonds traded during the day.
4. Price paid in the last or closing transaction during the day. Newspaper quote is stated as a percent of the face value.
5. Difference between the price paid for the last bond today and the price paid for the last bond on the previous day.

Source: *The Wall Street Journal,* June 29, 1994, p. C18.

READING BOND QUOTATIONS

Not all local newspapers contain bond quotations, but *The Wall Street Journal* and *Barron's* publish complete and thorough information on this subject. Purchases and sales of corporate bonds are reported in tables like that shown at the top of Figure 74–1. The third row of Figure 74–1 gives detailed information about an ATT bond that pays $7\frac{1}{2}$ percent interest and matures in the year 2006. (The numbers above each column in Figure 74–1 refer to the explanatory notes below the illustration in the figure.)

All bonds are issued with a stated face value, which is usually $1,000. This is the amount the bondholder will receive if the bond is held until it matures. But once the bond has been issued, its price may be higher or lower than its face value. When a bond is selling for less than its face value, it is said to be selling at a discount. When a bond is selling for more than its face value, it is said to be selling at a premium.

In bond quotations, prices are given as a percent of the face value, which is usually $1,000. Thus, to find the actual price paid, you must multiply the face value ($1,000) by the newspaper quote. For example, the market value of the third bond from the top of Figure 74–1 (the ATT bond that pays $7\frac{1}{2}$ percent) is $985.

EXAMPLE 3   Determine the Current Market Value for an ATT Bond with a Newspaper Quote of $98\frac{1}{2}$

CALCULATOR TIP

Depress clear key
Enter 1,000   Depress ×
Enter 0.985   Depress =
Read the answer $985

$1,000 × 0.985 = $985   Multiply the face value of the bond ($1,000) by the newspaper quote ($98\frac{1}{2}$ percent). Note that the percent has been converted to a decimal (0.985).

EXAMPLE 4   Determine the Current Market Value for a Cigna Bond with a Newspaper Quote of 109

$1,000 × 1.09 = $1,090   Multiply the face value of the bond ($1,000) by the newspaper quote (109 percent).

CALCULATING THE CURRENT YIELD FOR A BOND

Because a bond's market value may increase or decrease, good investors often calculate the current yield on a bond investment. The current yield for a bond investment is determined by dividing the annual dollar amount of interest by the bond's current price. For example, let's assume that you purchase a Southwestern Bell corporate bond that pays 6 percent interest. Let's also assume that the current market price for the bond is $880. The current yield is 6.82 percent as illustrated in Example 5.

EXAMPLE 5   Calculate the Current Yield for a Southwestern Bell Corporate Bond that Pays 6 Percent Interest and Has a Current Market Price of $880

CALCULATOR TIP

Depress clear key
Enter 1,000 Depress ×
Enter 0.06 Depress =
Read the answer $60
Enter 60 Depress ÷
Enter 880 Depress =
Read the answer 0.0682

$1,000 × 0.06 = $60   **STEP** 1   Multiply the face value ($1,000) by the interest rate (6 percent).

$60 ÷ $880 = 0.0682 = 6.82%   **STEP** 2   To determine the current yield, divide the interest amount from step 1 ($60) by the current market value ($880).

The current yield calculation allows investors to compare bond investments with the yields offered by other investment alternatives, which include certificates of deposit, common stocks, preferred stocks, and mutual funds.

DETERMINING THE COST OF BUYING AND SELLING BONDS

Brokerage firms are free to set their own commission charge for bond transactions, but the typical charge is $5 to buy *or* sell a $1,000 bond. The steps illustrated in Example 6 can be used to determine the total cost when buying bonds.

EXAMPLE 6   Find the Total Cost for a Bond Purchase

CALCULATOR TIP

Depress clear key
Enter 1,020 Depress ×
Enter 7 Depress =
Read the answer $7,140
Enter 7 Depress ×
Enter 5 Depress =
Read the answer $35
Enter 7,140 Depress +
Enter 35 Depress =
Read the answer $7,175

Charles Matlock purchased seven Home Depot corporate bonds. The current market price for each bond is $1,020. If the commission charged by the brokerage firm is $5 per bond, what is the total cost for this transaction?

$1,020 × 7 = $7,140   **STEP** 1   Multiply the number of bonds purchased (7) by the market price for one bond ($1,020).

7 × $5 = $35   **STEP** 2   Multiply the number of bonds purchased (7) by the commission charged by the brokerage firm for each bond ($5).

458

CHAPTER FOURTEEN

$7,140
+   35
$7,175

STEP 3   Add the cost of the bonds from step 1 ($7,140) and the amount of commission from step 2 ($35).

To determine the proceeds or the amount the investor receives when selling bonds, the commission amount is subtracted from the total selling price.

EXAMPLE 7   Find the Proceeds When an Investor Sells Bonds

At the end of two years, Charles Matlock decides to sell his seven Home Depot bonds. At that time, the market price for one bond was $1,150. If the commission charged by the brokerage firm is $5 per bond, what is the dollar amount of the proceeds that Mr. Matlock will receive?

CALCULATOR TIP

Depress clear key
Enter 1,150   Depress   ×
Enter 7   Depress   =
Read the answer $8,050
Enter 7   Depress   ×
Enter 5   Depress   =
Read the answer $35
Enter 8,050   Depress   −
Enter 35   Depress   =
Read the answer $8,015

$1,150 × 7 = $8,050   **STEP** 1   Multiply the number of bonds sold (7) by the market price for one bond ($1,150).

7 × $5 = $35   **STEP** 2   Multiply the number of bonds sold (7) by the commission charged by the brokerage firm for each bond ($5).

$8,050
−   35
$8,015

STEP 3   Subtract the amount of commission from step 2 ($35) from the cost of the bonds from step 1 ($8,050).

To determine the amount of profit or loss for a bond transaction, one additional step is necessary. The difference between the total cost of purchasing bonds and the proceeds received when selling bonds must be determined.

EXAMPLE 8   Find the Total Amount of Profit or Loss for Charles Matlock's Bond Transaction

CALCULATOR TIP

Depress clear key
Enter 8,015   Depress   −
Enter 7,175   Depress   =
Read the answer $840

$8,015
−7,175
$ 840

To determine the amount of profit or loss, find the difference between the total cost of purchasing bonds and the proceeds when selling bonds.

Because Mr. Matlock sold his Home Depot bonds for more than he paid for them, he made a total profit of $840, as illustrated in Example 8. If he had sold his bonds for less than he paid for them, he would have experienced a loss.

Complete the Following Problems

Unless otherwise indicated, round your answer to two decimal places.

T   F   **1.**   A mortgage bond is backed only by the reputation of the issuing corporation.

BONDS **459**

T F **2.** A convertible bond can be exchanged for a specified number of shares of a corporation's stock.

T F **3.** Most investors prefer corporate bonds because they are safer than bonds issued by the U.S. government.

T F **4.** The usual face value of a bond is $500.

For problems 5–12, determine the dollar amount of interest.

Bond	Face Value	Interest Rate	Dollar Amount of Interest
Ballys	$1,000	6%	**5.** _____
Kroger	1,000	9%	**6.** _____
USX	1,000	7%	**7.** _____
Ametek	1,000	$9\frac{3}{4}$%	**8.** _____
Mobil	1,000	$8\frac{3}{8}$%	**9.** _____
Mead	1,000	$6\frac{3}{4}$%	**10.** _____
Tenneco	1,000	$9\frac{1}{4}$%	**11.** _____
Sears	1,000	$9\frac{1}{2}$%	**12.** _____

13. Bill Barnes owns five $1,000 IBM bonds that pay 9 percent interest. How much interest will he receive each year?

$1,000 × .09 = $90 $90 × 5 = $450

14. Nancy Malone owns one $1,000 Time Warner corporate bond that pays 7.45 percent interest. The bond matures in 1998. If interest payments are made to bondholders semiannually or every six months, what is the dollar amount of interest that she will receive each six months?

Use the following information to answer questions 15–20.

Bonds	Cur Yld	Vol	Close	Net Chg
Mobil 8⅜01	7.9	25	105⅜	− ¼
Motrla zr13	...	80	63	− 1
NJBTI 7⅜12	7.8	25	94	...
NRut 8½98	8.3	10	102⅞	− ⅝
NMed 12⅛95	11.7	30	103½	...
Navstr 9s04	9.2	10	98	...
NETelTel 6⅛06	7.2	5	85⅝	+ ⅛
NETelTel 7⅜07	7.8	10	95	− ¼

15. What is the interest rate for the Mobil bond?

16. What is the current yield for the NJBTI bond?

17. What is the closing price for the Motorola bond?

18. What is the net change for the NMed bond?

19. What is the maturity date for the Navstr bond?

20. What is the volume for the Mobil bond?

For problems 21–26, determine the current market value for each bond.

Bond	Face Value	Newspaper Quotation		Current Market Value
Pacific Bell	$1,000	99	**21.**	_____
Revlon	1,000	97	**22.**	_____
Southern Bell	1,000	88	**23.**	_____
Pennzoil	1,000	$86\frac{3}{4}$	**24.**	_____
Champion	1,000	$103\frac{3}{4}$	**25.**	_____
Pier One	1,000	$102\frac{1}{4}$	**26.**	_____

 27. In your own words, describe how to find the current yield for a bond.

_____

_____

For problems 28–35, calculate the current yield for each problem. (Carry your answer to three decimal places.)

Bond	Face Value	Interest Rate	Current Market Price	Current Yield
Fieldcrest	$1,000	6%	82	**28.** _____

Bond	Face Value	Interest Rate	Current Market Price	Current Yield
RJR Nabisco	1,000	$8\frac{5}{8}\%$	87	29. _____
Detroit Edison	1,000	6.4%	$98\frac{1}{2}$	30. _____
Xerox	1,000	$13\frac{1}{4}\%$	108	31. _____
Brown Shoe	1,000	$9\frac{1}{4}\%$	$97\frac{1}{2}$	32. _____
Dole	1,000	7%	$92\frac{1}{4}$	33. _____
Ralston	1,000	$9\frac{1}{2}\%$	$104\frac{1}{2}$	34. _____
Safeway	1,000	10%	106	35. _____

36. Patsy Martinez purchased eight Chevron bonds. If the commission charged by the brokerage firm is $5 per bond, what is the total commission for this transaction?

37. Latoya Smith has decided to invest in Tennessee Gas corporate bonds. The newspaper quote for each bond is 103. If the commission charged by her brokerage firm is $10 per bond, what is the total cost for four bonds?

38. Tom Highland wants to sell nine Shell Oil bonds. The newspaper quote for each Shell bond is 98. If the commission charged by his brokerage firm is $5 per bond, what is the dollar amount of the proceeds that Mr. Highland will receive?

39. Matt Rancho paid a total including commissions of $4,275 for five AMR bonds. When he sold the bonds, he received $5,210. Did Mr. Rancho have a profit or loss for this transaction?

40. In problem 39, what is the amount of his profit or loss for this transaction?

41. Back in 1989, Wanda and Hugh McNichols purchased three Trump corporate bonds for $1,040 each plus $5 commission per bond. In 1995, they decided to sell the three bonds at the current market price of $785. When they sold the bonds, the brokerage firm charged them $5 per bond. What is the amount of their profit or loss for this transaction?

UNIT 75	MUTUAL FUNDS

LEARNING OBJECTIVES   *After completing this unit, you will be able to:*

1. calculate the net asset value, the offer price, and the contingent deferred sales fee for a mutual fund.
2. read mutual fund price quotations.
3. determine the costs involved when buying and selling mutual funds.

A **mutual fund** is an investment alternative chosen by individuals who pool their money to buy stocks, bonds, and other securities selected by professional managers who work for an investment company. Today, investors can choose from more than 3,400 mutual funds that range from ultraconservative to very speculative investments. And while the numbers are impressive, *mutual funds—like all potential investments—must be evaluated.*

There are two major reasons why investors purchase mutual funds. First, mutual funds have a professional who manages the mutual fund. In fact, most investment companies do everything possible to convince individual investors that they can do a better job of picking securities than individual investors can. Second, mutual fund investments are diversified. This diversification spells safety, because an occasional loss incurred with one investment is usually offset by gains from other investments.

CALCULATING NET ASSET VALUE

The **net asset value** (NAV) per share for a mutual fund is equal to the current market value of the mutual fund's portfolio minus the mutual fund's liabilities divided by the number of shares outstanding.

| EXAMPLE 1 | Determine the Net Asset Value for the Global Management Fund |

The market value of stocks, bonds, and other investments for the Global Management fund is $15,430,000. Liabilities for the fund are $1,234,000. If there are 2 million shares outstanding, what is the net asset value?

$15,430,000
– 1,234,000
$14,196,000

STEP 1 Subtract the fund's liabilities ($1,234,000) from the value of the fund's portfolio ($15,430,000).

$14,196,000 ÷ 2,000,000 = $7.10

STEP 2 Divide the answer from step 1 ($14,196,000) by the number of shares outstanding (2,000,000).

Changes in net asset value allow investors to monitor the value of their investments. The net asset value is *also* important because it is the price that investors receive when they sell shares in a mutual fund.

READING MUTUAL FUND PRICE QUOTATIONS

Most local newspapers, *The Wall Street Journal,* and *Barron's* provide information about mutual funds. The net asset value, offer price, and change in net asset value are reported in tables like that shown in Figure 75–1. The last row of Figure 75–1 gives detailed information about the Aim Charter Growth and Income Fund. (The numbers above each column in Figure 75–1 refer to the explanatory notes below the illustration in the figure.)

FIGURE 75–1 How to Read Mutual Fund Price Quotations in the Newspaper

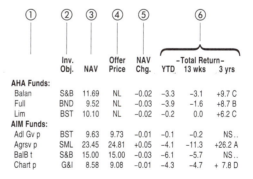

1. The name of the mutual fund appears in column 1.
2. The investment objective for this fund is abbreviated in column 2.
3. The net asset value (NAV) of one share is listed in column 3.
4. The offer price for one share is listed in column 4.
5. The difference between the net asset value today and the net asset value on the previous trading day is listed in column 5.
6. The last three columns (YTD, 13 wks, and 3 yrs) give the total return for this mutual fund for different time periods.

Source: *The Wall Street Journal,* June 29, 1994, p. C21.

The letters used in mutual fund quotations can be very informative. You can find out what they mean by looking at the footnotes that accompany the newspaper's mutual fund quotations. The two most common letters used with mutual fund quotations are *NL* and *r*. The letters *NL* mean no load and the letter *r* means that a redemption charge may be made.

THE COST OF PURCHASING MUTUAL FUND SHARES

With regard to cost, mutual funds are classified as load funds, low-load funds, and no-load funds. A **load fund** is a mutual fund in which investors pay a commission every time they purchase shares. The load charge, sometimes referred to as a sales fee or simply as a commission, may be as high as $8\frac{1}{2}$ percent for investments under $10,000.

A **low-load fund,** as the name implies, charges a lower commission than a load fund. This load charge usually ranges between 1 and 3 percent of the purchase price for investments under $10,000. (Typically, the commission declines for investments over $10,000 for both types of funds.)

A **no-load fund** is a mutual fund in which no sales charge is paid by the individual investor. No-load funds don't charge commissions when you buy shares because they have no sales force. If you want to buy shares of a no-load fund, you must deal directly with the investment company.

For load and low-load mutual funds, the load charge is added to the net asset value to determine the **offer price.** For example, the New York Mutual Small Business fund has a net asset value of $16.40. To invest in this fund, investors must pay a load charge (commission) of 5 percent. The portion formula ($P = B \times R$) presented in Unit 26 can be used to determine the load charge. The load charge (portion) is found by multiplying the net asset value (base) by the load charge stated as a percent (rate).

EXAMPLE 2   Calculating the Offer Price

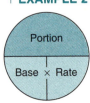

$16.40 × 0.05 = $0.82   **STEP** ☐1   To determine the dollar amount of load charge, multiply the net asset value ($16.40) by the load charge stated as a percent (5 percent).

$16.40
+ 0.82
――――
$17.22

STEP ☐2   Add the net asset value ($16.40) and the commission ($0.82).

CALCULATOR TIP

Depress clear key
Enter 16.40   Depress ×
Enter 0.05   Depress =
Read the answer $0.82
Enter 16.40   Depress +
Enter 0.82   Depress =
Read the answer $17.22

As an investor, you must decide whether to invest in a load fund, low-load fund, or a no-load fund. It should be pointed out that many financial analysts suggest that there is no significant difference in financial performance between mutual funds that charge commissions and those that do not. Although the sales commission should not be the decisive factor, the possibility of saving up to $8\frac{1}{2}$ percent commission is a factor to consider.

To determine the total cost of a mutual fund investment, you must multiply the offer price by the number of shares that you want to purchase. For a no-load mutual fund, the offer price is equal to the net asset value *because* there is no load charge (commission).

| EXAMPLE 3 | Find the Total Cost for a Mutual Fund Transaction |

CALCULATOR TIP

Depress clear key
Enter 200   Depress ×
Enter 12.24 Depress =
Read the answer $2,448

Debra Bally purchased 200 shares of the American Federal Stock Fund. The offer price for this fund was $12.24. What is the total cost for this transaction?

200 × $12.24 = $2,448   Multiply the number of shares (200) by the offer price ($12.24).

THE COST OF SELLING MUTUAL FUND SHARES

To determine the proceeds or the amount the investor receives when selling shares in a mutual fund, the contingent deferred sales fee, if there is one, must be subtracted from the total selling price. A **contingent deferred sales fee** (sometimes referred to as a back-end load) is a fee that ranges from 1 to 5 percent on redemptions or withdrawals during the first year that you own the fund. Generally, the deferred sales fee declines each year until there is no charge if you own the shares for more than five to seven years. Again, the portion formula can be used to determine the dollar amount of a contingent deferred sales fee.

| EXAMPLE 4 | Calculating a Contingent Deferred Sales Fee and Proceeds for a Mutual Fund Transaction |

CALCULATOR TIP

Depress clear key
Enter 200   Depress ×
Enter 16.82 Depress =
Read the answer $3,364
Enter 3,364 Depress ×
Enter 0.03   Depress =
Read the answer $100.92
Enter 3,364 Depress −
Enter 100.92 Depress =
Read the answer $3,263.08

After almost three years, Ms. Bally decided to sell her 200 shares of the American Federal stock fund. At that time, the net asset value for a share of the American Federal Stock Fund was $16.82. This fund charges a contingent deferred sales fee of 3 percent of the selling price for redemptions made during the third year of ownership. What is the dollar amount of proceeds that Ms. Bally will receive?

200 × $16.82 = $3,364   STEP 1   Multiply the number of shares sold (200) by the net asset value ($16.82).

$3,364 × 0.03 = $100.92   STEP 2   Multiply the answer from step 1 ($3,364) by the contingent deferred sales fee stated as a percent (3 percent).

$3,364.00
− 100.92
$3,263.08   STEP 3   To determine the proceeds, subtract the contingent deferred sales fee from step 2 ($100.92) from the answer from step 1 ($3,364).

For mutual funds that do not charge a contingent deferred sales fee, simply multiply the net asset value by the number of shares that are sold.

DETERMINING PROFIT OR LOSS FOR A MUTUAL FUND TRANSACTION

To determine the amount of profit or loss for a mutual fund transaction, one additional step is necessary. The difference between the total cost of purchasing shares in a mutual fund and the proceeds received when selling shares in a mutual fund must be determined.

| EXAMPLE 5 | Find the Total Amount of Profit or Loss for Debra Bally's Mutual Fund Transaction |

$3,263.08
−2,448.00
$ 815.08

To determine the amount of profit or loss, find the difference between the total cost of purchasing shares in a mutual fund and the proceeds when selling the shares.

CALCULATOR TIP

Depress clear key
Enter 3,263.08 Depress −
Enter 2,448   Depress =
Read the answer $815.08

Because Ms. Bally sold her shares in the American Federal Stock Fund for more than she paid for them, she made a total profit of $815.08, as illustrated in Example 5. If she had sold her shares for less than she paid for them, she would have experienced a loss.

Complete the Following Problems

Unless otherwise indicated, round your answer to two decimal places.

1. T F   Since mutual funds have professional management, there is no need to evaluate this type of investment.

2. T F   With a load fund, commissions cannot exceed $6\frac{1}{2}$ percent.

3. T F   Although investors who purchase no-load funds don't have to pay commissions, no-load funds do not perform as well as load and low-load funds.

4. T F   To determine the net asset value, an investor must add the dollar amount of commission to the offer price.

5. T F   Generally, contingent deferred sales fees range between 1 and 5 percent during the first year.

 6. In your own words, describe how to calculate the net asset value for a mutual fund.

_____

_____

For problems 7–10, determine the net asset value for each problem.

Portfolio Value	Fund's Liabilities	Number of Shares Outstanding	Net Asset Value
$ 8,900,000	$ 900,000	400,000	7. _____ $20 _____
			$8,900,000 − $900,000 = $8,000,000
			$8,000,000 ÷ 400,000 = $20
$12,450,000	650,000	800,000	8. _____
$24,560,000	1,400,000	1,500,000	9. _____

Portfolio Value	Fund's Liabilities	Number of Shares Outstanding	Net Asset Value
$33,100,000	332,000	3,000,000	**10.** _____

11. The Boston Pilgrim Fund contains the following investments: (a) government securities—$5,000,000; (b) certificates of deposit—$1,200,000 (c) corporate bonds—$12,400,000; and stocks—$21,750,000. The fund also has liabilities that total $480,000. If the fund has 900,000 shares outstanding, what is the fund's net asset value?

Use the following information to answer questions 12–15.

	Inv. Obj.	NAV	Offer Price	NAV Chg.	YTD	13 wks	3 yrs R
First Omaha:							
Equity	G&I	10.52	NL	−0.03	−0.6	−0.4	NS..
FxdInc	BND	9.62	NL	−0.04	−4.7	−2.4	NS..
SI FxIn	BST	9.64	NL	−0.02	−1.9	−1.0	NS..
FPDvAst p	S&B	12.26	12.84	−0.05	−4.3	−2.0	+9.1 C
FP Mu Bd	IDM	11.84	12.21	−0.02	−1.9	−0.3	+7.6 A
First Priority Fds:							
Equity Tr	G&I	10.09	NL	−0.04	−3.2	−4.1	NS..
FxdIncTr	BIN	9.71	NL	−0.05	−4.8	−2.2	NS..
LtdMGv	BST	9.70	9.90	−0.02	−0.9	−0.1	NS..
First Union:							
BalB p	S&B	11.51	11.99	−0.02	−3.1	−3.1	+9.3 C

(Total Return spans 13 wks and 3 yrs R columns.)

12. What is the NAV for the First Omaha Equity Fund?

13. What is the offer price for the First Union Balanced Fund?

14. What is the year-to-date (YTD) return for the First Priority Equity Fund?

15. What is the three-year return for the First Union Balanced Fund?

16. The net asset value for the GT Global Japan Fund is $13.49. The load charge (commission) is $0.67. What is the offer price?

For problems 17–26, determine (1) the load charge and (2) the offer price.

Net Asset Value	Load Charge Stated as a Percent	Dollar Amount of Load Charge		Offer Price	
$15.00	5%	17. _____		18. _____	
8.50	3%	19. _____		20. _____	
22.50	$8\frac{1}{2}$%	21. _____		22. _____	
10.54	$5\frac{1}{2}$%	23. _____		24. _____	
8.70	4%	25. _____		26. _____	

27. John Blackburn purchased 250 shares of the Franklin Group Growth Fund. At the time of purchase, the offer price was $14.84. What is the total cost of this transaction?

28. Martha Evens purchased 120 shares of First Investors High Yield Fund. The net asset value for this fund was $5.07. The offer price was $5.47. What is the total cost of this transaction?

29. When Mary Cunningham decided to sell her shares in a mutual fund, her account executive told her there would be a 4 percent contingent deferred sales fee on the total selling price of her shares. If the value of the shares she sold was $5,600, how much is the contingent deferred sales fee?

30. In problem 29, what are the proceeds that Ms. Cunningham will receive?

31. After almost two years, Franklin Snyder decided to sell his 310 shares in the Spartan Growth Fund. At the time of the sale, a share of Spartan Growth Fund had a net asset value of $21.15. If Spartan charges a contingent deferred sales fee of 5 percent of the selling price for redemptions made during the second year of ownership, what is the dollar amount of the contingent sales fee?

32. In problem 31, what are the proceeds that Mr. Snyder will receive?

For problems 33–37, determine the profit or loss for each problem.

Total Purchase Cost	Proceeds When Sold		Dollar Amount of Profit or Loss
$ 5,670	$ 7,210	**33.**	_____
$11,200	9,050	**34.**	_____
$14,500	16,730	**35.**	_____
$ 4,420	3,910	**36.**	_____
$21,780	32,510	**37.**	_____

Use the following information to answer problems 38–40.

Margo Washington bought 220 shares of the National Mutual Trust Fund that specializes in the Pacific Rim countries. At the time of the purchase, she paid $21.40 a share for this no-load fund. Five years later, Ms. Washington decided to sell 130 shares of the 220 shares that she owns in this fund. At the time of her decision to sell, the net asset value for a share was $33.

38. What is the total cost of Ms. Washington's investment?

39. Assuming that National Mutual Trust does not charge a contingent deferred sales fee, what is the amount of the proceeds that Ms. Washington will receive for the 130 shares that she sold?

40. Based on the above information, what is the current dollar value of her remaining investment in the National Mutual Trust fund?

INSTANT REPLAY

UNIT	IMPORTANT POINTS TO REMEMBER	EXAMPLES
UNIT 72 Determining the Future Value of an Annuity	The basic difference between an ordinary annuity and an annuity due is whether the deposit is made at the beginning of the investment period or at the end of the investment period. With both types, the simple interest formula can be used to determine the value at the end of a specific investment period. For most annuity problems, the easiest way to determine the future value of an annuity is to use an ordinary annuity table. If adjustments are made, it is also possible to use an ordinary annuity table to determine the value of an annuity due problem.	Find the value of an ordinary annuity at the end of five years when the interest rate is 6% and the annual deposit is $500. The table factor is 5.637. **$500 × 5.637 = $2,818.50**
UNIT 73 Common and Preferred Stock	There are two types of stocks: common and preferred. To calculate the dividend amount, multiply the dollar amount of dividend per share by the number of shares. To calculate the current yield, divide the annual dividend for one share by the market value of one share. To calculate the PE ratio, divide the price of a share of stock by the amount of earnings per share. To calculate the amount of commission, multiply either the total cost of buying stock or the total sales price when selling stock by the commission stated as a percent. To calculate the amount of profit or loss, find the difference between the purchase cost and the proceeds when selling stock.	100 shares × $1.10 per share dividend = $110 $1.10 dividend ÷ $27.50 market value = 0.04 = 4% $60 price per share ÷ $3 earnings per share = 20 PE ratio $6,800 cost × 0.02 commission = $136 commission amount $9,200 proceeds −7,000 purchase cost $2,200 profit

	IMPORTANT POINTS	
continued from page 470 **UNIT**	**TO REMEMBER**	**EXAMPLES**
UNIT 74 Bonds	Corporations, the U.S. government, and state and local governments issue bonds.	
	To calculate the amount of bond interest, multiply the face value by the interest rate.	$1,000 face value × 0.06 interest rate = $60 interest
	To determine the current market value for a bond, multiply the face value by the newspaper quotation.	$1,000 face value × 95 newspaper quote = $950 current market value
	To calculate the current yield, divide the interest amount by the current market value.	$60 annual interest ÷ $800 current market value = 0.075 = 7.5%
	To calculate the amount of profit or loss, find the difference between the purchase cost and the proceeds when selling bonds.	**$1,000** proceeds **– 850** purchase cost **$ 150** profit
UNIT 75 Mutual Funds	There are two reasons why investors purchase mutual funds: professional management and diversification.	
	To determine the net asset value, subtract the fund's liabilities from the value of the portfolio and divide by the number of shares outstanding.	$15 million portfolio value – $1 million liabilities ÷ 1 million shares = $14 NAV per share
	To determine the offer price, multiply the net asset value by the load charge. Then add the load charge to the net asset value.	$15 NAV × 0.04 load charge = $0.60 $15 NAV + $0.60 load charge = $15.60 offer price
	To determine the total cost for a mutual fund, multiply the number of shares by the offer price.	300 shares × $8.90 offer price = $2,670 total purchase price
	To calculate the contingent deferred sales fee, multiply the value of shares when sold by the fee stated as a percent. Then subtract the fee from the value of the shares.	$5,200 value when sold × 0.04 contingent deferred sales fee = $208 **$5,200** **– 208** **$4,992** proceeds
	To calculate the amount of profit or loss, find the difference between the purchase cost and the proceeds when selling mutual funds.	**$4,992** proceeds **–2,670** purchase cost **$2,322** profit

NOTES

CHAPTER 14	MASTERY QUIZ

DIRECTIONS   *Round each answer to two decimal places or hundredths. (Each answer counts 5 points.)*

Use Table 72–1 to answer problems 1–3. Each problem in this group is an *ordinary annuity.*

Annual Deposit Amount	Interest Rate	Number of Years	Dollar Value at the End of the Last Year
$1,000	6%	10	**1.** _____
$ 600	8	15	**2.** _____
$2,000	7	5	**3.** _____

4. What is the dollar value of an ordinary annuity that earns 4 percent interest when quarterly deposits of $300 are made for 10 years? (Use Table 72–1.)

5. Find the value of a five-year *annuity due* that pays 6 percent interest when annual deposits are $800. (Use Table 72–1.)

6. Michelle Thurman owns 125 shares of Disney stock. If Disney pays an annual dividend of $0.30 per share, what is the total dividend that Ms. Thurman will receive each year?

7. A preferred stock issue has a par value of $100. If the dividend is $7\frac{1}{2}$ percent of the par value, what is the annual dividend amount for one share of stock?

8. Robert Barker receives a dividend of $1.80 for each share of National Manufacturing common stock that he owned. If the market value of the stock is $30, what is the current yield?

9. The price for one share of stock for Microsoft stock is $51. If the company earns $1.76, what is the price-earnings ratio?

10. Chris Craft common stock is selling for 35\frac{1}{2}$ a share. If an investor purchases 200 shares and the brokerage firm charges 2 percent commission, what is the total cost for this stock transaction?

Bond	Face Value	Interest Rate	Dollar Amount of Interest
TWA	$1,000	8%	11. _____
Texaco	1,000	9	12. _____
Revlon	1,000	$9\frac{3}{4}$	13. _____

14. New York Telephone has issued a bond with a face value of $1,000. The newspaper quotation for this bond is $92\frac{1}{4}$. What is the current market value for this bond?

15. Owens Illinois has issued a $1,000 bond that pays 10 percent. If the current market value for this bond is $1,070, what is the current yield?

16. Alvia Jackson purchased seven Borden bonds. If each bond cost $920, and the commission charged by the brokerage firm is $5 per bond, what is the total cost for the seven bonds?

17. The Washington Liquidity Fund has a portfolio value of $10,500,000. This fund also has liabilities of $230,000. If the fund has 500,000 shares outstanding, what is the fund's net asset value?

18. Mattie Washington purchased 340 shares of Vision Growth and Income Fund. At the time of purchase, the offer price was $10.13. What is the total cost of this transaction?

19. When Bill Nelson decided to sell his shares in a mutual fund, his account executive told him there would be a 3 percent contingent deferred sales fee on the total selling price of his shares. If the value of the shares he sold was $4,900, how much is the contingent deferred sales fee?

20. Peter Lawson paid a total of $7,240 for 400 shares of the Noel High-Yield Income Fund. After four years, he decided to sell his investment. When he sold, his proceeds were $9,114. How much profit did Mr. Lawson make on this transaction?

Business Statistics

The final chapter of *Business Math* covers business statistics. The chapter begins by describing the steps used to calculate the arithmetic mean, the median, and the mode. These three statistical measures help managers, business owners, and individuals make decisions based on facts. Then, the chapter describes the steps required to build frequency distributions, graphs, bar charts, and pie charts.

STATISTICS, STATISTICS, STATISTICS

Question: What do the following statements have in common?

∎ The Dow Jones Industrial Average drops 34.88 points to finish at 3741.90[1]

∎ Seventeen percent of all U.S. households spend more than $100 per week on groceries[2]

Answer: Both statements contain statistics. While unrelated to each other, the statistics in each statement are important to someone. For example, it

---

[1]*The Wall Street Journal,* Tuesday, June 21, 1994, p. A1.
[2]"USA Snapshots," Life section, Section D, *USA Today,* Tuesday, June 21, 1994, p. 1D.

would be impossible for stock brokers and investors to gauge the movement of stocks listed on the New York Stock Exchange without the Dow Jones Industrial Average. It is the most publicized statistic about the financial markets in the United States and is reported in *The Wall Street Journal, Barron's,* and the financial section of most local newspapers.

The second statistic—17 percent of U.S. households spend more than $100 on groceries each week—is useful for grocery store managers and owners. With this type of information along with other vital statistics, people in the grocery industry can identify customers, plan inventory, develop marketing plans, and even price merchandise. Like the Dow Jones Industrial Average, this information was easy to find. The statistic was reported on the front page of the Life section of *USA Today.*

To be useful, statistics—like the two facts above—should always be presented in the form in which they have the most informational value. In some cases, a simple listing of numbers may be all that is needed. In other cases, it may be necessary to construct a graph, bar chart, or pie chart. For example, *The Wall Street Journal* could report that the Dow Jones Average was 3756.23 on Monday, 3741.90 on Tuesday, 3747.35 on Wednesday, 3,751.45 on Thursday, and 3,759.32 on Friday. Most investors and financial analysts would find it difficult to interpret changes if they just read the list of numbers. In this case, a graph could be used to illustrate the same statistical information.

Dow Jones Industrial Average

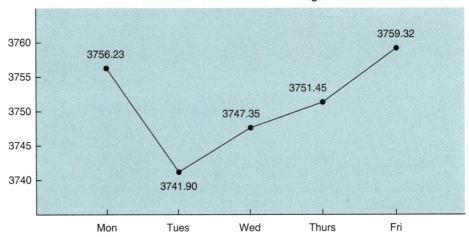

This graph quickly illustrates the changes in the Dow Jones Averages that occurred over a five-day period. While providing the same information, the graph is easier to interpret and ultimately more meaningful than if the numbers were just part of a written report.

UNIT 76   MEAN, MEDIAN, AND MODE

LEARNING OBJECTIVES   *After completing this unit, you will be able to:*

1. calculate the mean for a set of numbers.
2. determine the median for a set of numbers.
3. determine the mode for a set of numbers.

TABLE 76–1   Employee Salaries for the Stars & Stripes Corporation

STARS & STRIPES CORPORATION
Employee Salaries (for the month of November, 19XX)

Employees	Monthly Salary
Tammy Witherspoon	$4,000
Bill Hartford	4,000
Jane Haley	3,800
June Sanderson	3,700
Bob Wolf	3,500
Andy Laurels	3,500
William Long	3,500
Cindy Katz	3,400
Ursula Jackson	3,300
Sandra Waters	3,300
Lance Grandy	2,850
Al Culpepper	2,500
Ann Justin	2,500
Nancy Malone	2,150
Glen Haskins	2,000
Total	$48,000

Today, business statistics are so common that it is impossible to even thumb through a newspaper or financial magazine without finding percents, averages, bar graphs, and pie charts. But what is a statistic? A **statistic** is a measure that summarizes a particular characteristic for an entire group of numbers. Average miles per gallon, the number of people who earn in excess of $50,000 a year, or the most frequently purchased movie video during the month of December are all statistics. In this unit, we use the information in Table 76–1 to describe the three most common statistical measures: the arithmetic mean, the median, and the mode.

THE ARITHMETIC MEAN

Perhaps the most familiar statistic is the arithmetic mean, commonly called the **average.** The **arithmetic mean** for a set of numbers is the total of all number values divided by the number of items in the set. Using the information in Table 76–1, the average is $3,200 as illustrated in Example 1.

EXAMPLE 1   Calculate the Average (based on the salaries in Table 76–1)

Total Salaries = $48,000   STEP **1**   Determine the total for all employee salaries.

Total employees = 15   STEP **2**   Determine the total number of employees.

$48,000 ÷ 15 = $3,200   STEP **3**   Divide the total salaries found in step 1 ($48,000) by the total number of employees found in step 2 (15).

CALCULATOR TIP

Depress clear key
Enter 48,000   Depress ÷
Enter 15   Depress =
Read the answer $3,200

Although the arithmetic mean summarizes a large number of salaries into one number, it can be interpreted incorrectly. In Example 1, the average is $3,200, yet not one of the employee salaries is exactly equal to that amount. In this case, the distinction between actual salaries and average salaries is an important one that should never be disregarded.

THE MEDIAN

The **median** of a set of numbers is the value that appears at the exact middle of the data when they are arranged in order from the largest value to the smallest value. The data in Table 76–1 are already arranged from the largest value to the smallest value. Their median is thus $3,400, which is exactly halfway between the top and bottom values.

EXAMPLE 2   Determine the Median (based on the salaries in Table 76–1)

$4,000	**STEP** 1	Arrange all salaries from the largest value to the smallest value.
4,000		
3,800		
3,700	**STEP** 2	Find the number that is in the exact middle. Since there are a total of 15 numbers, the median ($3,400) is the eighth salary from the top.
3,500		
3,500		
3,500		
3,400 ← The Median		
3,300		
3,300		
2,850		
2,500		
2,500		
2,150		
2,000		

In Example 2, there are seven salaries above the median ($3,400) and seven salaries below the median. When there is an even number of values, the median is the average of the two middle values.

EXAMPLE 3   Determine the Median When There Is an Even Number of Values

During a six-day period, Beefmasters Feed & Supply Store had the following sales: Monday—$2,410; Tuesday—$2,200; Wednesday—$2,175; Thursday—$2,600; Friday—$2,800; Saturday—$1,480.

CALCULATOR TIP

Depress clear key
Enter 2,410   Depress +
Enter 2,200   Depress =
Read the answer $4,610
Enter 4,610   Depress ÷
Enter 2   Depress =
Read the answer $2,305

$2,800 Friday	**STEP** 1	Arrange all sales figures from the largest value to the smallest value.
2,600 Thursday		
2,410 Monday		
2,200 Tuesday		
2,175 Wednesday		
1,480 Saturday		

$2,410 + 2,200 = $4,610   **STEP** 2   Find the number that is in the exact middle. Since there are a total of six numbers, the median is the average of the two middle values ($2,410 and $2,200).

$4,610 ÷ 2 = $2,305 Median   **STEP** | 3 | To determine the median, divide the total of sales for Monday and Tuesday ($4,610) by two.

THE MODE

The **mode** of a set of numbers is the value that appears most frequently in the set. In Table 76–1, the $3,500 salary appears three times, which is more often than any other salary amount appears. Thus, $3,500 is the mode for this set of numbers.

EXAMPLE 4   Determine the Mode (based on the salaries in Table 76–1)

$4,000   **STEP** | 1 | Arrange all salaries from the largest
 4,000   value to the smallest value.
 3,800
 3,700
 3,500 ⎫
 3,500 ⎬ The mode   **STEP** | 2 | To determine the mode, find the salary
 3,500 ⎭   that appears most frequently.
 3,400
 3,300
 3,300
 2,850
 2,500
 2,500
 2,150
 2,000

In Example 4, the mode is $3,500 because it appears three times. It is possible to have more than one mode in a set of numbers. If two or more salaries had appeared three times, there would have been more than one mode in Example 4. And in some circumstances, there may be no mode. If all the salaries in Example 4 had been different, there would have been no mode for this set of numbers.

Complete the Following Problems

If necessary, round your answer to two decimal places.

 1. In your own words, define *arithmetic mean.* _____

_____

_____

 2. In your own words, define *median.* _____

_____

_____

 3. In your own words, define *mode.* _____

_____

For problems 4–15, calculate the arithmetic mean, median, and mode for each set of numbers below.

89   82   76   66   66   54   88

4. Arithmetic mean _____   **5.** Median _____   **6.** Mode _____

$1,124 $994 $765 $665 $665 $555 $540

7. Arithmetic mean _____   **8.** Median _____   **9.** Mode _____

100   88   88   76

10. Arithmetic mean _____   **11.** Median _____   **12.** Mode _____

$2,456 $2,356 $2,250 $2,200 $2,200 $1,900

13. Arithmetic mean _____   **14.** Median _____   **15.** Mode _____

16. During January, Atlanta Job Placement Services placed 96 workers. During February, the firm placed 76 workers. During March, the firm placed 108 workers. How many workers did Atlanta Job Placement Service place during the three-month period?

96 + 76 + 108 = 280

17. In problem 16, what is the arithmetic mean for workers placed during this three-month period?

18. Daily sales for the Koslow Furniture Store were as follows: Monday—$2,405; Tuesday—$1,985; Wednesday—$3,422; Thursday—$2,950, and Friday—$4,820: What is the total of sales?

19. In problem 18, what is the arithmetic mean for daily sales for Koslows?

20. During the first five months of the year, Best-Chicken-to-Go had the following electric bills:

January	$345.67
February	245.20
March	190.26
April	410.56
May	520.35

What is the average electric bill for Best-Chicken-to-Go?

21. Salaries for Tulsa Pizza-to-Go are as follows: $4,800; $3,100; $2,200; $2,200; $2,000; $1,500; $1,300. What is the median for these salaries?

22. All hourly workers for Nettles Manufacturing are grouped into one of the following seven wage groups.

Wage Group	Hourly Wage
Class 1	$ 5.00
Class 2	6.25
Class 3	7.50
Class 4	9.00
Class 5	10.50
Class 6	11.00
Class 7	11.75

What is the median hourly wage group for Nettles Manufacturing?

Critical Thinking Problem

You are the manager for a clothing store in a mall. You have been told by customers that your store's prices are higher than other stores in the mall. You decide to compare prices on one of your most popular products—a pair of Levi blue jeans that four other stores in the mall also carry. The prices below summarize your findings.

Your Store	$29.95
Jeans "R" Us	27.00
Baston's Clothing Store	31.00
All-American Clothing	27.00
Mayfair Department Store	24.95

23. What is the arithmetic mean price for jeans?

24. What is the median price for jeans?

25. What is the mode price for jeans?

26. Based on your calculations, would you raise the price, keep the price the same, or lower the price of jeans sold by your store? Why? _____

_____

_____

_____

_____

_____

UNIT 77 | **FREQUENCY DISTRIBUTIONS, GRAPHS, BAR CHARTS, AND PIE CHARTS**

LEARNING OBJECTIVES   *After completing this unit, you will be able to:*

1. construct a frequency distribution.
2. calculate the arithmetic mean for information contained in a frequency distribution.
3. construct graphs, bar charts, and pie charts.

As mentioned in the opening case, statistical information should always be presented in the form in which it has the most informational value. The material in this unit—frequency distributions, graphs, bar charts, and pie charts—can make comparisons easier or reflect trends among different statistical values.

FREQUENCY DISTRIBUTIONS

The number of items in a set of numbers can be reduced by developing a frequency distribution. A **frequency distribution** is a listing of the number of times each value appears in a set of numbers. We use the information in Table 77–1 to illustrate the steps required to construct a frequency distribution.

TABLE 77–1   Individual Responses to a Regional Home Buyer's Survey

When asked what price home that they were going to purchase, the following individual responses were obtained:

$ 95,000	$ 90,000	$110,000	$ 85,000	$100,000
$ 85,000	$115,000	$ 95,000	$115,000	$110,000
$100,000	$ 95,000	$110,000	$ 85,000	$ 95,000
$105,000	$120,000	$ 95,000	$110,000	$ 95,000

EXAMPLE 1   Steps Required to Construct a Frequency Distribution

$ 85,000 90,000 95,000 100,000 105,000 110,000 115,000 120,000	**STEP** 1 Arrange all possible responses in order from the lowest value to the highest value.

$ 85,000 ||| 90,000 | 95,000 ⊥⊥⊥⊥ | 100,000 || 105,000 | 110,000 |||| 115,000 || 120,000 |	**STEP** 2 Place tally marks for each response beside the appropriate dollar value.

$ 85,000 ||| = 3 90,000 | = 1 95,000 ⊥⊥⊥⊥ | = 6 100,000 || = 2 105,000 | = 1 110,000 |||| = 4 115,000 || = 2 120,000 | = 1	**STEP** 3 To complete the frequency distribution, add all tally marks and convert to numerical form.

It is also possible to use the information in a frequency distribution to calculate an arithmetic mean. Based on the frequency distribution constructed in Example 1, the arithmetic mean is $100,500 as illustrated in Example 2.

EXAMPLE 2   Determine the Arithmetic Mean for the Information Contained in a Frequency Distribution

CALCULATOR TIP

Depress clear key
Enter 85,000   Depress ×
Enter 3   Depress =
Read the answer $255,000
This step is repeated for each value in the frequency distribution.
Enter 255,000   Depress +
Enter 90,000   Depress +
Enter 570,000   Depress +
Enter 200,000   Depress +
Enter 105,000   Depress +
Enter 440,000   Depress +
Enter 230,000   Depress +
Enter 120,000   Depress =
Read the answer $2,010,000
Note: Step 3 should be a mental calculation.
Enter 2,010,000   Depress ÷
Enter 20   Depress =
Read the answer $100,500

$ 85,000 × 3 = $255,000 90,000 × 1 = 90,000 95,000 × 6 = 570,000 100,000 × 2 = 200,000 105,000 × 1 = 105,000 110,000 × 4 = 440,000 115,000 × 2 = 230,000 120,000 × 1 = 120,000	**STEP** 1 Multiply each dollar value in the frequency distribution by the number of frequencies for that value.

$ 255,000 90,000 570,000 200,000 105,000 440,000 230,000 120,000 $2,010,000	**STEP** 2 Add the answers obtained in step 1.

3 + 1 + 6 + 2 + 1 + 4 + 2 + 1 = 20	**STEP** 3 Determine the total number of frequencies in the distribution.

$2,010,000 ÷ 20 = $100,500   **STEP** 4   To determine the arithmetic mean, divide the answer from step 2 ($2,010,000) by the total frequencies determined in step 3 (20).

GRAPHS

A visual display is a diagram that represents several items of information in a manner that makes comparison easier or reflects trends among the items. The most accurate visual display is a **graph,** in which values are plotted to scale on a set of axes. All graphs should have a title. This title should clearly indicate what facts are presented in the graph. The steps required to construct a graph are illustrated in Example 3.

EXAMPLE 3   Construct a Graph to Illustrate Sales Trends

Total sales figures for Reynolds Truck Lines are as follows:

1991	$176,500
1992	$192,000
1993	$181,000
1994	$205,000
1995	$195,000

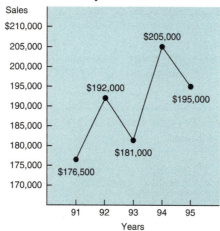

Yearly Sales Figures for Reynold's Truck Line

STEP 1   Draw the basic graph outline with vertical and horizontal axes.

STEP 2   Place the dates across the bottom of the horizontal axis.

STEP 3   Place dollar values along the vertical axis. (Given the sales figures to be graphed in Example 3, a logical starting point is $170,000. Then increments of $5,000 can be used to show sales increases and decreases.)

STEP 4   Place a graph point on the graph for each year's sales total. Then connect all graph points.

Graphs are most effective for presenting information about a single variable that changes with time (such as variations in total sales over the last five years). Graphs tend to emphasize trends as well as peaks and low points in the value of the variable that is being graphed.

BAR CHARTS

A **bar chart** is a chart where each value is represented with a vertical *or* horizontal bar. The longer the bar, the greater the value. Bar charts are useful for presenting values that are to be compared. The eye can quickly pick out the longest or shortest bar, or even those that seem to be of average size. The steps required to construct a bar chart are illustrated in Example 4.

EXAMPLE 4   Construct a Bar Chart to Illustrate Profits over a Seven-Year Period

The profits for Dillon Hardware Stores are as follows:

1989	$135,000
1990	156,000
1991	160,000
1992	148,500
1993	151,000
1994	140,000
1995	155,000

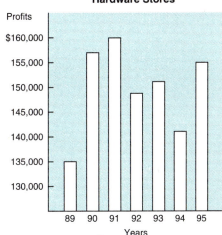

Yearly Profits for Dillon Hardware Stores

STEP 1   Draw the basic outline with vertical and horizontal axes.

STEP 2   Place the dates across the bottom of the horizontal axis.

STEP 3   Place dollar values along the vertical axis. (Given the profits to be graphed in Example 4, a logical starting point is $130,000. Then increments of $5,000 can be used to show profit increases and decreases.)

STEP 4   Using a ruler, draw a bar on the chart for each year's profit amount.

PIE CHARTS

A **pie chart** is a circle ("the pie") that is divided into "slices," each of which represents a different item. The circle represents the whole—for example, total expenses for a particular calendar year. The size of each slice shows the contribution of each expense item to the whole. The larger the slice, the larger the contribution. By their nature, pie charts are most effective in displaying the relative size or importance of various items of information. In fact, the government often uses pie charts to illustrate statistical information.

To construct a pie chart, you must know the following four basic facts.

1. You must know what percent of the total each individual part represents. For example, 25 percent of *all* college seniors have a job before they graduate.
2. If you do not know the percent for each part of the pie, you must use the rate formula (R = P − B) presented in Unit 27 and the available statistical information to find the percent (rate).
3. You must know that a circle has a total of 360 degrees.
4. You must know how to use a protractor to construct a pie chart. It is also possible to use a computer and a graphics software program to construct a pie chart.

Now you can follow the steps in Example 5 to construct a pie chart.

EXAMPLE 5   Construct a Pie Chart to Illustrate the Breakdown of Sales for the Month of May

Portland Wholesale Electronics sells computers, software programs, and telephone equipment. For May, total sales were $350,000. During the same month, total sales were broken down as follows:

Computer Sales	$210,000
Software Programs	87,500
Telephone Equipment	52,500
Total	$350,000

CALCULATOR TIP

Depress clear key
Enter 210,000 Depress ÷
Enter 350,000 Depress =
Read the answer 0.60
Enter 87,500 Depress ÷
Enter 350,000 Depress =
Read the answer 0.25
Enter 52,500 Depress ÷
Enter 350,000 Depress =
Read the answer 0.15
Enter 360   Depress ×
Enter 0.60   Depress =
Read the answer 216°
Enter 360   Depress ×
Enter 0.25   Depress =
Read the answer 90°
Enter 360   Depress ×
Enter 0.15   Depress =
Read the answer 54°

$P \div B = R$
$210,000 \div 350,000 = 0.60 = 60\%$
$87,500 \div 350,000 = 0.25 = 25\%$
$52,500 \div 350,000 = 0.15 = 15\%$

STEP 1   Use the rate formula (R = P ÷ B) to find the percent (rate) of the total sales (base) for each product sold (portion).

$360° \times 0.60 = 216°$
$360° \times 0.25 = 90°$
$360° \times 0.15 = 54°$

STEP 2   Multiply 360 degrees by the answers from step 1. (Remember a circle or pie has 360 degrees.)

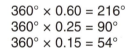

Breakdown of Sales by Product Type

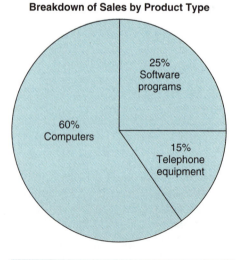

25%
Software programs

60%
Computers

15%
Telephone equipment

STEP 3   Draw a circle and then use a protractor to determine the size of each piece of the pie.

Complete the Following Problems

If necessary, round your answer to two decimal places.

HINT: Before completing this unit, you may want to review the material on arithmetic mean in Unit 76 and the material on rate in Unit 27.

For problems 1–3, construct a frequency distribution for each set of numbers below.

1. 76   80   80   54   76   82   54   76   86

     54   II   = 2
     76   III = 3
     80   II   = 2
     82   I   = 1
     86   I   = 1

2. $24,500 $30,000 $31,500 $39,000 $24,500 $30,000

3. $76,400 $48,700 $70,000 $48,700 $65,300 $76,400
   $45,000 $70,000 $80,000 $76,400 $48,700 $71,000

4. The statistical listing below represents the prices that were paid for homes constructed by Alliance Homes during the month of May. Using this information, construct a frequency distribution.

$110,000	$120,000	$ 94,000	$105,000
$120,000	$ 94,000	$110,000	$120,000
$ 94,000	$ 98,000	$110,000	$ 94,000
$ 94,000	$120,000	$105,000	$ 94,000

5. In problem 4, what are the total sales for Alliance Homes during the month of May?

6. In problem 4, how many homes were sold during the month of May?

7. In problem 4, what is the arithmetic mean for homes sold during the month of May? (Note: this calculation is similar to the calculation used to calculate average inventory in Unit 47 of Chapter 8.)

Use the following information to answer questions 8–11.

You are a work-study student at your college. As part of your job responsibilities, you help a history professor calculate test grades. Test scores for the last unit test are listed below.

88	76	82	66	92	96	72	88	66	92	72	84	76
76	80	92	88	90	88	70	72	76	90	74	90	88

8. In the space below, construct a frequency distribution for the test scores listed above.

9. What is the total number of points earned by students on this test?

10. How many students took this test?

11. What is the arithmetic mean for this unit test?

12. Your boss asks you to construct a graph to illustrate sales growth for the last three years. During 1993, sales totaled $765,000. During 1994, sales totaled $810,000. During 1995, sales totaled $710,000.

13. Construct a graph that will chart monthly telephone expenses for Ft. Worth–based Sammons Corporation.

January	$1,090	July	$1,240
February	950	August	1,030
March	1,150	September	970
April	1,240	October	860
May	910	November	900
June	1,080	December	1,150

14. As the owner of a small business, you are trying to get a handle on operating expenses. The accountant has provided total monthly expenses for the last six months. Total operating expenses are as follows:

July	$4,850	October	$5,500
August	3,900	November	5,120
September	3,750	December	5,375

In the space below, construct a graph that illustrates total operating expenses for the last six months.

15. Annual sales figures for the last five years for the Morton Interior design group are listed below.

1991	$240,600
1992	345,100
1993	320,900
1994	236,000
1995	310,566

Based on this information, construct a bar chart that illustrates sales trends.

Critical Thinking Problem

Ms. Watkins, the vice president of finance, seems perplexed. It seems that although the firm's sales have increased over the last four years, profits have gone down. She gives you the following information and asks that you construct bar charts to illustrate trends for both sales and profits.

	Sales	Profits
1992	$475,000	$92,300
1993	510,000	94,500
1994	562,000	90,300
1995	608,000	91,550

16. In the space provided, construct a bar chart that illustrates sales trends for this company.

17. In the space provided, construct a bar chart that illustrates profit trends for this company.

18. Based on the statistical information and your two bar charts, what actions, if any, would you take if you were Ms. Watkins? Why? _____

_____

_____

_____

19. Departmental sales for last month for Handy Man Hardware Center are broken down as follows: electrical—40 percent; paint—25%; plumbing—20%; and lumber 15%. Construct a pie chart that illustrates this breakdown.

20. Regal Outdoor Furniture has four retail stores in the state of Florida. Total sales for the firm are broken down as follows:

Orlando—40 percent of sales
Miami—35 percent of sales
Ft. Myers—15 percent of sales
Key West—10 percent of sales

Based on the above information, construct a pie chart that illustrates total sales broken by retail store location.

21. Last year, All-Star Products had total sales of $710,000. During this same year, total sales were broken down as follows:

Consumer products—$497,000

Industrial products—$213,000

Based on this information, construct a pie chart that illustrates the breakdown of sales by type of product for last year.

22. Last year, the management for Nationwide Paper Products decided to divide the United States into three different sales territories. At the end of the first year, the firm's sales totaled $900,000. Territorial sales were as follows: Eastern—$306,000; Midwest—$180,000; and Western—$414,000. Construct a pie chart that illustrates the regional sales figures for Nationwide Paper Products.

INSTANT REPLAY

UNIT	IMPORTANT POINTS TO REMEMBER	EXAMPLES
UNIT 76 Mean, Median, and Mode	The *arithmetic mean* (average) for a set of numbers is the total of all numbers in a set divided by the number of items in the set.	During 1995, A-1 Stores paid the following salaries:

$ 55,300
45,700
42,100
37,400
31,200
31,200
27,300
———
$270,200

The *median* of a set of numbers is the value that appears in the exact middle of the data when they are arranged from the largest value to the smallest value.

The *mode* of a set of numbers is the value that appears most frequently in the set.

The mean is
$270,200 ÷ 7 = $38,600.

The median is
$37,400.

The mode is
$31,200.

UNIT 77
Frequency Distributions, Graphs, Bar Charts, and Pie Charts

A frequency distribution is a listing of the number of times each value appears in a set of numbers. It is also possible to determine the arithmetic mean for the values included in a frequency distribution.

Number of VCRs sold at the following prices

$169 ||| = 3
199 |||| = 5
249 || = 2

$169 × 3 = $ 507
199 × 5 =   995
249 × 2 =   498
Total $2,000

$2,000 ÷ 10 = $200

A *graph* is a visual display in which values are plotted to scale on a set of axes.

Total Employees 1990–1995

continued from page 494

UNIT	IMPORTANT POINTS TO REMEMBER	EXAMPLES
UNIT 77 Frequency Distributions, Graphs, Bar Charts, and Pie Charts—*Concluded*	A *bar chart* is a chart where each value is represented with a vertical or horizontal bar.	
	A *pie chart* is a circle that is divided into slices, each of which represents a different item. (Note that a circle has a total of 360 degrees.)	

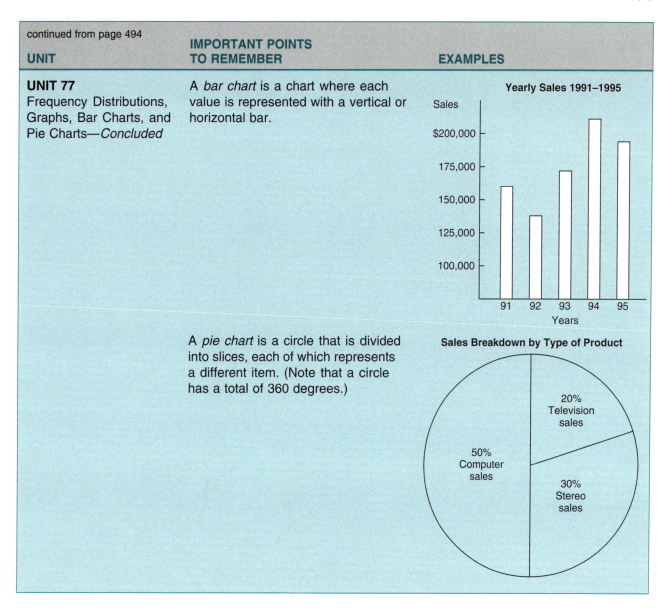

Yearly Sales 1991–1995

Sales

$200,000
175,000
150,000
125,000
100,000

91   92   93   94   95
Years

Sales Breakdown by Type of Product

20% Television sales

50% Computer sales

30% Stereo sales

NOTES

MASTERY QUIZ   CHAPTER 15

DIRECTIONS: *Round each answer to two decimal places or hundredths. (Each answer counts 5 points.)*

For problems 1–6, calculate the arithmetic mean, median, and mode for each set of numbers below.

| 80 | 80 | 76 | 56 | 80 | 84 | 62 | 56 | 72 |

1. Arithmetic mean _____   **2.** Median _____   **3.** Mode _____

| $2,345 | $2,210 | $4,560 | $5,665 | $1,900 | $3,324 | $7,250 |

4. Arithmetic mean _____   **5.** Median _____   **6.** Mode _____

7. The annual sales figures for Christy Appliance Stores are 1993—$756,245; 1994—$790,435; and 1995—$665,420. What is the total sales for the three-year period?

8. In problem 7, what is the arithmetic mean for sales for Christy's?

Use the information below to answer questions 9–11.

The number of years of experience for employees in the Accounting Department at KDDS TV station are listed below.

T. Appleby	8 years
C. Cooper	12 years
H. Franklin	9 years
S. Goldsborough	2 years
K. Justin	5 years
W. Miller	14 years
E. Everett	8 years

9. What is the average number of years of service for employees?

10. What is the median number of years of service?

11. What is the mode for years of service?

12. The statistical listing below represents the prices that were paid for color televisions at one retailer in the San Francisco area. Using this information, construct a frequency distribution.

$495	$999	$ 750	$399	$995	$1,695	$1,695
$999	$399	$1,695	$750	$495	$1,150	$ 495
$750	$495	$1,150	$995	$495	$1,695	$1,695

13. In problem 12, what is the total sales amount for color televisions?

14. In problem 12, how many color televisions were sold during this period?

15. In problem 12, what is the arithmetic mean for color televisions sold during this period?

16. Construct a graph that illustrates the following sales figures.

Year	Sales Totals
1991	$450,000
1992	375,000
1993	355,000
1994	378,000
1995	360,000

17. You are trying to determine how fast your electric bills are increasing. Over the last eight months, the monthly bills were: January—$87; February—$76; March—$69; April—$94; May—$99; June—$134; July—$178; and August—$128. Construct a graph to illustrate these electric utility bills.

18. Construct a bar chart to illustrate the following profit amounts.

Year	Profits
1992	$78,400
1993	55,200
1994	65,300
1995	81,256
1996 projected	94,000

500

CHAPTER FIFTEEN

19. Spring Creek Book Stores has three retail outlets in the St. Louis area. Total sales for the firm are broken down as follows:

Downtown—20 percent of sales
Riverfront—50 percent of sales
Cathedral Square—30 percent of sales

Based on the above information, construct a pie chart that illustrates total sales broken down by retail store location.

20. Last year, Goddard Furniture had total sales of $3,400,000. During this same period, total sales were broken down as follows: Eastern Division—$1,360,000; Western Division—$2,040,000.

Based on this information, construct a pie chart that illustrates the breakdown of sales by geographic region for last year.

| CHAPTERS 13–15 | CUMULATIVE REVIEW |

SPECIAL NOTE TO STUDENTS: To help you build a foundation for success in math, we have included a cumulative review at the end of every three or four chapters in the text. If you do not score at least 80 on this cumulative review, you may want to review the type of problems that you missed.

DIRECTIONS: *If necessary, round each answer to two decimal places or hundredths. (Each answer counts 5 points.)*

1. On January 1, Econo Plastics, Inc., had a beginning inventory of $36,300. During the year, the firm purchased merchandise valued at $432,700. On December 31, the firm's ending inventory was valued at $45,000. What was the firm's cost of goods sold?

2. In 1995, Denton Auto Repair had net sales of $870,000. In 1994, the firm's net sales were $750,000. What is the amount of change?

3. In problem 2, what is the percent of increase or decrease?

4. Top Drawer Menswear has assets valued at $310,000. The firm also has liabilities that total $60,000. What is the amount of owners' equity?

5. Last year, total assets for Moody's Home Furnishings were $312,000. During the same accounting period, cash totaled $28,080. What is the percent of total assets for the cash account?

6. In 1995, Antonio's Builder's Supply had net income after taxes totaling $56,700. If net sales were $708,750, what is the firm's net profit margin?

7. If current assets total $77,000 and current liabilities total $35,000, what is the current ratio?

8. What is the dollar value of an ordinary annuity that earns 6 percent interest when annual deposits of $500 are made for 12 years? (Use Table 71–1.)

9. Pat Cisneros owns 170 shares of Skyline Manufacturing. If Skyline pays an annual dividend of $0.45 per share, what is the total dividend that Ms. Cisneros will receive this year?

10. A preferred stock issue has a par value of $50. If the dividend is $5\frac{1}{4}$ percent of the par value, what is the annual dividend amount for one share of stock?

11. Art Bentley receives a dividend of $1.40 for each share of Montania High-Tech Manufacturers. If the market value of the stock is $35, what is the current yield?

12. World Publishing common stock is selling for $42 a share. If an investor purchases 150 shares and the brokerage firm charges 2 percent commission, what is the total cost for this stock transaction?

13. New Mexico Telephone has issued a bond with a face value of $1,000. The interest rate for this bond is $7\frac{1}{4}$ percent. What is the dollar amount of interest for each bond?

14. In problem 13, what is the current yield for the New Mexico bond if the current market value for the bond is $800?

15. The Thriftmasters Mutual Fund has a portfolio value of $12,700,000. This fund also has liabilities that total $300,000. If the fund has 500,000 shares outstanding, what is the fund's net asset value?

16. Barry O'Grady purchased 170 shares of Vitality Health Select Fund. At the time of purchase, the offer price was $12.10. What is the total cost of this transaction?

Use the following sales information for Philadelphia Home Furnishing to answer questions 17–20. **Total sales:** 1991—$500,000; 1992—500,000; 1993—600,000; 1994—700,000; 1995—800,000.

17. What is the total sales for the five-year period?

18. What is the arithmetic mean for sales for Philadelphia Home Furnishings?

19. What is the median sales amount for Philadelphia Home Furnishings?

20. What is the mode sales figure for Philadelphia Home Furnishings?

Standard Weights and Measures

There are two commonly used systems of measurement in the world today. Most of the world uses the International System (sometimes called the metric system). The United States continues to use the U.S. customary measure system. Some of the most commonly used U.S. measurements and their approximate International equivalents are presented below.

Length or Distance

1 foot (ft.) = 12 inches = 30.48 centimeters
1 yard (yd.) = 3 feet = 0.914 meter
1 rod (rd.) = $5\frac{1}{2}$ yards = 5.029 meters
1 mile (mi.) = 5,280 feet = 1.609 kilometers

Surface Area

1 square foot (sq. ft.) = 144 inches = 0.093 square meters
1 square yard (sq. yd.) = 9 square feet = 0.836 square meters
1 square rod (sq. rd.) = $30\frac{1}{4}$ square yards = 25.293 square meters
1 acre (A.) = 4,840 square yards = 4,047 square meters
1 square mile (sq. mi.) = 640 acres = 2.590 square kilometers

Cubic Measure

1 cubit foot (cu. ft.) = 1,728 cubic inches = 0.028 cubic meters
1 cubit yard (cu. yd.) = 27 cubic feet = 0.765 cubic meters

Dry Measure

1 pint (pt.) = 33.600 cubic inches = 0.550 liters
1 quart (qt.) = 2 pints = 1.101 liters
1 peck (pk.) = 8 quarts = 8.809 liters
1 bushel (bu.) = 4 pecks = 35.238 liters

Liquid Measure

1 cup = 8 fl. ounces = 0.237 liters
1 pint (pt.) = 2 cups = 0.473 liters
1 quart (qt.) = 2 pints = 0.946 liters
1 gallon (gal.) = 4 quarts = 3.785 liters

Weight

1 grain (gr.) = 0.036 drams = 0.065 grams
1 dram (dr.) = 27.343 grains = 1.771 grams
1 ounce (oz.) = 16 drams = 28.349 grams
1 pound (lb.) = 16 ounces = 453.592 grams
1 short ton = 2,000 pounds = 0.907 metric tons
1 long ton = 2,240 pounds = 1.016 metric tons

Selected Employment Problems

The following problems are representative of those used by both private and government employers to screen job applicants. Your ability to solve problems like these may help you score higher on various employment tests.

1. Write $2,456.70 in words.

2. Write ninety-three dollars and twenty-three cents in numeral form.

3. Round 2,346 to the nearest 10.

4. Round 56.744 to the nearest hundredth.

5. Add $2.94 + $32.56 + $56 + $789.28.

6. During the first week of April, Denise Tran worked the following hours: Monday, 7.5 hours; Tuesday, 8 hours; Wednesday, off; Thursday, 5 hours; Friday, 9.5 hours; and Saturday, 4.25 hours. How many hours did she work during this week?

7. On October 1, Bob Gonzales had $3,456.70 in his checking account. During the month, he made a deposit for $2,456.50 and wrote checks for $564.23, $119.30, $56.70, and $23.45. What is his checking account balance at the end of the month?

8. Subtract $56.34 from $489.92.

9. On Saturday, Peter Chang spent $34.50 at the hardware store, $56.78 at the grocery store, and $19.50 at the dry cleaners. Assuming that Peter had $150 when he started this shopping trip, how much money does he have now?

10. A customer purchases merchandise valued at $35.68. If the customer gives you a $100 bill, how much change should you give the customer?

11. An employee earns $564.00 a week. The following deductions are withheld from the employee's paycheck: federal withholding, $86; Social Security, $43.15; and Voluntary Savings Plan, $25. What is the amount of the employee's take-home pay?

12. Multiply 67.52 × 4.59.

13. A customer purchased 18 coffee mugs. If the price of each mug is $1.95, how much does the customer owe?

14. On Monday, a retail sales clerk sold four shirts priced at $21.95, two suits priced at $349, and two ties priced at $19.50. What were the employee's total sales for the day?

15. Peggy Olga worked 40 hours during the second week of November. If Peggy is paid $7.85 an hour, how much did she earn?

16. Divide 6,840 by 15.

17. Calculate the average miles per gallon if it takes 16.4 gallons of gasoline to fill the car's gasoline tank and the car has been driven 432 miles.

18. A clerk purchased 12 typewriter ribbons for $40.80. What was the cost of each ribbon?

19. Reduce the fraction $\frac{54}{108}$ to its lowest terms.

20. Add $\frac{5}{8} + \frac{3}{4} + \frac{1}{2}$.

21. Subtract $4\frac{4}{5}$ from $8\frac{1}{3}$.

22. Multiply $2\frac{3}{5} \times 1\frac{2}{3}$.

23. Divide $\frac{5}{12} \div \frac{1}{4}$.

24. Divide $8\frac{1}{2} \div 2\frac{3}{4}$.

25. Jack Johnson earns $8.20 an hour. If his firm pays time and a half for overtime, what is his overtime pay rate?

26. According to a partnership agreement, Sandra Day is to receive $\frac{1}{5}$ of all profits that the business earns. If the firm earns $148,360, how much is Sandra's share?

27. Divide 20 by $3\frac{1}{5}$.

28. Convert 43.5% to a decimal.

29. Convert $\frac{3}{4}$ to a percent.

30. Convert 45% to a fraction.

31. Find 3.5% of 1240.

32. 210 is what percent of 4,200?

33. 440 is 20 percent of what number?

34. Jane Gomez earns $2,368 a month. If the employer withholds 24.3% of her salary for deductions, what is her take-home pay?

35. A manufacturer produced 12,800 computer monitors during the month of May. If 256 monitors were defective, what percent of the monitors were defective?

36. A real estate salesperson receives $2\frac{1}{2}$ percent commission for all sales. If the salesperson sells real estate valued at $856,000, how much is the commission?

37. Sales for 1994 were $234,000. Sales for 1993 were $195,000. What is the percent of increase for sales?

38. Last month, Ed Oberg wrote the following checks that have not been presented to the bank for payment: $15.40, $1,200.40, $69.80, and $12.25. What is the total of Mr. Oberg's outstanding checks?

39. At the end of October, Jill Clayburn's credit card balance was $210. If the bank charges a monthly rate of $1\frac{1}{2}$ percent calculated on the previous month's balance, what is the finance charge?

40. Milton Auburn's gross pay is $710 a week. His Social Security deduction is $54 a week and his federal withholding is $80 a week. What is his net pay amount?

41. Assume that Jackson Cole earns $6.70 an hour. Also assume that he worked $44\frac{1}{2}$ hours during one week. What is his gross pay?

42. A salesperson for Watley Chemical Company is paid 5 percent commission on all sales up to $18,000 and 8 percent commission on all sales in excess of $18,000. If this salesperson sold merchandise valued at $67,400, what is his gross pay?

43. Lynch's Hardware originally priced an electric drill at $29.95. To increase sales, management decided to lower the price of the drill to $24.50. What is the dollar amount of discount?

44. A 12-volt battery has a suggested retail price of $58.50. If the manufacturer offers retailers a trade discount of 40 percent, what is the discount amount?

45. A desk-top calculator has a suggested retail price of $64.95. The manufacturer offers retailers series discounts of 20 percent, 15 percent, and 10 percent. What is the wholesale price for one of the calculators?

46. O'Brien Construction Company received an invoice dated June 10 for $34,580 from one of their suppliers. The supplier offers the following cash discount terms: $\frac{2}{10}$, $\frac{N}{30}$. If the firm pays the invoice on June 15, what is the cash discount amount?

47. Mills Computer Service bought the following merchandise from Tri-Lakes Office Supply: (a) 48 printer ribbons at $4.50 each; (b) 6 desk lamps at $19.80 each; and (c) 10 boxes of computer paper at $9.95 a box. What was the total value of the merchandise purchased from Tri-Lakes Office Supply?

48. East Town Ford sold a Ford Crown Victoria for $18,750. If the car cost the dealership $15,990, what is the markup amount?

49. General Nutrition pays $4.50 for a jar of Vitamin E. If the markup is 30 percent on cost, what is the selling price?

50. A new telephone cost U.S. Telephone Supply $19.45. If the markup amount is $5.54, what is the selling price?

51. If a retailer pays $129 for a coffee table that is later sold at a markup of 40 percent on the selling price, what is the selling price?

52. A yard of carpet originally priced at $19.95 is marked down 20 percent. What is the reduced selling price?

53. Find the interest for a $5,000 loan at 9 percent for one year.

54. Find the interest for a $10,400 loan at $8\frac{1}{2}$ percent for 60 days if an exact year is used.

55. T F The due date for an 85-day loan that originates on March 3 is May 26.

56. T F The number of days between June 10 and August 15 is 61 days.

57. Julian Jewelry borrowed $15,000 from Red Oak Federal Bank. The owners of Julian Jewelry signed a 90-day simple discount note with a rate of 9 percent. What is the dollar amount of the bank discount? (Use a banker's year.)

58. What is the monthly escrow payment for real estate taxes if annual taxes for a home are $1,800?

59. Find the total repayment amount for a 15-year home mortgage if the monthly payment is $710.

60. If the straight-line method of depreciation is used, what is the annual depreciation amount for a desk that cost $750, has no salvage value, and has a useful life of five years?

61. If the first-year depreciation expense is $375, what is the book value at the end of year one for a piece of equipment that cost $3,000?

62. Southern Plastics purchased a new forklift for $19,800. It has a salvage value of $1,000 and a useful life of eight years. What is the first-year depreciation amount if the sum-of-the-years'-digits method of depreciation is used?

63. Janice Mooney purchased an automobile for business use. The car cost $17,600 with a salvage value of $1,500 and an estimated life of five years. What is the first-year depreciation amount if the declining-balance method is used?

64. On Wednesday, Charlie Newbury purchased one shirt at $15.99 and one tie for 11.99. What is the sales tax amount if the sales tax rate is 7 percent?

65. If the property tax rate is $1.21 per $100 of assessed value, what is the dollar amount of property tax for a piece of real estate with an assessed value of $71,000?

66. During the year, Sandra Columbo earned $22,900. Her husband earned $21,500. What is their total income?

67. During 1995, All Star Sportswear had gross sales of $1,240,000. The firm also had sales returns that total $25,000 and sales allowances that total $12,500. What is the firm's net sales?

68. According to the union contract, Northwood Manufacturing must pay 80 percent of the health care cost for workers. If each worker's health care coverage cost $260 a month, what is Northwood's share?

69. A health care policy requires that the insurance company pay 70 percent of all approved medical expenditures. If a policyholder has approved medical expenditures of $5,600, what is the amount the insurance company will pay?

70. A brick building has a replacement cost of $240,000. If there is a coinsurance clause that requires that the property be insured for 80 percent of its replacement value, what is the dollar amount of required coverage?

71. During the month of January, The Office Store had net sales of $245,600. The firm's cost of goods sold totaled $120,000. Also, the firm had expenses that total $82,500. What is the firm's net income before taxes?

72. Metroplex Electrical Supply had assets that total $425,000. If the firm has liabilities that total $120,000, what is the dollar amount of owners' equity?

73. In 1994, total assets for the Christmas Store were $3,450,000. In 1995, total assets were $3,795,000. What is the percent of increase or decrease for total assets?

74. Shawn Jackson owns 180 shares of Disney stock. If Disney pays an annual dividend of $0.30 a share, what is the total dividend that Mr. Jackson will receive?

75. Morgan Claus receives a dividend of $1.20 for each share of East Coast Fiberglass Corporation that he owns. If the market value of the stock is $24, what is the current yield?

76. Rob North purchased stock valued at $4,200. If the brokerage firm charges 2 percent commission, what is the total cost for this stock transaction?

77. New York Electric has issued a $1,000 bond with an interest rate of $9\frac{1}{4}$ percent. What is the dollar amount of interest?

78. Mathew Gold purchased 200 shares of the Vision High Income Mutual Fund. At the time of purchase, the offer price was $13.40. What is the total cost of this transaction?

79. Annual sales figures for Newport Appliances are 1993—$234,000; 1994—$210,000; 1995—$315,200. What is the arithmetic mean for sales for Newport Appliances?

80. Monthly utility expenses for the first five months of this year for Harmon Insurance are: January—$155; February—$110; March—$120; April—$95; and May—$105. What is the arithmetic mean for utility expenses for Harmon Insurance?

Answers to Selected Problems and Mastery Quizzes

CHAPTER 1—*Number Values, Addition, and Subtraction*

Unit 1

1a. 1
1c. 2
1e. 3
3a. 193
3c. 12,246
3e. 269,784
3g. 70,433

Unit 2

1a. 9
1c. 3
3a. 0.39
3c. 143.2
3e. 0.046
3g. $355.21

Unit 3

1. 560
3. 2,700
5. 800
7. 500,000
9. 46,000
11. 0.4
13. 1,777
15. 567,595
17. $1,356; $1,355.62
19. $6,780; $6,780.12
21. 20
23. $62
25. $133,000
27. two thousand, five hundred dollars
29. four hundred fifty-two thousandths

Unit 4

1a. 0
1c. 2
1e. 4

1g. 6
1i. 8
3a. 2
3c. 4
3e. 6
3g. 8
3i. 10
5a. 4
5c. 6
5e. 8
5g. 10
5i. 12
7a. 6
7c. 8
7e. 10
7g. 12
7i. 14
9a. 8
9c. 10
9e. 12
9g. 14
9i. 16
11a. 3
11c. 2
11e. 10
11g. 11
11i. 12

Unit 5

1a. 76
1c. 80
1e. 177
1g. 89
3a. 5,015
3c. 5,991
3e. 11,471
5. 436
7. 5,943
9. 2,685
11. $962
13. $5,775
15. 123
17. $7,700

Unit 6

1a. 24.68
1c. 280.60
1e. 2,933.21
3. 18.0562
5. 1,401.003
7. $225.64
9. 37.75
11. $264,238.82
13. $79,866.27

Unit 7

1a. 1
1c. 2
1e. 0
1g. 2
1i. 0
3a. 6
3c. 4
3e. 2
3g. 0
3i. 6
5a. 3
5c. 1
5e. 9
5g. 7
5i. 5
7a. 15
7c. 8
7e. 33
7g. 48
9a. 22,257
9c. 12,139; 12,700
11. $1,692
13. $38,456
15. $343
17. 4,574,000

CHAPTER 1—*Concluded*

Unit 8

1a.	0.13
1c.	1.35
1e.	26.89
3.	1.865

5.	2,228.185
7.	$5,568.20
9.	1,369.3
11.	$223.71
13.	$331.73

15.	$1,294.40
17.	$80,017.58
19.	$507.95

CHAPTER 1—*Mastery Quiz*

1.	4
2.	9
3.	1
4.	7
5.	forty-five hundredths
6.	twenty-nine and six tenths
7.	fifty-six dollars and seventy cents

8.	fifty-six thousand, seven hundred eighty-nine dollars and four cents
9.	126
10.	108.05
11.	353.21
12.	$2,882.55
13.	528.32

14.	3,571.46
15.	$64.75
16.	$6,972.69
17.	36.0
18.	1,065
19.	$69.55
20.	$926.51

CHAPTER 2—*Multiplication and Division*

Unit 9

1a.	0
1c.	0
1e.	0
1g.	0
1i.	0
3a.	2
3c.	6
3e.	10
3g.	14
3i.	18
5a.	4
5c.	12
5e.	20
5g.	28
5i.	36
7a.	6
7c.	18
7e.	30
7g.	42
7i.	54
9a.	8
9c.	24
9e.	40
9g.	56
9i.	72
11a.	6
11c.	9

Unit 10

1a.	54
1c.	360
1e.	182
1g.	215
3a.	5,828
3c.	14,490
3e.	19,908
3g.	54,080
5.	$4,950
7.	$1,242
9.	$3,902

Unit 11

1a.	163.8
1c.	2.070
1e.	5,453.3
3.	$219.45
5.	$8.45
7.	$81.00
9.	$283.76
11.	$35,787.50

Unit 12

1a.	34
1c.	12,410

1e.	45,600
3a.	900
3c.	182,480
3e.	2,050,400
5a.	5,760 5,000
5c.	6,715 7,200
7.	297
9.	$480
11.	$4,800
13.	$439

Unit 13

1a.	0
1c.	0
1e.	0
1g.	0
1i.	0
3a.	1
3c.	6
3e.	5
3g.	3
3i.	9
5a.	1
5c.	4
5e.	6
5g.	5
5i.	7
7a.	1

CHAPTER 2—*Concluded*

Unit 13—*Concluded*

7c.	7
7e.	8
7g.	9
7i.	5
9a.	1
9c.	8
9e.	3
9g.	4
9i.	5
11a.	4
11c.	8

Unit 14

1a.	78
1c.	87

1e.	45
3a.	36
3c.	39
3e.	37
5.	$13
7.	$1,441
9.	$270
11.	120

Unit 15

1a.	12.25
1c.	4.84
3a.	38.79
3c.	2.06
5.	$.25
7.	$293.95

9.	$1.40
11.	943

Unit 16

1a.	51.14
1c.	3.452
1e.	2.3451
3a.	23.400
3c.	22
3e.	13
5.	124.656
7.	0.45621
9a.	9; 10 (estimate)
11.	$3.15
13.	$.1495 or $.15
15.	$78.15

CHAPTER 2—*Mastery Quiz*

1.	4,788
2.	15,219
3.	168.00
4.	8.84
5.	2,885.09
6.	23,459.80
7.	2,456,200

8.	28
9.	132
10.	6.75
11.	23.5
12.	47.10
13.	9.2
14.	$25,220

15.	$1,286.04
16.	$241.88
17.	$1,795.50
18.	$72,350
19.	$10.50
20.	$58

CHAPTER 3—*Fractions*

Unit 17

1.	numerator
3.	denominator
5.	mixed number
7.	improper fraction
9.	improper fraction
11a.	$5\frac{1}{4}$
13a.	$13\frac{1}{2}$
15a.	$7\frac{2}{15}$
17a.	$\frac{21}{8}$
19a.	$\frac{11}{3}$

Unit 18

1a.	$\frac{1}{3}$
1c.	$\frac{1}{4}$
3a.	$\frac{1}{2}$
3c.	$\frac{1}{3}$

5a.	$\frac{10}{21}$
5c.	$\frac{14}{43}$
7a.	$\frac{1}{4}$
7c.	$\frac{2}{3}$
9a.	$\frac{9}{21}$
9c.	$\frac{7}{28}$
11a.	$\frac{15}{42}$
11c.	$\frac{16}{136}$
13a.	$\frac{225}{275}$
13c.	$\frac{112}{152}$

Unit 19

1a.	1
1c.	$\frac{6}{7}$
3a.	$\frac{3}{5}$
3c.	$1\frac{1}{2}$

5a.	$\frac{2}{9}$
5c.	$\frac{3}{7}$
7a.	$\frac{1}{3}$
7c.	$\frac{2}{7}$
9.	$\frac{2}{5}$
11.	$\frac{2}{5}$

Unit 20

1a.	24
3a.	30
5a.	24
7a.	72
9a.	60

Unit 21

1a.	$\frac{11}{12}$
3a.	$\frac{17}{20}$

CHAPTER 3—*Concluded*

Unit 21—*Concluded*

5a. $\frac{31}{40}$

7a. $22\frac{19}{20}$

9a. $16\frac{7}{44}$

11. $41\frac{3}{4}$

13. $75\frac{11}{24}$

15. $7\frac{5}{6}$

Unit 22

1a. $\frac{7}{20}$

3a. $\frac{1}{2}$

5a. $4\frac{3}{8}$

7a. $8\frac{3}{8}$

9. $\frac{5}{8}$

11. $32\frac{7}{8}$

13. $22\frac{1}{6}$

Unit 23

1a. $\frac{1}{2}$

3a. $\frac{8}{33}$

5a. $\frac{7}{96}$

7a. $4\frac{13}{35}$

9a. $18\frac{7}{11}$

11. Jack $27,000
Martha $18,000
Sam $9,000

13. 76

15. $66\frac{1}{2}$

17. $45

Unit 24

1a. $1\frac{1}{3}$

3a. $\frac{21}{40}$

5a. $1\frac{1}{2}$

7a. $2\frac{10}{19}$

9a. $162\frac{1}{2}$

11. .667

13. .067

15. $8

17. 20

19. $2,088

CHAPTER 3—*Mastery Quiz*

1. denominator

2. numerator

3. proper

4. mixed

5. improper

6. $\frac{3}{8}$

7. $\frac{42}{90}$

8. 54

9. $\frac{23}{40}$

10. $10\frac{7}{8}$

11. $\frac{2}{15}$

12. $\frac{59}{60}$

13. $\frac{1}{36}$

14. $7\frac{7}{34}$

15. $1\frac{1}{6}$

16. $4\frac{3}{10}$

17. .444

18. $41\frac{1}{4}$

19. $68

20. $12

APPENDIX A—*The Hand-Held Calculator*

1. 78

3. 10,571

5. $149.43

7. $2,387.98

9. 756

11. 80.44

13. $349.29

15. 288

17. 2,837,376

19. 10,806.29

21. $162.49

23. 53.6

25. 2.14

27. $13.01

29. 60

31. 120

33. 201.72

35. 27.06

37. 3,552

39. 519.90

41. 2,320

43. 2,850

45. 18.20

CHAPTER 4—*Percent: Portion, Rate, and Base*

Unit 25

1. .23

3. .10

5. .114

7. .031

9. .145

11. 33%

13. 12%

15. 40%

17. 11.5%

19. 60%

21. 25%

23. 70%

25. 37.5%

27. 75%

29. 30%

31. $\frac{1}{5}$

33. $\frac{3}{25}$

35. $\frac{9}{20}$

37. $\frac{11}{20}$

39. $\frac{4}{125}$

41. .45

43. 44%

45. 5%

47. 3.4%

49. 3%

CHAPTER 4—*Concluded*

Unit 26

1. 84
3. 195
5. 8.2
7. 229.46
9. 8,118
11. 168
13. $1,626.56
15. $21.12
17. $8.13
19. $2,652

Unit 27

1. 25%
3. 46.667%
5. 95.833%
7. 22.5%

9. −$34,742.40
11. +$12,104.60
13. 12.5%
15. 55.556%
17. $94,395
19. 9.114%

Unit 28

1. 55
3. 48.78
5. 307.5
7. 246.32
9. $95,600
11. $411.67
13. 160
15. $1,260,000

Unit 29

1. .297
3. .057
5. $\frac{1}{25}$
7. 38.33%
9. 533.33
11. 5,096.8
13. 26.67%
15. 4.95%
17. 3,720
19. 6.75%
21. $643,861.40
23. 44.94%
25. $12,306.25

CHAPTER 4—*Mastery Quiz*

1. .26
2. 85.4
3. 75%
4. $\frac{2}{5}$
5. $P = R \times B$
6. $R = P \div B$
7. $B = P \div R$

8. 117
9. 65.15%
10. 7,843.75
11. 18.29%
12. $800
13. $4,100
14. 6,750

15. 11.11 decrease
16. $671.22
17. $2,298.78
18. 4%
19. $141,900
20. $290,855.64

CHAPTERS 1–4—*Cumulative Test*

1. one-thousand, three hundred forty-six dollars and fifty-four cents
2. 239.43
3. $1,341.75
4. 2,118.65
5. $108,78
6. 386,400

7. $357.50
8. $24,568,000
9. 374
10. 7.11
11. $\frac{1}{4}$
12. $1\frac{23}{40}$
13. $3\frac{35}{36}$

14. $\frac{9}{32}$
15. $3\frac{9}{19}$
16. .22
17. 40%
18. 34.5%
19. 990
20. 405

CHAPTER 5—*Banking Records, Credit Cards, and Installment Purchases*

Unit 30

1. $4,149.46
3. $6,219.49
5. blank
7. blank

Unit 31

1. $2,985.93
3. $2,610.30
5. 4-15-95 $2,145.60
   Ck No 440 $1,235.43
   $910.17
7. $2,726.95
9. $5.80

Unit 32

1. decrease
3. decrease
5. decrease
7. decrease
9. $4,890.11
11. $2,538.73
13j. $2,587.18
15j. $12,355.72

CHAPTER 5—*Concluded*

Unit 33

3. T
5. $249.39
7. $94.50
9. $4.60
11. $43.59
13. $11.43
15. $498.75

17. $223.68
19. $28.14

Unit 34

3. annual percentage rate
5. $1,600

7. $242
9. $129
11. $188.89
13. $459
15. $33.75
17. 14.50%
19. 14.50%
21. 16%

CHAPTER 5—*Mastery Quiz*

1. $568.70
2. blank
3. special
4. restrictive
5. $645.45
6. $8,886.05
7. $2,590.98

8. $739.40
9. $1,973.98
10. $568.61
11. $1,405.37
12. $1,412.87
13. $1,405.37
14. $88.05

15. $4.28
16. $11.85
17. $530
18. $71
19. $185
20. 15.00%

CHAPTER 6—*Payroll*

Unit 35

3. $615.38
5. $1,333.33
7. $43\frac{1}{2}$
9. $244
11. $285.18
13. $280
15. $337.75

Unit 37

1. T
3. F
5. $22.32
7. $254.60
9. $3.69
11. $22.39
13. $27.63
15. $61.63
17. $72.14

Unit 39

1. F
3. Federal Unemployment Tax Act
5. $1,519
7. $347.20
9. $8,611.80
11. $465.45
13. $126.15
15. $3,197.25
17. $1,392.52
19. $325.67
21. $9,051.38
23. $164
25. $44.80
27. $1,073.60
29. $91.36
31. $56

Unit 36

1. $2,400
3. $5,052
5. $1,876
7. $469.35
9. $4,344.58
11. $1,929.52
13. $2,700

Unit 38

1. $62
3. $106
5. $236.11
7. $650
9. $9.43
11. $651.60
13. $9.45
15. $535.75

CHAPTER 6—*Mastery Quiz*

1. an employee's compensation before any payroll deductions have been made
2. an employee's compensation after all payroll deductions have been made
3. $42
4. $9.45
5. $276.90
6. $486.94

CHAPTER 6—*Mastery Quiz—Concluded*

7. $2,739.85
8. $2,603.25
9. $290.78
10. $34.51
11. $228

12. $64.60
13. $693.50
14. $43.00
15. $10.06
16. $65.00

17. $575.44
18. $12,362.76
19. $378
20. $56

CHAPTER 7—*Discounts and Invoices*

Unit 40

1. $74.75
3. $174.90
5. $40.24
7. 85%
9. 68%
11. $48.75
13. $43.61
15. $142.00
17. $93.24
19. $129.99
21. $276.75

3. 27.1%
5. $1,392.94
7. $1,137.02
9. $263.98
11. $14.83
13. $29.44
15. $17.21
17. $4,574.44
19. $193.02

Unit 42

1. $4.50
3. $14.61
5. $219
7. $25

9. $12.90
11. $123.50
13. No
15. $1,732.50
19. $12,152
21. $1,266.21

Unit 43

1. $622.55
3. $151.97
5. $5,400
7. $1,981.30
9. $1,155.29
11. $584.71
13. $11.69

Unit 41

1. 35.875%

CHAPTER 7—*Mastery Quiz*

1. T
2. F
3. 43.48%
4. 30.88%
5. $644
6. $2,576
7. $738.28

8. $481.72
9. $65.00
10. $3,185.10
11. $170.11
12. $5,500.29
13. $11,270
14. $7,170

15. $11,858
16. $875.85
17. $591.20
18. $284.65
19. $11.82
20. $579.38

CHAPTER 8—*Markup, Inventory, and Markdown*

Unit 44

1. $7
3. $1.63
5. $14.97
7. $15.62
9. $10.96
11. $2.88
13. $40.30
15. $1.60
17. $2,160
19. $15.13

Unit 45

1. $13.30
3. 85%
5. 89.5%
7. $15.00
9. $33.79
11. $15.43
13. $.60
15. $32.67
17. $81.67
19. $2.70
21. $16.88

Unit 46

1. $1.00
3. $2.50
5. $3.00
7. $2.50
9. $.79
11. $2.25
13. $4.80
15. $15.00
17. $3.75
19. 25.21%
21. 32.67%

CHAPTER 8—*Concluded*

Unit 47

1. F
3. T
5. T
7. 115
9. $144.38
11. 905
13. $357.50
15. $315.00
17. 50

19. $12,320.95
21. $10,423.10

Unit 48

1. $5.00
3. $12.00
5. $12.40
7. $11.25
9. $27.50

11. $10.00
13. $30.00
15. $2.45
17. $50.00
19. $20.45
21. $150
23. $125
25. $0.20

CHAPTER 8—*Mastery Quiz*

1. $5.60
2. $40.60
3. $1.38
4. $6.88
5. $1.47
6. $3.67
7. $42.86

8. $142.86
9. $11.95
10. $189.95
11. $40
12. 31.11%
13. $2,325
14. $2,560

15. $2,300
16. $2,381.28
17. $3.75
18. $16.25
19. $1.25
20. 7.69%

CHAPTERS 5–8—*Cumulative Review*

1. Special Endorsement
2. $2,255.08
3. $3,485.40
4. $16.20
5. $362
6. $3,853.95
7. $213.90

8. $50.03
9. $378
10. $23.76
11. $64.80
12. $3,175.20
13. $19.00
14. $1,881.00

15. $19.13
16. $29.40
17. $31.50
18. 28%
19. 109
20. 20%

CHAPTER 9—*Interest Calculations and Promissory Notes*

Unit 49

1. 360
3. $960
5. $62.50
7. $108.49
9. $794.56
11. $119.60
13. $446.50
15. $465.75
17. $900

Unit 50

1. $1,305
3. $192.50
5. 90 days
7. 9%

9. $63,000
11. $526.75
13. $8,800

Unit 51

3. 31
5. 28
7. November 11
9. November 22
11. June 5
13. 20
15. 128
17. 176
19. 139
21. $5,390.74
23. $285
25. $475

Unit 52

3. $141.67
5. $220
7. $151.67
9. $501.76
11. $1,419.56
13. 8.70%
15. 6%
17. 8.08%
19. 118 (days)
21. 62 (days)
23. $3,554.90
25. $22,078.93
27. 9%
29. 9.15%

CHAPTER 9—*Concluded*

Unit 53

1. July 17
3. August 2
5. 45
7. 100
9. $735

11. $347.92
13. $34,722.50
15. $14,680.65
17. $902
19. $629.63
21. $42,826.67
23. $12,628.91

25. $11,852
27. $25,331.06
29. 63
31. $24,754.78
33. $4,885
35. 62
37. $4,784.04

CHAPTER 9—*Mastery Quiz*

1. $I = P \times R \times T$
2. 365
3. $R = \frac{I}{P \times T}$
4. $2,450
5. $19,000
6. $136
7. 63 days

8. 7%
9. December 18
10. October 26
11. 114
12. $55.23
13. $2,944.77
14. $58.13

15. $3,158.13
16. June 25
17. 52
18. $52.46
19. $3,105.67

CHAPTER 10—*Compound Interest, Present Value, and Home Mortgages*

Unit 54

1. 1.311
3. 1.268
5. 2.427
7. $19,565.25
9. $7,663.60
11. $132,544.80
13. $58,198.28
15. 1.284
17. $8,148
19. $48,140
21. $23,804
23. 8.2%
25. $51,960
27. 1.356
29. $31,775
31. $54,150

Unit 55

3. 0.557
5. 0.820

7. 0.788
9. $3,500
11. $25,685
13. $4,532
15. $10,805.80
17. $13,277.80
19. $7,096
21. $346,500
23. $22,464
25. $2,600

Unit 56

3. $85,500
5. $102,340
7. $7.16
9. $7.39
11. $8.78
13. $990
15. $1,153.50
17. $700.25
19. $1,317.30
21. $845.90

23. $175
25. $6,250
27. $955.94
29. $83.33

Unit 57

1. 180
3. 300
5. $216,000
7. $240,000
9. $504,000
11. $134,700
13. $175,720
15. $360.00
17. $874.08
19. $730.71
21. $487.50
23. $67,016.35
25. $101,827
27. $101,477.54
29. $1,130.04
31. $835.91

CHAPTER 10—*Mastery Quiz*

1. $28,150
2. $9,811.20
3. $36,425.69
4. $9,895.69
5. $16,050
6. $8,112
7. $4,418.40
8. $37,015
9. $11,160
10. $12,460
11. $21,300
12. $120,700
13. $709.16
14. $116.67
15. $198,000
16. $101,220
17. $473.33
18. $70,879.33
19. $472.53
20. $70,757.86

CHAPTER 11—*Depreciation*

Unit 58

5. $350
7. $240
9. $2,688
11. $105
13. $765
15. $18,900
17. $345,500
19. $285
21. $1,865

Unit 59

1. $.02
3. $.11
5. $.128
7. $4,582.76
9. $18,867.24
11. $12,452.16
13. $5,646.20
15. $5,351.64
17. $21,000.00
19. $2,357.30

Unit 60

1. $\frac{4}{10}$
3. $\frac{8}{55}$
5. $720
7. $2,672.73
9. $500
11. $4,000
13. $9,800
15. $7,350
17. $4,900
19. $2,450

Unit 61

1. 20%
3. 40%
5. 50%
7. $970
9. $1,320
11. $900
13. $6,600
15. $9,600

17. $1,485
19. $7,250
21. $3,625
23. $1,812.50
25. $1,500

Unit 62

1. T
3. T
5. 27.5-year class
7. 5-year class
9. 33.33%
11. 8.93%
13. $2,200
15. $435.00
17. $4,476.92
19. $2,181.82
21. $10,000
23. $2,400
25. $1,440

CHAPTER 11—*Mastery Quiz*

1. F
2. T
3. $143.75
4. $1,106.25
5. $143.75
6. $962.50
7. $.11
8. $4,757.50
9. $\frac{7}{28}$
10. $33,000
11. $91,000
12. $27,500
13. $63,500
14. 10%
15. $6,960
16. $10,440
17. $4,176
18. $2,680
19. $10,720
20. $4,288

CHAPTER 12—*Taxes and Insurance*

Unit 63

1. $.06
3. $.32
5. $.59
7. $.74
9. $.67
11. $1.68
13. $2.58
15. $7.74
17. $10.50
19. $15.56
21. $4.53
23. $82.10
25. $21.33
27. $575
29. $1,046.50
31. $1,669.80
33. $2,763.45
35. $4,092
37. $264

Unit 64

1. T
3. T
5. F
7. $5,014
9. $6,449
11. $6,094
13. $6,452
15. $7,261
17. $83,050
21. $8,558
23. $6,368

25. $7,564
27. $4,954
29. $5,542
31. $5,284

Unit 65

3. T
5. $41,430
7. $2,309,580
11. $525,000
13. $52,900
15. $285,370
17. $100,250
19. $26,150
21. $1,275,000
23. $374,500
25. $795,000
27. $238,940

Unit 66

1. T
3. F
5. $60
7. $14,006.88
9. $303.60
11. $3,440
13. $1,340
15. $750
17. $2,750
19. $1,050
21. $8,796
23. T

25. $1.37
27. $15.16
29. $1.53
31. $136
33. $260
35. $185
37. $143

Unit 67

1. F
3. T
5. $224
7. $864
9. $441
11. $608
13. $661.50
15. $359.68
17. $312.74
19. $721.23
21. $1,666.29
23. $168,000
25. $82,500
27. F
29. $101.33
31. $111.46
33. $227.14
35. $291.90
37. $149.18
39. $98
41. $145
43. $139
45. $126
47. $211

CHAPTER 12—*Mastery Quiz*

1. $6.63
2. $117.07
3. $731.60
4. $1,810.12
5. $1,347.56
6. $19,790
7. $6,990

8. $5,696
9. $7,242
10. $633,550
11. $56,960
12. $124,440
13. $291.60
14. $5,136

15. $143
16. $485
17. $849.60
18. $105,000
19. $378.51
20. $146

CHAPTERS 9–12—*Cumulative Review*

1. $246.67
2. October 10
3. 131 days
4. $312

5. $15,912
6. 47 days
7. $228.51
8. $27,874

9. $27,080
10. $742.50
11. $327,600
12. $2,880

CHAPTERS 9–12—*Cumulative Review—Concluded*

13. $777.78
14. $6,080
15. $3,040

16. $11.53
17. $1,785.60
18. $5,344

19. $148
20. $75,000

CHAPTER 13—*Evaluation of Financial Statements*

Unit 68

3. T
5. $796,040
7. $6,092.42
9. $29,119
11. $1,023 decrease
13. $21,000   25.93%
15. 1.93%
17. $9,750
19. 12.95% decrease

9. $370,000
11. $59,700
13. 16.75%
15. 23.14% decrease
17. 4.60%
19. 14.11%, 22.18%, 48.39%, 84.68%, 10.89%, 4.44%, 15.32%; 21.17%, 21.17%, 56.45%, 22.38%, 78.83%, 100%

15. $0.82
17. 22.77%

Unit 71

5. 1.18
7. 1.0
9. 0.91
15. 12.47
17. 33%
19. 24%
21. 2.76
23. 38.48%

Unit 69

3. F
5. $104,500
7. $156,900

Unit 70

5. 4.2%
7. Between 12% and 15%
9. 17.06%

CHAPTER 13—*Mastery Quiz*

1. F
2. T
3. F
4. F
5. F
6. $221,660
7. $41,060

8. $224,500 decrease
9. 19.79% decrease
10. 25.78%
11. $149,800
12. $10,200
13. 15.64%
14. 8.83%

15. 8.00%
16. 14.20%
17. $1.68
18. 1.86
19. 1.11
20. 35.06%

CHAPTER 14—*Annuities, Stocks, Bonds, and Mutual Funds*

Unit 72

3. $3,000
5. $6,120
7. $120
9. $247.20
11. $13,064
13. $6,289
15. $13,923
17. $86,540
19. $364

21. $753.48
23. $1,170.22
25. No. They were $2,112.30 short. $20,000 − $17,887.70 = $2,112.30
27. $6,603.50
29. $104,850
31. $48,670
33. $13,979.25
35. $222,300

Unit 73

1. T
3. F
5. F
7. $16
9. $75
11. $49
13. $8
17. 3.897%
19. 2.605%

CHAPTER 14—*Concluded*

Unit 73—*Concluded*

21. 0.548%
25. 27.273
27. $55\frac{7}{8}$ or $55.875
29. $39\frac{1}{4}$ or $39.25
31. $20\frac{3}{4}$ or $20.75 per share
33. KM
35. $100
37. $206
39. $346.50
41. $4,692
43. $466.38
45. $4,388 profit
47. $3,492
49. $996 loss

Unit 74

1. F
3. F
5. $60
7. $70
9. $83.75
11. $92.50
13. $450
15. 8.375%
17. 63 or $630
19. 2004
21. $990
23. $880
25. $1,037.50
29. 9.914%
31. 12.269%
33. 7.588%
35. 9.434%
37. $4,160
39. profit
41. $795 loss

Unit 75

1. F
3. F
5. T
7. $20
9. $15.44
11. $44.30
13. $11.99
15. +9.3 percent
17. $.75
19. $.26
21. $1.91
23. $.58
25. $.35
27. $3,710
29. $224
31. $327.83
33. $1,540 profit
35. $2,230 profit
37. $10,730 profit
39. $4,290

CHAPTER 14—Mastery Quiz

1. $13,181
2. $16,291.20
3. $11,502
4. $14,665.80
5. $4,780
6. $37.50
7. $7.50
8. 6%
9. 28.98
10. $7,242
11. $80
12. $90
13. $97.50
14. $922.50
15. 9.35%
16. $6,475
17. $20.54 NAV
18. $3,444.20
19. $147
20. $1,874 profit

CHAPTER 15—Business Statistics

Unit 76

5. 76
7. $758.29
9. $665
11. 88
13. $2,227
15. 2,200
17. 93.33
19. $3,116.40
21. $2,200
23. $27.98
25. $27.00

Unit 77

1. 54 ΙΙ = 2
   76 ΙΙΙ = 3
   80 ΙΙ = 2
   82 Ι = 1
   86 Ι = 1

3. $45,000 Ι = 1
   $48,700 ΙΙΙ = 3
   $65,300 Ι = 1
   $70,000 ΙΙ = 2
   $71,000 Ι = 1
   $76,400 ΙΙΙ = 3
   $80,000 Ι = 1
5. $1,682,000
7. $105,125
9. 2,124
11. 81.69

CHAPTER 15—*Concluded*

Unit 77—*Concluded*

13.

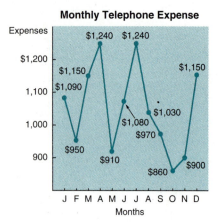

Monthly Telephone Expense

15.

Yearly Sales for
Morton Interior Design

17.

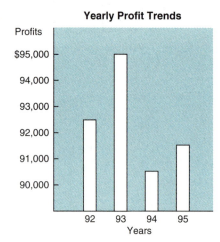

Yearly Profit Trends

19.

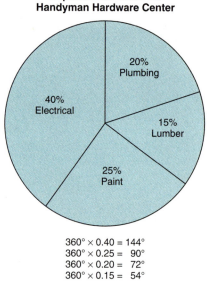

Departmental Sales for
Handyman Hardware Center

$360° \times 0.40 = 144°$
$360° \times 0.25 = 90°$
$360° \times 0.20 = 72°$
$360° \times 0.15 = 54°$

21.

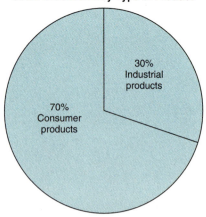

Sales Breakdown by Type of Product

$497,000 \div 710,000 = .070$
$213,000 \div 710,000 = .030$

$360° \times 0.70 = 252°$
$360° \times 0.30 = 108°$

CHAPTER 15—*Mastery Quiz*

1. 71.78
2. 76
3. 80
4. $3,893.43
5. $3,324
6. No mode
7. $2,212,100
8. $737,366.67
9. 8.29 years
10. 8—The middle value
11. 8—The most common value

12. $399   II = 2
    $495   HIt = 5
    $750   III = 3
    $995   II = 2
    $999   II = 2
    $1,150   II = 2
    $1,695   HIt = 5
13. $20,286
14. 21
15. $966

16.

17.
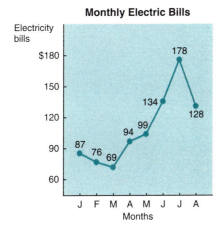

18.
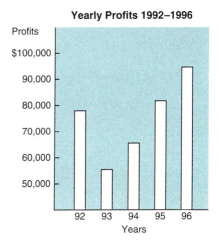

19.
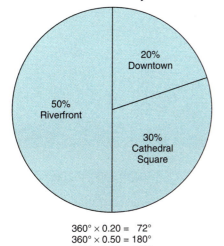

$$360° × 0.20 = 72°$$
$$360° × 0.50 = 180°$$
$$360° × 0.30 = 108°$$

20.
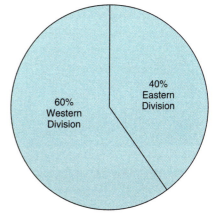

$$\$1,360,000 ÷ \$3,400,000 = 0.40$$
$$\$2,040,000 ÷ \$3,400,000 = 0.60$$

$$360° × 0.40 = 144°$$
$$360° × 0.60 = 216°$$

CHAPTERS 13–15—*Cumulative Review*

1. $424,000
2. $120,000
3. 16% increase
4. $250,000
5. 9%
6. 8%
7. 2.2
8. $8,435
9. $76.50
10. $2.63
11. 4%
12. $6,426
13. $72.50
14. 9%
15. $24.80
16. $2,057
17. $3,100,000
18. $620,000
19. $600,000
20. $500,000

APPENDIX C—*Selected Employment Problems*

1. Two thousand, four hundred fifty-six dollars and seventy cents.
2. $93.23
3. 2,350
4. 56.74
5. $880.78
6. 34.25 hours
7. $5,149.52
8. $433.58
9. $39.22
10. $64.32
11. $409.85
12. 309.92
13. $35.10
14. $824.80
15. $314
16. 456
17. 26.3
18. $3.40
19. $\frac{1}{2}$
20. $1\frac{7}{8}$
21. $3\frac{8}{15}$
22. $4\frac{1}{3}$
23. $1\frac{2}{3}$
24. $3\frac{1}{11}$
25. $12.30
26. $29,672
27. $6\frac{1}{4}$
28. 0.435
29. 75%
30. $\frac{9}{20}$
31. 43.40
32. 5%
33. 2,200
34. $1,792.58
35. 2%
36. $21,400
37. 20%
38. $1,297.85
39. $3.15
40. $576
41. $313.23
42. $4,852
43. $5.45
44. $23.40
45. $39.75
46. $691.60
47. $434.30
48. $2,760
49. $5.85
50. $24.99
51. $215
52. $15.96
53. $450
54. $145.32
55. F
56. F
57. $337.50
58. $150
59. $127,800
60. $150
61. $2,625
62. $4,177.78
63. $7,040
64. $1.96
65. $859.10
66. $44,400
67. $1,202,500
68. $208
69. $3,920
70. $192,000
71. $43,100
72. $305,000
73. 10%
74. $54
75. 5%
76. $4,284
77. $92.50
78. $2,680
79. $253,066.67
80. $117

INDEX